TRANSACTIONS
OF THE
MOSCOW MATHEMATICAL SOCIETY
FOR THE YEAR 1973

ТРУДЫ
МОСКОВСКОГО
МАТЕМАТИЧЕСКОГО
ОБЩЕСТВА
ТОМ 28 (1973)

VOLUME 28

This translation was prepared jointly by the
American Mathematical Society
and the
London Mathematical Society

Published by the American Mathematical Society
Providence, Rhode Island
1975

AMS (MOS) subject classifications (1970)
Primary 31A05, 31A25, 31A35, 35A05, 82A30, 35B40, 34C35, 35D05,
76D10, 76D99, 46F05, 47F05, 58F99, 60F05, 47G05, 35J25,60J50, 35P20, 35Q99, 35S15;
Secondary 28A40, 30A24, 35A35, 31B15, 45B05,
47B25, 46E30, 46E35, 35J10, 60J45, 60J65, 35R15

International Standard Book Number 0-8218-1628-4

Library of Congress Card Number 65-4713

Printed in the United States of America

TRANSACTIONS OF THE MOSCOW MATHEMATICAL SOCIETY

(Труды Московского Математического Общества, Том 28, 1973)

TABLE OF CONTENTS

Page

Trudy Moskov. Mat. Obšč.
Tom 28 (1973)

Trans. Moscow Math. Soc.
Vol. 28 (1973)

SPECTRAL ASYMPTOTICS OF NONSMOOTH ELLIPTIC OPERATORS. II

M. Š. BIRMAN and M. Z. SOLOMJAK

Contents

Introduction

The present work is a continuation of [1]. The basic aim of this work is to establish a formula for the spectral asymptotics of the "nonsmooth" Dirichlet problem under the same broad assumptions which we used in obtaining bounds for the spectrum in [1].

We will recall the statement of the basic problem and we will then formulate the fundamental result of the present work. It will not be necessary to refer to [1] for the formulation of the result. The proof, however, depends heavily on the material in [1]. We will quote the necessary results from [1] without additional explanation, and when we make references to [1] we will preface the number of the formula or theorem with the symbol I (for instance, Lemma I.8.1, formula (I.3.1), and so forth). In the proof of Theorem 2 we will make use of the basic concepts and notation in [1].

Let $\mathbf{C}^k$ denote a k-dimensional unitary space, and let $f\overline{g}$ denote the scalar product of two elements $f, g \in \mathbf{C}^k$. If w is a $k \times k$ matrix, we let $\bar{w}$ denote the Hermitian-conjugate matrix and let $|w|$ denote the norm of the corresponding linear transformation.

Let $\Omega \subset \mathbf{R}^m$ be an open set, and suppose that we are given two differential quadratic forms defined on the vector-functions $u : \Omega \to \mathbf{C}^k$. The "higher-order" form can be written

$$A[u, u] = \int_\Omega \langle a(x) \nabla_l u, \nabla_l u \rangle \, dx \overset{\text{def}}{=} \sum_{|\sigma_1| = |\sigma_2| = l} \int_\Omega a_{\sigma_1, \sigma_2}(x) D^{\sigma_2} u \overline{D^{\sigma_1} u} \, dx. \tag{0.1}$$

AMS (MOS) subject classifications (1970). Primary 35P20, 47F05; Secondary 47B25, 46E35, 35J10.

Here $a(x)=\{a_{\sigma_1\sigma_2}(x)\}$ is a "block" matrix; each block $a_{\sigma_1\sigma_2}$ is a square matrix of order k, and $\overline{a_{\sigma_1\sigma_2}}=a_{\sigma_2\sigma_1}$.

We impose the following conditions on the set Ω and the matrix $a(x)$.

CONDITION I.[1] *The open set* $\Omega\subset\mathbf{R}^m$ *is bounded;* $a\in L_{1,\mathrm{loc}}\Omega$; *for almost all* $x\in\Omega$ *the matrix* $a(x)$ *is positive definite and* $a^{-1}\in L_\alpha\Omega$ *for* $1\leqq\alpha\leqq\infty$, $a^{-1}<2lm^{-1}$.

From Condition I it follows that in the space $L_2\Omega$ the form A considered on $C_0^\infty(\Omega)$ is positive definite and admits a closure. The domain of definition of the corresponding closed form will be denoted by $\overset{\circ}{H}(a)$.

In the Hilbert space $\overset{\circ}{H}(a)$ we will consider operators which are determined by forms of lower order. In these forms the coefficients for the derivatives may in some cases be derivatives of measures.

We say that a complex measure η is in the class $\mathfrak{M}_\beta=\mathfrak{M}_\beta\Omega$, $1<\beta\leqq\infty$, if η is absolutely continuous and $d\eta/dx\in L_\beta\Omega$; here $\|\eta\|_{(\beta)}\overset{\text{def}}{=}\|d\eta/dx\|_\beta$, where the symbol $\|\cdot\|_\beta$ denotes the norm in $L_\beta\Omega$. We let $\mathfrak{M}_1$ denote the class of all finite Borel measures with the norm $\|\eta\|_{(1)}=|\eta|(\Omega)$. A matrix measure μ belongs to the class $\mathfrak{M}_\beta$, $1\leqq\beta\leqq\infty$, if each of its entries is in that class.

Let r be an integer or half-integer, $0\leqq r<l$; let τ_1 and τ_2 be tuples such that $|\tau_1|,|\tau_2|\leqq l$ and $|\tau_1|+|\tau_2|=2r$. Suppose that to each such pair of tuples there corresponds a matrix measure $\mu_{\tau_1\tau_2}$, where $\overline{\mu_{\tau_1\tau_2}}=\mu_{\tau_2\tau_1}$. The set of measures $\mu=\{\mu_{\tau_1\tau_2}\}$ will also be called a measure. The set of matrices $b=\{b_{\tau_1\tau_2}\}=\{d\mu_{\tau_1\tau_2}/dx\}$ will also be called a matrix.

A measure μ of the form just described generates a quadratic form

$$B[u,\ u]=\sum_{\tau_1\tau_2}\int_\Omega \mu_{\tau_1\tau_2}(dx)\,D^{\tau_2}u\,\overline{D^{\tau_1}u} \qquad (|\tau_1|,\ |\tau_2|\leqslant l,\quad |\tau_1|+|\tau_2|=2r,\quad r<l), \tag{0.2}$$

which for absolutely continuous measures can be written

$$B[u,\ u]=\sum_{\tau_1\tau_2}\int_\Omega b_{\tau_1\tau_2}(x)\,D^{\tau_2}u\,\overline{D^{\tau_1}u}\,dx. \tag{0.3}$$

We note that in [1] our main concern was with forms of a more restricted type:

$$B[u,\ u]=\sum_{|\tau_1|=|\tau_2|=r}\int_\Omega b_{\tau_1\tau_2}(x)\,D^{\tau_2}u\,\overline{D^{\tau_1}u}\,dx \quad (r\text{ an integer}). \tag{0.4}$$

The following condition connecting the matrix a and the measure μ is of fundamental importance.

CONDITION II. *Each measure* $\mu_{\tau_1\tau_2}$ *in* (0.2) *belongs to the class* $\mathfrak{M}_{\beta_{\tau_1\tau_2}}$, *where*

$$\alpha^{-1}+(\beta_{\tau_1\tau_2})^{-1}<2(l-r)m^{-1}; \tag{0.5}$$

$$\alpha^{-1}+2(\beta_{\tau_1\tau_2})^{-1}<2(l-|\tau_i|)m^{-1}+1,\quad i=1,\ 2. \tag{0.5a}$$

---

[1] Condition I is somewhat changed from its formulation in [1]. Here it includes an assumption about the matrix a^{-1} which, however, follows from Condition II (see (0.5)).

In a number of cases (0.5a) follows from (0.5) or is satisfied automatically. This is true, for instance, for forms of the type (0.4), and also when $\beta_{\tau_1\tau_2} = \infty$.

If Conditions I and II are both satisfied, the form (0.2) is completely continuous in the space $\overset{\circ}{H}(a)$ and consequently it determines in $\overset{\circ}{H}(a)$ a completely continuous operator, namely, the "operator of the problem $\mathscr{D}(A,B;\Omega)$" (the Dirichlet problem). The positive (negative) eigenvalues $\lambda_n^+(A,B,\Omega;\mathscr{D})$ $(-\lambda_n^-(A,B,\Omega;\mathscr{D}))$ of this problem coincide with the successive maxima (minima) of the ratio

$$B[u,u]/A[u,u], \qquad u \in \overset{\circ}{H}(a). \tag{0.6}$$

In certain cases we will speak of the "spectrum of the ratio (0.6)". The distribution functions of the sequences $\lambda_n^{\pm}(A,B,\Omega;\mathscr{D})$ are denoted by $n_{\pm}(\lambda;A,B,\Omega;\mathscr{D})$.

THEOREM 1. *Suppose that Conditions* I *and* II *are satisfied. Then*[2)]

$$\lambda_n^{\pm}(A,\ B,\ \Omega;\ \mathscr{D}) \leqslant C_1 n^{-q} \| a^{-1} \|_{\alpha} \sum_{\tau_1\tau_2} \| \mu_{\tau_1\tau_2} \|_{(\beta_{\tau_1\tau_2})}, \quad q = 2(l-r)m^{-1}. \tag{0.7}$$

The constant C_1 depends on m, k, l, r, α, $\beta_{\tau_1\tau_2}$ and the diameter of the set Ω, but not on a or μ.

If the measure μ is singular, then along with (0.7) *we have*

$$\lambda_n^{\pm}(A,B,\Omega;\mathscr{D}) = o(n^{-q}).$$

This theorem coincides with the assertion in part a) of Theorem I.3.1. The proof is given in §4 of [**1**].

We proceed to results on asymptotics. We let $\mathbf{a}(x,\xi)$ and $\mathbf{b}(x,\xi)$ denote the symbols of the forms A and B:

$$\mathbf{a}(x,\ \xi) = \sum_{\sigma_1\sigma_2} a_{\sigma_1\sigma_2}(x)\,\xi^{\sigma_1+\sigma_2},$$

$$\mathbf{b}(x,\ \xi) = \sum_{\tau_1\tau_2} b_{\tau_1\tau_2}(x)\,\xi^{\tau_1+\tau_2}.$$

The symbols are square matrices of order k. The singular component of the measure μ does not influence the symbol of B.

THEOREM 2. *Under the hypothesis of Theorem* 1, *for the spectrum of the problem $\mathscr{D}(A,B;\Omega)$ we have the asymptotic formula*

$$\lim_{\lambda\to 0} \lambda^{\theta} n_{\pm}(\lambda) = \omega_{\pm}(a,\ b;\ \Omega), \tag{0.8}$$

where

$$2\theta = 2q^{-1} = m(l-r)^{-1}, \tag{0.9}$$

[2)] The enumeration of "essential" constants, denoted by C_i, is different from the enumeration in [**1**]. Here c will denote nonessential constants whose values are not related to each other.

$$\omega_\pm(a,\ b;\ \Omega)$$

$$(0.10)\qquad = \frac{(2\pi)^{-m}}{m} \int\limits_{\Omega} \int\limits_{|\xi|=1} \operatorname{tr}\{[\mathrm{a}^{-1/2}(x,\ \xi)\,\mathrm{b}(x,\ \xi)\,\mathrm{a}^{-1/2}(x,\ \xi)]^0_\pm\}\,dS(\xi)\,dx.$$

The inner integral in (0.10) is taken with respect to $(m-1)$-dimensional Lebesgue measure on a sphere. Formula (0.8) – (0.10) shows that under the conditions being considered the singular component of μ makes no contribution to the principal term of the asymptotics.

The proof of Theorem 2 is the main content of the present work. This theorem significantly strengthens the result of Theorems I.3.2 – I.3.4 by removing all of the additional (as compared to the theorem on bounds of the spectrum) restrictions.

Let us recall the content of those restrictions. In Theorem I.3.2 we assumed that B was a form of the type (0.4) and that for almost all $x \in \Omega$ the matrix $b(x)$ was definite. And in the hypothesis of Theorem I.3.2 there were additional restrictions on the matrix a; we assumed that $a \in L_\kappa\Omega$ and

$$(0.11)\qquad \alpha^{-1} + \kappa^{-1} < 2(l-r)m^{-1}, \qquad \alpha^{-1} < (l-r)m^{-1}.$$

In Theorem I.3.3 the above restrictions on a were replaced by an assumption on the structure of the set of its "singular points". In Theorem I.3.4 we removed the assumption that b was definite; instead we required a to be continuous except on some closed (relative to Ω) set of measure zero.

All of these restrictions are done away with by Theorem 2. Unfortunately the proof of this theorem contains many steps. In §1 we establish some facts in the abstract theory of operators. Especially useful is Lemma 1.2, by virtue of which we overcome some basic difficulties. We also need a series of preparatory results. First we artificially raise the order of the forms A and B, replacing u in (0.1) and (0.3) by $\Delta^p u$, which is related to the loss of positive definiteness of the corresponding matrix. This change enables us to remove the restriction of type (0.11). The relevant material is discussed in §2. Second, some technical difficulties force us to consider operators which are generated by forms of the type (0.1), (0.3) in Sobolev spaces H^s, where s is not in general an integer. This set of questions is treated in §3. These considerations could be shortened if we could manage to extend the assertions in Lemmas 1 and 2 to all $\eta > 0$.[3] In §4 the proof of Theorem 2 is completed. Finally, §5 contains applications of the fundamental result to the Neumann problem, to problems in unbounded domains, and also to problems containing a small parameter.

We have already mentioned that the technical apparatus developed in [1] will be used repeatedly in the present work. In particular, the results

---

[3] *Added in proof*. This extension has now been carried out. The resulting simplifications in the proof of Theorem 2 will be discussed elsewhere.

in §§5–7 of [1] are very important. The results in §8 of [1] form the basis of §2 in the present work. On the other hand, the complicated constructions in §9 of [1] can be ignored.

A report on the results in this work appeared in [2].

§1. On fractional powers of selfadjoint operators

1. We begin with assertions concerning the spectrum of the operator $\sqrt{T_1}-\sqrt{T_2}$ under the assumption that we have information on the spectrum of T_1-T_2.

LEMMA 1.1. *Let T_1 and T_2 be nonnegative selfadjoint operators in a Hilbert space $\mathfrak{H}$, and let $T_1-T_2\in\mathfrak{S}_\infty$. Then*

a)

$$\sqrt{T_1}-\sqrt{T_2}\in\mathfrak{S}_\infty; \tag{1.1}$$

b)

$$\sum_{k=1}^{n}\left[\lambda_k^{\pm}\left(\sqrt{T_1}-\sqrt{T_2}\right)\right]^2\leqslant\sum_{k=1}^{n}\lambda_k^{\pm}(T_1-T_2),\quad n=1,\,2,\,\ldots. \tag{1.2}$$

PROOF. a) First suppose that $T_1, T_2\geqq\gamma I$ with $\gamma>0$. Then we can use an integral representation for fractional powers of the operator (see [3], formula (11.41)), and thus (1.1) is clear. In the general case we have

$$\sqrt{T_1}-\sqrt{T_2}=\left(\sqrt{T_1+\gamma I}-\sqrt{T_2+\gamma I}\right)-\left(\sqrt{T_1+\gamma I}-\sqrt{T_1}\right)$$
$$+\left(\sqrt{T_2+\gamma I}-\sqrt{T_2}\right),\quad \gamma>0.$$

The first term on the right side is an operator in $\mathfrak{S}_\infty$, and the second and third terms have norms which do not exceed $\sqrt{\gamma}$. If we let $\gamma\to0$ we obtain (1.1).

b) We will use the identity

$$\left(\sqrt{T_1}-\sqrt{T_2}\right)^2=(T_1-T_2)-\sqrt{T_2}\left(\sqrt{T_1}-\sqrt{T_2}\right)$$
$$-\left(\sqrt{T_1}-\sqrt{T_2}\right)\sqrt{T_2}. \tag{1.3}$$

We let P_n^+ denote the projection onto the linear span of the eigenvectors of $\sqrt{T_1}-\sqrt{T_2}$ corresponding to the eigenvalues $\lambda_k^+(\sqrt{T_1}-\sqrt{T_2})$, $k=1,\cdots,n$. If we multiply both sides of (1.3) on the left and the right by P_n^+, we obtain

$$P_n^+\left(\sqrt{T_1}-\sqrt{T_2}\right)^2P_n^+=P_n^+(T_1-T_2)P_n^+$$
$$-P_n^+\sqrt{T_2}P_n^+\cdot\left(\sqrt{T_1}-\sqrt{T_2}\right)P_n^+-\left(\sqrt{T_1}-\sqrt{T_2}\right)P_n^+\cdot P_n^+\sqrt{T_2}P_n^+. \tag{1.4}$$

Since the operators $P_n^+\sqrt{T_2}P_n^+$ and $(\sqrt{T_1}-\sqrt{T_2})P_n^+$ are nonnegative, their product has a nonnegative trace. If we calculate the trace of both sides of (1.4), we find that

$$\sum_{k=1}^{n}\left[\lambda_k^{+}\left(\sqrt{T_1}-\sqrt{T_2}\right)\right]^2=\operatorname{tr}P_n^+\left(\sqrt{T_1}-\sqrt{T_2}\right)^2P_n^+\leqq$$

$$\leqslant \operatorname{tr} P_n^+ (T_1 - T_2) P_n^+ \leqslant \sum_{k=1}^{n} \lambda_k^+ (T_1 - T_2).$$

If we interchange the roles of T_1 and T_2 we obtain an analogous inequality for λ_k^-. The lemma is proved.

LEMMA 1.2. *Suppose the hypothesis of Lemma* 1.1 *is satisfied and for some* η_+, $0 < \eta_+ < 1$, *we have the asymptotic bound*

$$\limsup_{n\to\infty} n^{\eta_+} \lambda_n^+ (T_1 - T_2) \leqslant M_+.$$

Then

$$\limsup_{n\to\infty} n^{\eta_+} \left[\lambda_n^+ \left(\sqrt{T_1} - \sqrt{T_2}\right)\right]^2 \leqslant M_+ (1 - \eta_+)^{-1}. \tag{1.5}$$

An analogous assertion holds for λ_n^-.

PROOF. Let c denote the quantity $\sup_n n^{\eta_+} \lambda_n^+ (T_1 - T_2)$. We fix $\epsilon > 0$ and choose n_ϵ such that for $n > n_\epsilon$ we have

$$\lambda_n^+ (T_1 - T_2) \leqq (M_+ + \epsilon) n^{-\eta_+}.$$

From (1.2) we find that for $n > n_\epsilon$

$$n \left[\lambda_n^+ \left(\sqrt{T_1} - \sqrt{T_2}\right)\right]^2 \leqslant \sum_{k=1}^{n} \left[\lambda_k^+ \left(\sqrt{T_1} - \sqrt{T_2}\right)\right]^2$$

$$\leqslant \sum_{k=1}^{n} \lambda_k^+ (T_1 - T_2) \leqslant c n_\varepsilon + (M_+ + \varepsilon)(1 - \eta_+)^{-1} n^{1-\eta_+},$$

which implies (1.5).

Along with Lemmas 1.1 and 1.2 we will find the following assertion to be useful.

LEMMA 1.3. *Let* Γ *be a selfadjoint operator in* $\mathfrak{H}$ *such that* $0 \leqq \Gamma \leqq I$, *and let* K *be a completely continuous operator. Then*

$$s_n\left[\left(I - \sqrt{\Gamma}\right)K\right] \leqslant s_n\left(\sqrt{I - \Gamma}\, K\right), \quad n = 1, 2, \ldots.$$

PROOF. It is clear that $\|(I - \sqrt{\Gamma})y\| \leqq \|\sqrt{I - \Gamma}\, y\|$, $y \in \mathfrak{H}$. It remains to set $y = Kx$ and to apply the minimax principle for singular numbers.

2. In this subsection we will establish some interpolation inequalities for singular numbers of operators acting in a Hilbert scale of spaces. The first inequality, pertaining to the finite-dimensional case, is obtained by applying the Heinz inequality to associated matrices of various orders. A useful form of the Heinz inequality is contained in [**3**] (Theorem 12.5 in Chapter 3). Necessary information on associated matrices can be found in Chapter 1 of [**4**]. We note that associated matrices were used by Weyl [**5**] for obtaining inequalities involving eigenvalues and singular numbers.

Lemma 1.4. *Let J_1 and J_2 be nonnegative selfadjoint operators in $\mathfrak{H}$. Let T be a completely continuous operator, and let J_1TJ_2 be defined on $\mathfrak{D}(J_2)$ and bounded in $\mathfrak{H}$. Then for each $\kappa \in (0,1)$ the operator $J_1^\kappa TJ_2^\kappa$ is completely continuous and its singular numbers satisfy*

$$\prod_{k=1}^{n} s_k(J_1^\kappa TJ_2^\kappa) \leqslant \prod_{k=1}^{n} s_k^{1-\kappa}(T)\,\|J_1TJ_2\|^{n\kappa}, \quad n = 1, 2, \ldots. \tag{1.6}$$

If $J_1TJ_2 \in \mathfrak{S}_\infty$, then also

$$\prod_{k=1}^{n} s_k(J_1^\kappa TJ_2^\kappa) \leqslant \prod_{k=1}^{n} s_k^{1-\kappa}(T) \cdot \prod_{k=1}^{n} s_k^{\kappa}(J_1TJ_2), \quad n = 1, 2, \ldots. \tag{1.7}$$

We will just present the proof of (1.7), since the proof of (1.6) involves simpler calculations.

a) Suppose that J_1, J_2 and T are operators in a finite-dimensional space $\mathbf{C}^N$. We will use the same letters to denote the matrices corresponding to these operators in the standard basis. If K is a square matrix of order N, by $K_{[n]}$, $n \leqq N$, we denote its associated matrix of degree n. From the properties of associated matrices it follows that

$$(J_1^\kappa TJ_2^\kappa)_{[n]} = (J_{1[n]})^\kappa T_{[n]}(J_{2[n]})^\kappa. \tag{1.8}$$

From the Heinz inequality (see (12.21) in Chapter 3 of [**3**]) we have

$$\|(J_{1[n]})^\kappa T_{[n]}(J_{2[n]})^\kappa\| \leqslant \|T_{[n]}\|^{1-\kappa}\|J_{1[n]}T_{[n]}J_{2[n]}\|^\kappa. \tag{1.9}$$

From Kronecker's theorem on the spectrum of an associated matrix (see [**4**]) we find that

$$\|T_{[n]}\| = s_1(T_{[n]}) = \prod_{k=1}^{n} s_k(T) \tag{1.10}$$

and we get analogous results for the other terms in (1.9). If we compare (1.8) and (1.9) with relations of the form (1.10), we can see that (1.7) is true in the finite-dimensional case.

b) Suppose that J_1 and J_2 are bounded operators in $\mathfrak{H}$. Let P_N be a sequence of orthoprojectors of rank N which converge strongly to the identity operator in $\mathfrak{H}$. We use the notation $J_{iN} = P_N J_i P_N$, $i = 1, 2$. We have the limits

$$\begin{gathered}(u)\text{-}\lim T_N = T;\\ (s)\text{-}\lim J_{iN} = J_i, \qquad (s)\text{-}\lim J_{iN}^\kappa = J_i^\kappa, \quad i = 1, 2.\end{gathered} \tag{1.11}$$

Thus it is easy to see that

$$(u)\text{-}\lim(J_{1N}T_NJ_{2N}) = J_1TJ_2, \qquad (u)\text{-}\lim(J_{1N}^\kappa T_NJ_{2N}^\kappa) = J_1^\kappa TJ_2^\kappa. \tag{1.12}$$

We fix n and apply the result of part a) to the operators J_{1N}, T and J_{2N} restricted to the space $P_N\mathfrak{H}$, $N \geqq n$. We obtain

$$(1.13)\qquad \prod_{k=1}^{n} s_k\,(J_{1N}^{\varkappa} T_N J_{2N}^{\varkappa}) \leqslant \prod_{k=1}^{n} s_k^{1-\varkappa}\,(T_N) \prod_{k=1}^{n} s_k^{\varkappa}\,(J_{1N} T_N J_{2N}).$$

We take the limit as $N \to \infty$ and remember that (u)-convergence of a sequence of completely continuous operators implies convergence of all of the singular numbers. Thus from (1.13) and (1.11), (1.12) we obtain (1.7).

c) We have to eliminate the assumption that J_1 and J_2 are bounded. Without this assumption it is difficult to prove that $J_1^{\varkappa} T J_2^{\varkappa} \in \mathfrak{S}_\infty$. This fact will be established later on the basis of inequalities (1.6) and (1.7). For the proof of these inequalities we make use of the properties of s-numbers of arbitrary bounded operators (see §7 in Chapter II of [**6**]).

We begin with the following elementary remarks.

Suppose $K \in \mathfrak{R}$, $K_j \in \mathfrak{S}_\infty$ *for* $j = 1, 2, \cdots$, *and* $K = (s)\text{-}\lim K_j$. *Then*

$$(1.14)\qquad s_n\,(K) \leqslant \liminf_{j\to\infty} s_n\,(K_j), \quad n = 1,\ 2,\ \ldots .$$

To see this, we fix $\epsilon > 0$ and find an n-dimensional projector P such that $s_n(KP) \geqq s_n(K) - \epsilon$. Since $KP = (u)\text{-}\lim K_j P$, we have

$$s_n\,(K) - \varepsilon \leqslant s_n\,(KP) = \lim_{j\to\infty} s_n\,(K_j P) \leqslant \liminf_{j\to\infty} s_n\,(K_j),$$

which implies that (1.14) is true.

Now assume that in the above terminology $K_j = Q_j^{(1)} K Q_j^{(2)}$, *where* $\| Q_j^{(i)} \| \leqq 1$, $i = 1, 2$. *Then*

$$s_n\,(K) = \lim_{j\to\infty} s_n\,(K_j), \quad n = 1,\ 2,\ \ldots .$$

In fact, it suffices to establish (1.14) and the inequality

$$s_n(K_j) \leqq \| Q_j^{(1)} \|\ \| Q_j^{(2)} \|\, s_n(K) \leqq s_n(K).$$

After these remarks the proof of the lemma can be completed without difficulty. We fix $\epsilon > 0$ and consider the bounded operators $J_{i\epsilon} = J_i(I + \epsilon J_i)$, $i = 1, 2$. We have

$$J_{1\varepsilon} T J_{2\varepsilon} = (I + \varepsilon J_1)^{-1} J_1 T J_2 (I + \varepsilon J_2)^{-1};$$

$$J_{1\varepsilon}^{\varkappa} T J_{2\varepsilon}^{\varkappa} = (I + \varepsilon J_1)^{-\varkappa} J_1^{\varkappa} T J_2^{\varkappa} (I + \varepsilon J_2)^{-\varkappa}.$$

The operators $(I + \epsilon J_i)^{-1}$ and $(I + \epsilon J_i)^{-\kappa}$ $(i = 1, 2)$ converge strongly to I as $\epsilon \to 0$, and their norms do not exceed unity. The operator $J_1^{\kappa} T J_2^{\kappa}$ is bounded by virtue of the Heinz inequality. Thus we can make use of the preceding remarks and conclude that for all $n = 1, 2, \cdots$

$$s_n\,(J_1 T J_2) = \lim_{\varepsilon\to 0} s_n\,(J_{1\varepsilon} T J_{2\varepsilon}), \quad s_n\,(J_1^{\varkappa} T J_2^{\varkappa}) = \lim_{\varepsilon\to 0} s_n\,(J_{1\varepsilon}^{\varkappa} T J_{2\varepsilon}^{\varkappa}).$$

Now it is clear that inequality (1.7) for the unbounded operators J_1 and J_2 can be obtained from the corresponding inequality for the bounded operators $J_{1\epsilon}$ and $J_{2\epsilon}$ by taking the limit as $\epsilon \to 0$.

From (1.7) it is clear that $s_n(J_1^\kappa T J_2^\kappa) \to 0$ as $n \to \infty$. This is equivalent to $J_1^\kappa T J_2^\kappa \in \mathfrak{S}_\infty$. Thus Lemma 1.4 is proved.

Lemma 1.5. *Suppose that in the terminology of Lemma* 1.4 *we have the asymptotic bounds*[4]

$$\limsup_{n\to\infty} n^{\eta_0} s_n(T) \leqslant M_0, \quad \limsup_{n\to\infty} n^{\eta_1} s_n(J_1 T J_2) \leqslant M_1, \tag{1.15}$$

$$\eta_0 > 0, \quad \eta_1 \geqslant 0.$$

Then for each $\kappa \in (0,1)$

$$\limsup_{n\to\infty} n^{\eta} s_n(J_1^\varkappa T J_2^\varkappa) \leqslant e^\eta M_0^{1-\varkappa} M_1^\varkappa, \tag{1.16}$$

$$\eta = \eta(\varkappa) = (1-\varkappa)\eta_0 + \varkappa\eta_1.$$

Proof. Suppose $\sup_n n^{\eta_0} s_n(T) \leqq c$, $\sup_n n^{\eta_1} s_n(J_1 T J_2) \leqq c$. We fix $\epsilon > 0$ and find n_ϵ such that for $n > n_\epsilon$

$$s_n(T) \leqq (M_0+\epsilon) n^{-\eta_0}, \qquad s_n(J_1 T J_2) \leqq (M_1+\epsilon) n^{-\eta_1}.$$

By Lemma 1.4 we know that for $n > n_\epsilon$

$$[s_n(J_1^\varkappa T J_2^\varkappa)]^n \leqslant \prod_{k=1}^n s_k(J_1^\varkappa T J_2^\varkappa)$$

$$\leqslant c^{n_\varepsilon} (M_0+\varepsilon)^{(1-\varkappa)(n-n_\varepsilon)} (M_1+\varepsilon)^{\varkappa(n-n_\varepsilon)} (n!)^{-\eta}.$$

Since $n! \geqq n^n e^{-n}$, we have

$$n^\eta s_n(J_1^\varkappa T J_2^\varkappa) \leqslant c^{n_\varepsilon/n} (M_0+\varepsilon)^{(1-\varkappa)(1-n_\varepsilon/n)} (M_1+\varepsilon)^{\varkappa(1-n_\varepsilon/n)} e^\eta.$$

If we take the upper limit as $n \to \infty$ and then let $\epsilon \to 0$, we obtain (1.16).

§2. The forms $A_{(p)}$ and $B_{(p)}$

1. In the Introduction we mentioned that for the proof of Theorem 2 we can eliminate unnecessary restrictions on the matrix $a(x)$ by artificially raising the order of the forms A and B. In this section we will set forth the material related to this topic. It is not necessary to consider a form B of the general type (0.2); in accordance with Lemmas I.5.1 and I.5.2 it suffices to consider forms of the type (0.3) generated by matrices $b \in L_\infty\Omega$. We note that for $b \in L_\infty\Omega$ Condition II reduces to the single inequality

$$\alpha^{-1} < 2(l-r)m^{-1}. \tag{2.1}$$

2. Let $p \geqq 0$ be an integer. With the matrices $a(x)$ and $b(x)$ we associate the differential quadratic forms

$$A_{(p)}[u,u] = A[\Delta^p u, \Delta^p u], \tag{2.2}$$

$$B_{(p)}[u,u] = B[\Delta^p u, \Delta^p u], \tag{2.3}$$

---

[4] If the second inequality in (1.15) is satisfied for $\eta_1 = 0$ we set $M_1 = \| J_1 T J_2 \|$.

where Δ is the Laplace operator. Thus $A_{(p)}$ is a form of order $l+2p$, $A_{(0)}=A$, and analogous statements hold for $B_{(p)}$. We can write the quadratic form $A_{(p)}$ as follows, by analogy with (0.1):

$$A_{(p)}[u,\ u]=\int_{\Omega}\langle a_p(x)\nabla_{l+2p}u,\ \nabla_{l+2p}u\rangle\,dx,$$

where $a_p(x)$ is a nonnegative matrix of order $k\cdot\iota(m,l+2p)$. It is easy to see that $\det a_p(x)$ is identically zero. Therefore we can see that extension of the results in [1] to problems concerning the form $A_{(p)}$ is not automatic (but there will not be any fundamental difficulties).

If Condition I is satisfied we have inequalities of the form

$$A_{(p)}[u,\ u]\geqslant c\|\Delta^p u\|^2,\quad \|\Delta^p u\|\geqslant c\|u\|\quad (u\in C_0^\infty\Omega),$$

which imply that in $L_2\Omega$ the form $A_{(p)}$ is positive definite when considered on functions $u\in C_0^\infty\Omega$. It is easy to see that this form admits a closure in $L_2\Omega$. The domain of definition of the corresponding closed form will be denoted by $\overset{\circ}{H}(a,p)$; the operator in $L_2\Omega$ which is generated by this form will be denoted by $\mathbf{A}_{(p)}$.

In $\overset{\circ}{H}(a,p)$ we will consider operators generated by forms of lower order. For our purposes it suffices to consider forms of the type $B_{(p)}$. If $b\in L_\infty\Omega$, $a^{-1}\in L_\alpha\Omega$ and (2.1) is satisfied, then the operator determined by the form $B_{(p)}$ is completely continuous in $\overset{\circ}{H}(a,p)$. By analogy with [1] we will call this operator the operator of the problem $\mathscr{D}(A_{(p)},B_{(p)};\Omega)$, and we will use the notation $\lambda_n^\pm(A_{(p)},B_{(p)},\Omega;\mathscr{D})$ and $n_\pm(\lambda;A_{(p)},B_{(p)},\Omega;\mathscr{D})$ for the eigenvalues of this operator and their distribution functions.

We let Δ_p denote the selfadjoint p-harmonic operator of the Dirichlet problem in Ω. In other words, $(-1)^p\Delta_p$ is the selfadjoint operator in $L_2\Omega$ generated by the form $\int_\Omega|\nabla_p u|^2dx$ considered in $\overset{\circ}{H}{}^p\Omega$. We note that in the case $\partial\Omega\in C^\infty$ we have $\mathfrak{D}(\Delta_p)=H^{2p}\Omega\cap\overset{\circ}{H}{}^p\Omega$.

In some cases it will be necessary to consider the form $A_{(p)}$ on a domain of definition $\widetilde{H}(a,p)$ not equal to $\overset{\circ}{H}(a,p)$. Here $u\in\widetilde{H}(a,p)$ if $u\in\mathfrak{D}(\Delta_p)$ and $\Delta_p u\in\overset{\circ}{H}(a)$. The form $A_{(p)}$ defines a Hilbert metric in $\widetilde{H}(a,p)$. Clearly the operator Δ_p maps $\widetilde{H}(a,p)$ isometrically onto $\overset{\circ}{H}(a)$. We have the inclusion

$$\overset{\circ}{H}(a,p)\subset\widetilde{H}(a,p). \tag{2.4}$$

The operator determined by the form $B_{(p)}$ in the space $\widetilde{H}(a,p)$ will be called the operator of the problem $\widetilde{\mathscr{D}}$. Notation of the type $\lambda_n^\pm(\cdots;\widetilde{\mathscr{D}})$ needs no further explanation. By virtue of the above isometry between $\widetilde{H}(a,p)$ and $\overset{\circ}{H}(a)$ we know that

$$\lambda_n^\pm(A_{(p)},B_{(p)},\Omega;\widetilde{\mathscr{D}})=\lambda_n^\pm(A,B,\Omega;\mathscr{D}). \tag{2.5}$$

In the following lemma we extend the result of Theorem 1 to the forms $A_{(p)}$ and $B_{(p)}$. In the formulation of this lemma (and of most of the results in this section) we will not write down the most general case, but rather seek results in a form which is immediately applicable to the proof of Theorem 2.

LEMMA 2.1. a) *Suppose Condition* I *is satisfied, and that* $b \in L_\infty\Omega$ *and* (2.1) *is satisfied. Then*

$$\lambda_n^{\pm}(A_{(p)}, B_{(p)}, \Omega; \mathscr{D}) \leq C_1 n^{-q} \|a^{-1}\|_\alpha \|b\|_\infty, \tag{2.6}$$

where C_1 *is the constant in* (0.7) *and* $q = 2(l-r)m^{-1}$.

b) *Suppose that Condition* I *is satisfied. Then*

$$\lambda_n(\mathbf{A}_{(p)}^{-1}) \leqq C_2 n^{-2(l+2p)/m} \|a^{-1}\|_\alpha; \tag{2.7}$$

the constant C_2 *depends on* m, k, l, α, p *and the diameter of* Ω.

PROOF. a) By means of variational methods we can show that (2.4) and (2.5) imply

$$\lambda_n^{\pm}(A_{(p)}, B_{(p)}, \Omega; \mathscr{D}) \leqq \lambda_n^{\pm}(A, B, \Omega; \mathscr{D}). \tag{2.8}$$

If we bound the right side by means of Theorem 1 we obtain (2.6).

b) By means of the variational methods used in a) we find that

$$\lambda_n(\mathbf{A}_{(p)}^{-1}) \leqslant \lambda_n(\mathbf{A}^{-1/2}\boldsymbol{\Delta}_p^{-2}\mathbf{A}^{-1/2}) = \lambda_n(\boldsymbol{\Delta}_p^{-1}\mathbf{A}^{-1}\boldsymbol{\Delta}_p^{-1}).$$

If we use (I.1.10) and the bound for $\lambda_n(\Delta_p^{-1})$ which follows from Theorem 1, we obtain (2.7).

3. We consider the case where the matrix $a(x)$ is smooth and nondegenerate (that is, a and a^{-1} are in $C^\infty\overline{\Omega}$). Then integration by parts of the form $A_{(p)}$ leads to the representation

$$A_{(p)}[u,u] = A^{(p)}[u,u] + \cdots \qquad (u \in C_0^\infty\Omega),$$

where

$$A^{(p)}[u, u] = \sum_{\substack{|\sigma_1|=|\sigma_2|=l \\ |\omega|=2p}} \frac{(2p)!}{\omega!} \int_\Omega a_{\sigma_1\sigma_2}(x) D^{\sigma_2+\omega} u \overline{D^{\sigma_1+\omega} u}\, dx,$$

and the dots stand for lower terms. We write down a representation of the type (0.1) for the form $A^{(p)}$:

$$A^{(p)}[u, u] = \int_\Omega \langle a^{(p)}(x) \nabla_{l+2p} u, \nabla_{l+2p} u\rangle\, dx,$$

where

$$a^{(p)}(x) = \{a^{(p)}_{\varkappa_1\varkappa_2}(x)\}, \quad |\varkappa_1| = |\varkappa_2| = l + 2p,$$

$$a^{(p)}_{\varkappa_1\varkappa_2}(x) = \sum_{\substack{\sigma_1+\omega=\varkappa_1,\ \sigma_2+\omega=\varkappa_2 \\ |\omega|=2p}} \frac{(2p)!}{\omega!} a_{\sigma_1\sigma_2}(x).$$

In contrast to a_p the matrix $a^{(p)}$ is not degenerate. We can show that its lower bound coincides with the lower bound of the matrix a. Thus for the problem $\mathscr{D}(A^{(p)}, B_{(p)}; \Omega)$ we know that in each case the hypothesis of Lemma I.7.1 is satisfied and the spectral asymptotics for the problem can be calculated by means of (0.8) – (0.10). According to Remark I.2.2 the problems $\mathscr{D}(A_{(p)}, B_{(p)}; \Omega)$ and $\mathscr{D}(A^{(p)}, B_{(p)}; \Omega)$ have the same spectral asymptotics. We note that the symbolic matrix $\mathbf{a}^{(p)}(x, \xi)$ of the form $A^{(p)}$ can be written in terms of the symbolic matrix $\mathbf{a}(x, \xi)$ of the form A by means of the formula

$$\mathbf{a}^{(p)}(x, \xi) = \mathbf{a}(x, \xi) \cdot |\xi|^{4p}.$$

Consequently on the sphere $|\xi| = 1$ we have $\mathbf{a}^{(p)}(x, \xi) = \mathbf{a}(x, \xi)$. Analogously, for $|\xi| = 1$ the symbol of the form $B_{(p)}$ coincides with the symbol of the form B.

Thus we have the following result.

LEMMA 2.2. *Suppose $a, a^{-1} \in C^\infty \overline{\Omega}$ and $b \in L_\infty \Omega$. Then*

$$\lim_{\lambda \to 0} \lambda^\theta n_\pm (\lambda;\ A_{(p)},\ B_{(p)},\ \Omega;\ \mathscr{D}) = \omega_\pm (a, b; \Omega),$$

where θ and $\omega_\pm$ are defined by (0.9) *and* (0.10).

4. For the transition from smooth nondegenerate matrices to matrices of a general type we use analogues of Lemmas I.5.5 and I.8.1 applied to $A_{(p)}$. We will formulate the necessary assertions. Their proofs are almost identical to the corresponding proofs in [1] and we will only mention those portions of the proofs which are different.

The following lemma is analogous to Lemma I.8.1.

LEMMA 2.3. *Suppose $\partial\Omega \in C^\infty$ and $2(l+2p) > m$, and let $\hat{A}_{(p)}$ and $\tilde{A}_{(p)}$ be forms of the type* (2.2), *where $\hat{a} \in L_1 \Omega$, $\hat{a}^{-1} \in L_\infty \Omega$ and $\tilde{a}, \tilde{a}^{-1} \in C^\infty \overline{\Omega}$; let $\hat{\mathbf{A}}_{(p)}$ and $\tilde{\mathbf{A}}_{(p)}$ be the corresponding selfadjoint operators in $L_2 \Omega$. Then for arbitrary $f, g \in L_2 \Omega$*

$$(2.9) \qquad (\hat{\mathbf{A}}_{(p)}^{-1} f, g) - (\tilde{\mathbf{A}}_{(p)}^{-1} f, g) = \int_\Omega \langle (\tilde{a} - \hat{a})\ \nabla_l \Delta^p \tilde{\mathbf{A}}_{(p)}^{-1} f,\ \nabla_l \Delta^p \hat{\mathbf{A}}_{(p)}^{-1} g \rangle dx.$$

PROOF. The domain of definition of $\tilde{\mathbf{A}}_{(p)}$ is

$$(2.10) \qquad \mathfrak{D}(\tilde{\mathbf{A}}_{(p)}) = H^{2(l+2p)} \Omega \cap \overset{\circ}{H}{}^{l+2p} \Omega.$$

By an imbedding theorem (as a result of the condition on p) we have

$$(2.11) \qquad \mathfrak{D}(\tilde{\mathbf{A}}_{(p)}) \subset C_0^{l+2p} \overline{\Omega} \subset \mathfrak{d}[\hat{A}_{(p)}];$$

here $C_0^{l+2p} \overline{\Omega}$ denotes the closure of the set $C_0^\infty \Omega$ in the metric of the space $C^{l+2p} \overline{\Omega}$. Relation (2.11) replaces formula (I.8.3). The rest of the proof repeats the arguments in the proof of Lemma I.8.1.

Now we introduce an analogue of Lemma I.5.5.

LEMMA 2.4. *Suppose $\partial\Omega \in C^\infty$; let $\tilde{A}_{(p)}$ be a form of the type* (2.2), *where $\tilde{a}, \tilde{a}^{-1} \in C^\infty \overline{\Omega}$, and let $\tilde{\mathscr{A}}_{(p)}$ denote the corresponding differential expression. Let $p \geqq 0$ be an integer and let $\alpha \geqq 1$ satisfy the inequality*

$$(2.12) \qquad \alpha^{-1} + 1 < 2(l + 2p) m^{-1}.$$

Let $g(x)$ be a matrix in the class $L_2 \Omega$. Then for the eigenvalues λ_n of the ratio

$$\frac{\int_\Omega |g\nabla_l \Delta^p u|^2 dx}{\|\widetilde{\mathcal{A}}_{(p)} u\|^2}, \qquad u \in \mathfrak{D}(\widetilde{\mathbf{A}}_{(p)}),$$

we have the asymptotic bound

$$\limsup_{n\to\infty} n^{2(l+2p)/m} \lambda_n \leqslant C_3 \|\widetilde{a}^{-1}\|_\alpha \|g\widetilde{a}^{-1/2}\|_2^2. \tag{2.13}$$

The constant C_3 depends on m, k, l, p, α and the diameter of Ω, but not on the matrices $\widetilde{a}$ and g.

The proof is a copy of the proof of Lemma I.5.5. We note that the domain of definition $\mathfrak{D}(\widetilde{\mathbf{A}}_{(p)})$ is given by (2.10). On this domain we have the coercive inequality

$$\|\widetilde{\mathcal{A}}_{(p)} u\| \geqslant c\mathring{N}(u/H^{2(l+2p)}\Omega), \qquad u \in \mathfrak{D}(\widetilde{\mathbf{A}}_{(p)}), \quad c = c(a, p, \Omega).$$

This follows, for instance, from the analogous inequality for the strongly elliptic operator $\widetilde{\mathscr{A}}^{(p)}$ corresponding to the form $\widetilde{A}^{(p)}$ and from the fact that in $L_2\Omega$ the form $\widetilde{A}_{(p)}$ is positive definite on $\mathring{H}^{l+2p}\Omega$.

We stated that one of the fundamental steps in the proof of Lemma I.5.5 consisted in making an appropriate choice of a differential expression Q which satisfied (I.5.8). Now we replace this expression by

$$Q_{(p)} = \widetilde{\mathscr{A}}^2_{(p)} - \mathscr{R}^*_{(p)} \widetilde{\gamma} \mathbf{1} \mathscr{R}_{(p)},$$

where $\mathscr{R}_{(p)} = \nabla_l \Delta^p \widetilde{a}^{1/2} \nabla_l \Delta^p$; inequality (I.5.9) then becomes

$$\|\widetilde{\mathcal{A}}_{(p)} u\|^2 \geqslant \int_\Omega \widetilde{\gamma} |\nabla_l \Delta^p \widetilde{a}^{1/2} \nabla_l \Delta^p u|^2 dx - cN^2(u \mid H^{2l+4p-1/2}\Omega), \qquad u \in C_0^\infty\Omega.$$

Then we repeat the proof of Lemma I.5.5.

5. From Lemmas 2.3 and 2.4 we obtain asymptotic bounds for the eigenvalues of the operators $\mathbf{A}^{-1}_{(p)} - \widetilde{\mathbf{A}}^{-1}_{(p)}$ in terms of integral characteristics of the proximity of the matrices a and $\widetilde{a}$. The following lemma corresponds to the results in §8 of [1].

LEMMA 2.5. *Suppose $\partial\Omega \in C^\infty$ and the matrix a satisfies Condition* I *and $a \in L_1\Omega$; suppose $\widetilde{a}$ is a matrix such that $\widetilde{a}, \widetilde{a}^{-1} \in C^\infty\overline{\Omega}$. Let $p \geqq 0$ be an integer such that* (2.12) *is satisfied. Then*

$$\begin{aligned} &\limsup_{n\to\infty} n^{2(l+2p)/m} s_n(\mathbf{A}^{-1}_{(p)} - \widetilde{\mathbf{A}}^{-1}_{(p)}) \\ &\qquad \leqslant C_4 \|a^{-1}\|_\alpha^{1/2} \|\widetilde{a}^{-1}\|_\alpha^{1/2} \|a^{-1/2}(a - \widetilde{a})\widetilde{a}^{-1/2}\|_2, \end{aligned} \tag{2.14}$$

where the constant C_4 depends on m, k, l, p, α and the diameter of Ω, but not on the matrices a and $\widetilde{a}$.

The proof follows the pattern set in §8 of [1]. Let $a_j(x)$, $j > 0$, be the matrices described in [1], §8.3. We apply Lemma 2.3 to the matrices $\widetilde{a}$ and

$\hat{a} = a_j$; the form of the type (2.2) corresponding to a_j will be denoted by $A_{j(p)}$ and the corresponding selfadjoint operator in $L_2\Omega$ will be denoted by $\mathbf{A}_{j(p)}$. Equation (2.9) clearly leads to the representation

(2.15) $$\mathbf{A}_{j(p)}^{-1} - \widetilde{\mathbf{A}}_{(p)}^{-1} = \Phi_{jp}^{*}\Psi_{jp},$$

where Φ_{jp} and Ψ_{jp} are completely continuous operators acting from $L_2(\Omega; k)$ into $L_2(\Omega; k_l(m, l))$ according to the formula

$$\Phi_{jp} = a^{1/2}\nabla_l \Delta^p \mathbf{A}_{j(p)}^{-1}; \qquad \Psi_{jp} = a^{-1/2}(\tilde{a} - a_j)\nabla_l \Delta^p \widetilde{\mathbf{A}}_{(p)}^{-1}.$$

From (2.14) and (I.1.10) we find that

(2.16) $$s_{2n-1}(\mathbf{A}_{j(p)}^{-1} - \widetilde{\mathbf{A}}_{(p)}^{-1}) \leqslant s_n(\Phi_{jp})\, s_n(\Psi_{jp}).$$

For the numbers $s_n(\Phi_{jp})$ we have the following bound for all j:

$$s_n^2(\Phi_{jp}) \leqslant C_2 n^{-2(l+2p)/m} \| a^{-1} \|_\alpha,$$

which can be proved in the way that we proved (I.8.8), except that instead of Theorem I.3.1 we use part b) of Lemma 2.1. We substitute this bound into (2.16) and take the limit as $j \to \infty$; the justification for taking this limit is a copy of the argument in [1] which allowed us to take the limit in (I.8.9). We obtain

(2.17) $$s_{2n-1}(\mathbf{A}_{(p)}^{-1} - \widetilde{\mathbf{A}}_{(p)}^{-1}) \leqslant C_2^{1/2} \| a^{-1} \|_\alpha^{1/2} n^{-(l+2p)/m} \cdot s_n(\Psi_p),$$

where $\Psi_p = a^{-1/2}(\tilde{a} - a)\nabla_l \Delta^p \widetilde{\mathbf{A}}_{(p)}^{-1}$. The numbers $s_n^2(\Psi_p)$ are the successive minima of the ratio

$$\frac{\| a^{-1/2}(\tilde{a} - a)\nabla_l \Delta^p u \|^2}{\| \widetilde{\mathcal{A}}_{(p)} u \|^2}, \qquad u \in \mathfrak{D}(\widetilde{\mathbf{A}}_{(p)}).$$

Now we can use Lemma 2.4. We have $g \overset{\text{def}}{=} a^{-1/2}(\tilde{a} - a) \in L_2\Omega$, since a, $a^{-1} \in L_1\Omega$. If we bound the numbers $s_n(\Psi_p)$ by means of (2.13) and substitute the result into (2.17) we obtain (2.14). The lemma is proved.

LEMMA 2.6. *Suppose that Condition* I *is satisfied,* $\partial\Omega \in C^\infty$, $a \in L_1\Omega$, *and* $p \geqq 0$ *is an integer for which* (2.12) *is true. Then for each* $\epsilon > 0$ *there exist a matrix* $\tilde{a}(x)$ $(\tilde{a}, \tilde{a}^{-1} \in C^\infty\overline{\Omega})$ *and a matrix* $\hat{a}(x)$ $(\hat{a}(x) \geqq a(x)$ *for almost all* $x \in \Omega$; $\hat{a}^{-1} \in L_\infty\Omega)$ *such that for the corresponding operators* $\widetilde{\mathbf{A}}_{(p)}$ *and* $\hat{\mathbf{A}}_{(p)}$ *we have*

(2.18) $$\limsup_{n\to\infty} n^{2(l+2p)/m} s_n(\mathbf{A}_{(p)}^{-1} - \widetilde{\mathbf{A}}_{(p)}^{-1}) \leqslant \varepsilon;$$

(2.19) $$\limsup_{n\to\infty} n^{2(l+2p)/m} s_n(\mathbf{A}_{(p)}^{-1} - \hat{\mathbf{A}}_{(p)}^{-1}) \leqslant \varepsilon.$$

PROOF. We choose the matrices $\tilde{a}$ and $\hat{a}$ in accordance with Lemma I.5.6 and Remark I.5.7. Then (2.18) follows immediately from (2.14). Furthermore,

$$s_{2n-1}(\mathbf{A}_{(p)}^{-1} - \hat{\mathbf{A}}_{(p)}^{-1}) \leqslant s_n(\mathbf{A}_{(p)}^{-1} - \widetilde{\mathbf{A}}_{(p)}^{-1}) + s_n(\widetilde{\mathbf{A}}_{(p)}^{-1} - \hat{\mathbf{A}}_{(p)}^{-1});$$

we multiply both sides by $n^{2(l+2p)/m}$ and then bound the right side by means of (2.14). This leads to (2.19).

6. Now we will show that under some additional assumptions on $a(x)$ the spectral asymptotics of the ratio (0.6) are not changed if we replace the forms A and B by $A_{(p)}$ and $B_{(p)}$. We introduce the following notation (see §5.1 of [1]):

$$\Delta_{\pm}^{(p)}(a, b, \Omega; \mathscr{D}) = \limsup_{\lambda \to 0} \lambda^{\theta} n_{\pm}(\lambda; A_{(p)}, B_{(p)}, \Omega; \mathscr{D});$$
$$\delta_{\pm}^{(p)}(a, b, \Omega; \mathscr{D}) = \liminf_{\lambda \to 0} \lambda^{\theta} n_{\pm}(\lambda; A_{(p)}, B_{(p)}, \Omega; \mathscr{D}).$$

Here θ is given by (0.9). When $p = 0$ the symbol p will be omitted from $\Delta_{\pm}$ and $\delta_{\pm}$..

LEMMA 2.7. *Suppose that Condition* I *and* (2.1) *are satisfied and* $b \in C_0^{\infty}\Omega$. *Suppose further that there exists an open set* Ω_0 *such that* $\overline{\Omega}_0 \subset \Omega$ *and* $a, a^{-1} \in L_{\infty}(\Omega \backslash \overline{\Omega}_0)$. *Then for each* $p > 0$

$$\Delta_{\pm}^{(p)}(a, b, \Omega; \mathscr{D}) = \Delta_{\pm}(a, b, \Omega; \mathscr{D}),$$
$$\delta_{\pm}^{(p)}(a, b, \Omega; \mathscr{D}) = \delta_{\pm}(a, b, \Omega; \mathscr{D}). \tag{2.20}$$

PROOF. The inequalities

$$\Delta_{\pm}^{(p)} \leqq \Delta_{\pm}, \qquad \delta_{\pm}^{(p)} \leqq \delta_{\pm} \tag{2.21}$$

clearly follow from (2.8). We will prove the reverse inequalities. Instead of the problem $\mathscr{D}(A, B)$ we will consider the problem $\widetilde{\mathscr{D}}(A_{(p)}, B_{(p)})$, which according to (2.5) has the same spectrum.

Let Ω' and Ω'' be open sets with $\overline{\Omega}_0 \subset \Omega'$, $\overline{\Omega}' \subset \Omega''$ and $\overline{\Omega}'' \subset \Omega$; and let $b = 0$ outside Ω'. We select a function $\zeta \in C_0^{\infty}\Omega''$ such that $0 \leqq \zeta \leqq 1$ and $\zeta = 1$ in Ω'. We fix $\epsilon > 0$ and consider the inequality

$$A_{(p)}[u, u] \geqslant \varepsilon A_{(p)}[u, u] + (1 - \varepsilon) A_{(p)}[\zeta u, \zeta u]$$
$$+ (1 - \varepsilon) \int_{\Omega} [\zeta^2 \langle a\nabla_l \Delta^p u, \nabla_l \Delta^p u\rangle - \langle a\nabla_l \Delta^p(\zeta u), \nabla_l \Delta^p(\zeta u)\rangle] dx. \tag{2.22}$$

The last term on the right side is a form of lower differential order; here the integration is in fact done over the set $\Omega'' \backslash \overline{\Omega}'$ on which the matrix a is bounded. Consequently the form is completely continuous with respect to the form $N^2(u \,|\, H^{l+2p}(\Omega'' \backslash \overline{\Omega}'))$. We will show that for a function $u \in \widetilde{H}(a, p)$ we have

$$N^2(u \,|\, H^{l+2p}(\Omega'' \backslash \overline{\Omega}')) \leqq c A_{(p)}[u, u]. \tag{2.23}$$

Since $\Delta_p u \in \mathring{H}(a)$, known local bounds for solutions of elliptic equations lead to the inequality

$$N^2(u \,|\, H^{l+2p}(\Omega'' \setminus \overline{\Omega}')) \leqslant c\, \mathring{N}^2(\Delta^p u \,|\, H^l(\Omega \setminus \overline{\Omega}_0)) + c N^2(\Delta^p u \,|\, L_2\Omega).$$

Both terms on the right side are majorized by the form $A_{(p)}[u, u]$ (for $u \in \widetilde{H}(a, p)$); the first because the matrix a is bounded from below on the set $\Omega \backslash \overline{\Omega}_0$ and the second because the form A is positive definite on $\mathring{H}(a)$.

From (2.22) and (2.23) it follows that

$$A_{(p)}[u,u] \succ \varepsilon A_{(p)}[u,u] + (1-\varepsilon) A_{(p)}[\zeta u, \zeta u] \geqslant (1-\varepsilon) A_{(p)}[\zeta u,\zeta u], \quad u \in \widetilde{H}(a,p) \tag{2.24}$$

(the meaning of the symbol $\succ$ is given in [1], §2.5).

From (2.23) it follows that the equation $v = \zeta u$ defines a continuous mapping from $\widetilde{H}(a,p)$ into $\mathring{H}(a,p)$; here $B_{(p)}[v,v] = B_{(p)}[u,u]$. From (2.24), Lemma I.1.2 and Remark I.2.4 we obtain

$$\Delta_{\pm}(a,b,\Omega;\mathcal{D}) \leqslant \limsup_{\lambda\to 0} \lambda^{\theta} n_{\pm}(\lambda(1-\varepsilon);\ A_{(p)},\ B_{(p)},\ \Omega;\ \mathcal{D}) = (1-\varepsilon)^{-\theta} \Delta_{\pm}^{(p)}(a,b,\Omega;\mathcal{D}).$$

Analogous inequalities are true for the quantities $\delta_{\pm}$. If we take the limit as $\epsilon \to 0$ and use (2.21) we obtain (2.20).

§3. The operators $\mathbf{A}_{(p,\rho)}$ and $\mathbf{B}_{(p,\rho)}$

1. The second (after introducing the forms $A_{(p)}$ and $B_{(p)}$) preparatory step in the proof of Theorem 2 consists in studying the operators generated by these forms in the spaces of a certain Hilbert scale. This scale is similar to the Sobolev-Slobodeckiĭ scale $H^{\rho}\Omega$, where ρ is in general not an integer. Our main concern will be the spaces $\mathring{H}^{\rho}\Omega$.

We recall the appropriate definitions (for ρ not an integer). Let $\Omega \in \mathbf{R}^m$ be an open set; let $\rho > 0$ be a noninteger, and let $[\rho]$ be its integer part. The space $H^{\rho}\Omega$ is the set of all functions $u \in H^{[\rho]}\Omega$ for which the following quantity is finite:

$$G_{\rho}[u,u] = G_{\rho,\Omega}[u,u] = \sum_{|\omega|=[\rho]} \int_{\Omega}\int_{\Omega} \frac{|D^{\omega}u(x) - D^{\omega}u(y)|^2}{|x-y|^{m+2(\rho-[\rho])}}\, dx dy.$$

The norm in $H^{\rho}\Omega$ is given by

$$N^2(u \mid H^{\rho}\Omega) = \int_{\Omega} |u|^2 dx + G_{\rho,\Omega}[u,u].$$

For the sake of uniformity of notation the principal term of the metric form of the classes $H^{\rho}\Omega$ for integer ρ will also be denoted by G_{ρ}.

For the remainder of this section (except in Lemma 3.4) we will assume that the set Ω is bounded and $\partial\Omega \in C^{\infty}$. Then there exists a bounded linear operator which extends functions, with the operator acting from $H^{\rho}\Omega$ into $H^{\rho}\mathbf{R}^m$.

We let $\mathring{H}^{\rho}\Omega$ denote the closure in $H^{\rho}\Omega$ of the class $C_0^{\infty}\Omega$. If d is an integer, the space $\mathring{H}^{d}\Omega$ coincides with the domain of definition of the quadratic form of the operator $(-1)^d\Delta_d$. In the general case there is a close connection between the spaces $H^{\rho}\Omega$ and fractional powers of a polyharmonic operator. This connection can be formulated in terms of operators acting in $L_2\Omega$:

$$\mathbf{F}_d = \mathbf{F}_d(\Omega) = [(-1)^d \Delta_d]^{d/2}.$$

Let $0 \leqq \rho \leqq d$. The domain of definition of the operator $\mathbf{F}_d^\rho$ will be denoted by $E_d^\rho\Omega$. The class $E_d^\rho\Omega$, equipped with the norm $N(u\,|\,E_d^\rho\Omega) = \|\mathbf{F}_d^\rho u\|$, is a Hilbert space.

LEMMA 3.1. *Let $0 \leqq \rho \leqq d$, where d is an integer. Then we have a continuous imbedding $\overset{\circ}{H}{}^\rho\Omega \subset E_d^\rho\Omega$:*

$$N(u\,|\,H^\rho\Omega) \leqq C_5\|\mathbf{F}_d^\rho u\|, \qquad u \in \overset{\circ}{H}{}^\rho\Omega. \tag{3.1}$$

If $\rho - [\rho] \neq \frac{1}{2}$, then $H^\rho\Omega = E_d^\rho\Omega$ and

$$\|\mathbf{F}_d^\rho u\| \leqslant C_6 N(u\,|\,H^\rho\Omega), \quad \rho - [\rho] \neq 1/2; \quad u \in \overset{\circ}{H}{}^\rho\Omega. \tag{3.2}$$

The constants C_5 and C_6 depend on Ω, d and ρ.

The above lemma is contained in Chapter 1 of [7].

If $\rho \leqq d < d_1$, then $E_{d_1}^\rho\Omega = E_d^\rho\Omega$, and the corresponding norms are equivalent. We also mention the bound

$$\lambda_n(\mathbf{F}_d^{-1}) = O(n^{-1/m}), \tag{3.3}$$

which follows from Theorem 1 applied to the operator Δ_d.

2. Suppose that the matrix a satisfies Condition I in Ω and $A_{(p)}$ is the form (2.2) considered on $\overset{\circ}{H}(a,p)$. For $u \in C_0^\infty\Omega$ we have the inequality (see (I.2.15))

$$\begin{gathered}\|a^{-1}\|_\alpha A_{(p)}[u,u] \geqslant \overset{\circ}{N}{}^2(\Delta^p u\,|\,W_s^l\Omega) \geqslant c\overset{\circ}{N}{}^2(u\,|\,W_s^{l+2p}\Omega),\\ 2s^{-1} = 1 + \alpha^{-1}.\end{gathered} \tag{3.4}$$

Under the assumption that

$$\alpha^{-1} < 2(l + 2p - \rho)^{m-1} \tag{3.5}$$

the space $\overset{\circ}{W}{}_s^{l+2p}\Omega$ is (completely) continuously imbedded in $\overset{\circ}{H}{}^\rho\Omega$. Thus from (3.4) it follows that the form $A_{(p)}$ is positive definite in $\overset{\circ}{H}{}^\rho\Omega$:

$$A_{(p)}[u,u] \geqslant cN^2(u\,|\,H^\rho\Omega), \quad c > 0 \quad (u \in \overset{\circ}{H}(a,p)).$$

If $\rho - [\rho] \neq \frac{1}{2}$, then by Lemma 3.1 this is equivalent to positive definiteness in $E_d^\rho\Omega$, $d \geqq \rho$:

$$A_{(p)}[u,u] \geqslant c\|\mathbf{F}_d^\rho u\|^2, \quad c > 0 \quad (u \in \overset{\circ}{H}(a,p)).$$

In the case $\rho - [\rho] = \frac{1}{2}$ this relation still holds true. For suppose that a number $\rho_0 > \rho$ satisfies (3.5) and is not a half-integer. Then for $u \in \overset{\circ}{H}(a,p)$ we have

$$A_{(p)}[u,u] \geqslant c_0\|\mathbf{F}_d^{\rho_0} u\|^2 \geqslant c\|\mathbf{F}_d^\rho u\|^2, \qquad c > 0.$$

The selfadjoint operator generated in the space $E_d^\rho\Omega$ by the form $A_{(p)}$ will be denoted by $\mathbf{A}_{(p,\rho,d)}$. As a rule the symbol d will be omitted. If $\rho = 0$ the symbol ρ will also be omitted. Condition (3.5) guarantees complete continuity of $\mathbf{A}_{(p,\rho)}^{-1}$ in $E_d^\rho\Omega$. The operator $\mathbf{A}_{(p,\rho)}^{-1}$ can be expressed in terms of the operator $\mathbf{A}_{(p)}^{-1}$, which acts in $L_2\Omega$, by means of the formula

$$\mathbf{A}_{(p,\rho)}^{-1} = \mathbf{A}_{(p)}^{-1/2}(\mathbf{F}_d^\rho \mathbf{A}_{(p)}^{-1/2})^* \mathbf{F}_d^\rho.$$

Under the natural isometry between $E_d^\rho \Omega$ and $L_2\Omega$ realized by the operator $\mathbf{F}_d^\rho$ the operator $\mathbf{A}_{(p,\rho)}^{-1}$ is unitarily equivalent to $M_{p,\rho}$ in the space $L_2\Omega$, where

$$M_{p,\rho} = \mathbf{F}_d^\rho \mathbf{A}_{(p)}^{-1/2}(\mathbf{F}_d^\rho \mathbf{A}_{(p)}^{-1/2})^*.$$

For elements u in the set $E_d^\rho\Omega$ which is dense in $L_2\Omega$ the action of the operator $M_{p,\rho}$ is given by the formula

$$M_{p,\rho} u = \mathbf{F}_d^\rho \mathbf{A}_{(p)}^{-1} \mathbf{F}_d^\rho u.$$

Later we will be interested in the asymptotic behavior of the singular numbers of the operator $\mathbf{A}_{(p,\rho)}^{-1} - \widetilde{\mathbf{A}}_{(p,\rho)}^{-1}$, where $\widetilde{\mathbf{A}}_{(p,\rho)}$ corresponds to a smooth matrix $\tilde{a}$ which approximates the matrix a. It suffices to consider the nondegenerate case (that is, $a^{-1} \in L_\infty\Omega$).

Lemma 3.2. *Suppose $a \in L_1\Omega$ and $a^{-1} \in L_\infty\Omega$; let $p \geqq 0$ be an integer such that* (2.12) *is satisfied when $\alpha = \infty$, and let $d = l + 2p$ and $\rho \in [0, d]$. Then for each $\epsilon > 0$ there exists a matrix $\tilde{a}$ such that $\tilde{a}, \tilde{a}^{-1} \in C^\infty\overline{\Omega}$, $\|\tilde{a}^{-1}\|_\infty \leqq 4\|a^{-1}\|_\infty$, and for the corresponding operators $\mathbf{A}_{(p,\rho,d)}$ and $\widetilde{\mathbf{A}}_{(p,\rho,d)}$ we have the bounds*

$$\limsup_{n\to\infty} n^{2(l+2p-\rho)/m} s_n (\mathbf{A}_{(p,\rho)}^{-1} - \widetilde{\mathbf{A}}_{(p,\rho)}^{-1}) \leqslant \varepsilon. \tag{3.6}$$

Proof. We construct the matrix $\tilde{a}$ in accordance with Lemma I.5.6. If we replace $\mathbf{A}_{(p,\rho)}^{-1}$ and $\widetilde{\mathbf{A}}_{(p,\rho)}^{-1}$ by operators which are unitarily equivalent to them in $L_2\Omega$, we arrive at the equivalent problem of bounding the s-numbers of the operators

$$\widetilde{K}_\rho = M_{p,\rho} - \widetilde{M}_{p,\rho} = \mathbf{F}_d^\rho (\mathbf{A}_{(p)}^{-1} - \widetilde{\mathbf{A}}_{(p)}^{-1}) \mathbf{F}_d^\rho .$$

In order to bound $s_n(\widetilde{K}_\rho)$ we use Lemma 1.5 and introduce an interpolation between the cases $\rho = 0$ and $\rho = d$. For $\rho = 0$ the result which we need is given by (2.18). The operator $\widetilde{K}_d$ is bounded in $L_2\Omega$. For $u \in C_0^\infty\Omega$ we have

$$\| \mathbf{F}_d^d u \|^2 = \int_\Omega |\nabla_{l+2p} u|^2 \, dx = \int_\Omega | \nabla_l \Delta^p u |^2 dx \leqslant \| a^{-1} \|_\infty A_{(p)} [u, u] ;$$

$$\| \mathbf{F}_d^d u \|^2 \leqslant \| \tilde{a}^{-1} \|_\infty \widetilde{A}_{(p)} [u, u] \leqslant 4 \| a^{-1} \|_\infty \widetilde{A}_{(p)} [u, u] .$$

Therefore $\|M_{p,d}\| = \|\mathbf{F}_d^d \mathbf{A}_{(p)}^{-1/2}\|^2 \leqq \|a^{-1}\|_\infty$, $\|\widetilde{M}_{p,d}\| \leqq 4\|a^{-1}\|_\infty$ and $\|\widetilde{K}_d\| \leqq 5\|a^{-1}\|_\infty$.

Now we use Lemma 1.5, setting $J_1 = J_2 = \mathbf{F}_d^d$, $T = \widetilde{K}_0$ and $\kappa = \rho/d$. We obtain

$$\limsup_{n\to\infty} n^{2(d-\rho)/m} s_n (\widetilde{K}_\rho) \leqslant c \, \| a^{-1} \|_\infty^\varkappa \varepsilon^{1-\varkappa}.$$

If we adjust the notation for the constants we obtain (3.6).

3. If the matrix a is degenerate, the above method for obtaining bounds does not work since $\widetilde{K}_{l+2p}$ is not bounded in $L_2\Omega$. We can use a method based on Lemma 1.5 if for one of the limits of the interpolation we choose a suitable number $\rho_0 < l + 2p$. We cannot always choose ρ_0 to be an integer. In connection

with this, when we consider bounds of the type (0.7) we have to deal with spectra of ratios with numerators which are forms of fractional order. In obtaining the necessary bounds we encounter difficulties connected with the nonlocal character of the operator $\mathbf{F}_d$. These difficulties can be overcome by means of a technical lemma borrowed from [8].

In §4.1 of [1] we introduced piecewise polynomial projection operators $P_{\Sigma,l}$. We agree that for nonintegers ρ we will write $P_{\Sigma,\rho}$ instead of $P_{\Sigma,[\rho]+1}$.

LEMMA 3.3. *Suppose 2ρ is not an integer and Σ is an arbitrary partition of a cube Q^m into a finite number of cubes Δ. Then for each function $u \in H^\rho Q^m$ such that $P_{\Sigma,\rho}u = 0$ we have*

$$G_{\rho,Q^m}[u,u] \leqslant C_7 \sum_{\Delta\in\Sigma} G_{\rho,\Delta}[u,u]. \tag{3.7}$$

The constant $C_7 = C_7(m,\rho)$ does not depend on the partition Σ.

The proof of Lemma 3.3 is given in [8].

We will make use of the fact that Lemma I.4.6 is true for nonintegers $\bar{l}$.

LEMMA 3.4. *Suppose that Condition I is satisfied;*[5] *let $0 < \rho < l$, $\rho - [\rho] \neq \frac{1}{2}$ and $\alpha^{-1} < 2(l-\rho)m^{-1}$. Then the form $G_{\rho,\Omega}$ is completely continuous in the space $\overset{\circ}{H}(a)$, and for the eigenvalues $\lambda_n(A, G_\rho, \Omega; \mathscr{D})$ of the ratio*

$$G_\rho[u,u]/A[u,u], \qquad u \in \overset{\circ}{H}(a),$$

we have the bound

$$\lambda_n(A, G_\rho, \Omega; \mathscr{D}) \leqq C_8 n^{-2(l-\rho)/m} \|a^{-1}\|_\alpha. \tag{3.8}$$

The constant C_8 depends on m, k, l, ρ, α and the diameter of Ω.

PROOF. For integers ρ the bound (3.8) is a particular case of Theorem 1, and thus we only have to consider nonintegers.

a) Suppose $\Omega = Q^m$. By repeating the proof of Lemma I.4.1 (see §4.2 in [1]) we first find a partition Ξ of Q^m such that $|\Xi| \leqq n$ and

$$\sum_{\Delta\in\Xi} G_{\rho,\Delta}[v,v] \leqslant c\,A[u,u]\, \|a^{-1}\|_\alpha\, n^{-2(l-\rho)/m},$$
$$v = u - P_{\Xi,\rho}u.$$

If $P_{\Xi,\rho}u = 0$, then from (3.7) we obtain

$$G_{\rho,Q^m}[u,u] \leqslant C_7\cdot cA[u,u]\, \|a^{-1}\|_\alpha\, n^{-2(l-\rho)/m}.$$

From the last inequality and the minimax principle (see §4.3 in [1]) we immediately obtain (3.8).

b) Suppose Ω is an arbitrary bounded open set. Without loss of generality we can assume that $\Omega \subset Q^m$. We let a_η denote the matrix in Q^m which equals $a(x)$ for $x \in \Omega$ and equals $\eta\cdot\mathbf{1}$, $\eta > 0$, for $x \in Q^m\setminus\Omega$; we let A_η denote the

---

[5] In contrast to the other assertions in this section, here we do not assume that $\partial\Omega \in C^\infty$.

corresponding quadratic form in Q^m. From the inequality

$$\frac{G_{\rho,\Omega}[u,u]}{A[u,u]} \leqslant \frac{G_{\rho,Q^m}[u,u]}{A_\eta[u,u]}, \qquad u \in C_0^\infty \Omega,$$

and the bound (3.8) which has already been proved for the case of the cube we find that

$$\lambda_n(A, G_\rho, \Omega; \mathscr{D}) \leqslant \lambda_n(A_\eta, G_\rho, Q^m; \mathscr{D}) \leqslant C_8 n^{-2(l-\rho)/m} \| a_\eta^{-1} \|_\alpha.$$

If we take the limit as $\eta \to \infty$ we obtain the desired inequality. The lemma is proved.

The following result is analogous to part a) of Lemma 2.1.

LEMMA 3.5. *Suppose that Condition* I *and* (3.5) *are satisfied, and let* $\rho > 0$, $\rho \geqq 2p$ *and* $\rho - [\rho] \neq \frac{1}{2}$. *Then the form* G_ρ *is completely continuous in* $\mathring{H}(a,p)$, *and for the eigenvalues* $\lambda_n(A_{(p)}, G_\rho, \Omega; \mathscr{D})$ *of the ratio*

$$(3.9) \qquad G_{\rho,\Omega}[u,u]/A_{(p)}[u,u], \qquad u \in \mathring{H}(a,p),$$

we have

$$(3.10) \qquad \lambda_n(A_{(p)}, G_\rho, \Omega; \mathscr{D}) \leqq C_9 n^{-2(l+2p-\rho)/m} \| a^{-1} \|_\alpha,$$

where $C_9 = C_9(\Omega, k, l, p, \rho, \alpha)$.

PROOF. Well-known (see [7]) bounds for solutions of elliptic equations in the spaces H^ρ enable us to write the inequality

$$G_{\rho,\Omega}[u,u] \leqslant c\left[G_{\rho-2p}[\Delta^p u, \Delta^p u] + \int_\Omega |u|^2\,dx\right] \quad (u \in C_0^\infty \Omega).$$

The spectrum of the ratio

$$G_{\rho-2p}[\Delta^p u, \Delta^p u]/A_{(p)}[u,u], \qquad u \in \mathring{H}(a,p),$$

can be bounded by means of arguments used in the proof of part a) of Lemma 2.1. The spectrum of the ratio $\int_\Omega |u|^2 dx / A_{(p)}[u,u]$ was bounded in part b) of Lemma 2.1. As a result of these bounds we obtain (3.10).

4. By virtue of Lemmas 3.4 and 3.5 we can bound the spectrum of the operators $\mathbf{A}_{(p,\rho)}^{-1}$.

LEMMA 3.6. *Suppose that Condition* I *is satisfied; let* $\rho > 0$, *and let* $p \geqq 0$ *be an integer such that* (3.5) *is satisfied. Then for* $d = l + 2p$ *we have*

$$(3.11) \qquad \lambda_n(\mathbf{A}_{(p,\rho,d)}^{-1}) \leqslant C_{10} n^{-2(d-\rho)/m} \| a^{-1} \|_\alpha \qquad (d = l + 2p),$$

where $C_{10} = C_{10}(\Omega, k, l, p, \rho, \alpha)$.

PROOF. The numbers $\lambda_n(\mathbf{A}_{(p,\rho,d)}^{-1})$ are the successive maxima of the ratio

$$(3.12) \qquad \| \mathbf{F}_d^\rho u \|^2 / A_{(p)}[u,u], \qquad u \in \mathring{H}(a,p).$$

If $\rho \geqq 2p$ and $\rho - [\rho] \neq \frac{1}{2}$, then according to (3.2) the numerator in (3.12) is bounded by $G_\rho[u,u] + \|u\|^2$. Then (3.10) and (2.7) imply (3.11).

If $\rho > 2p$ and $\rho - [\rho] = \frac{1}{2}$, we choose $\rho_0 > \rho$ such that $2(d-\rho_0)m^{-1} > \alpha^{-1}$ and $\rho_0 - [\rho_0] \neq \frac{1}{2}$. If we switch to unitarily equivalent operators in $L_2\Omega$, we obtain

$$M_{p,\rho} = \mathbf{F}_d^{-(\rho_0-\rho)} M_{p,\rho_0} \mathbf{F}_d^{-(\rho_0-\rho)}.$$

From (I.1.10) we have

$$\lambda_{3n-2}(M_{p,\rho}) \leqslant \lambda_n^2(\mathbf{F}_d^{-(\rho_0-\rho)})\,\lambda_n(M_{p,\rho_0}),$$

and if we apply (3.3) and (3.11) we obtain the desired result.

Finally, suppose that $\rho < 2p$. We write (3.12) in the form

$$\frac{\|\mathbf{F}_d^{\rho}u\|^2}{\|\Delta^p u\|^2}\cdot\frac{\|v\|^2}{A[v,v]}, \qquad v = \Delta^p u, \qquad u \in \mathring{H}(a,p).$$

In the first factor the denominator can be transformed by integration by parts into $\int_\Omega |\nabla_{2p}u|^2 dx = \|\mathbf{F}_{2p}^{2p}u\|^2$; in the numerator we can replace $\|\mathbf{F}_d^{\rho}u\|$ by $\|\mathbf{F}_{2p}^{\rho}u\|$, which amounts to a change to an equivalent norm. Now elementary variational considerations (as in Lemma I.1.1) lead to the bound

$$\lambda_{2n-1}(\mathbf{A}_{(p,\rho)}^{-1}) \leqslant c\lambda_n^2(\mathbf{F}_{2p}^{-(2p-\rho)})\cdot\lambda_n(\mathbf{A}^{-1}).$$

To complete the proof we apply (3.3) and (0.7).

REMARK 3.7. If $\partial\Omega \in C^\infty$, the assumption that $\rho - [\rho] \neq \frac{1}{2}$ in Lemma 3.4 is unnecessary. If suffices to compare (3.11) and (3.1). It is clear that we can also eliminate the assumptions $\rho \geqq 2p$ and $\rho - [\rho] \neq \frac{1}{2}$ in Lemma 3.5.

Now we are in a position to prove an analogue of Lemma 3.2 for degenerate matrices a. The main difference in the formulation consists in the fact that here we will need a "truncation" matrix $\hat{a}$ instead of a "smoothing" matrix $\tilde{a}$.

LEMMA 3.8. *Suppose that Condition* I *is satisfied and* $a \in L_1\Omega$. *Let* $\rho \geqq 0$, *and let* $p \geqq 0$ *be an integer such that* (2.12) *and* (3.5) *are satisfied. Let* $d = l + 2p$. *Then for each* $\epsilon > 0$ *there exists a matrix* $\hat{a}$ *such that* $\hat{a}(x) \geqq a(x)$ *for almost all* $x \in \Omega$, *and* $\hat{a}^{-1} \in L_\infty$, *and for the corresponding operators* $\mathbf{A}_{(p,\rho,d)}$ *and* $\hat{\mathbf{A}}_{(p,\rho,d)}$ *we have*

$$\limsup_{n\to\infty} n^{2(l+2p-\rho)/m} s_n(\mathbf{A}_{(p,\rho)}^{-1} - \hat{\mathbf{A}}_{(p,\rho)}^{-1}) \leqslant \varepsilon. \tag{3.13}$$

PROOF. We construct the matrix $\hat{a}$ in accordance with Remark I.5.7. As in the proof of Lemma 3.2 we change to the equivalent problem of bounding the s-numbers of the operator

$$\hat{K}_\rho = M_{p,\rho} - \hat{M}_{p,\rho} = \mathbf{F}_d^{\rho}[\mathbf{A}_{(p)}^{-1} - \hat{\mathbf{A}}_{(p)}^{-1}]\mathbf{F}_d^{\rho}.$$

Suppose $\rho_0 > \rho$ and $2(d-\rho_0)m^{-1} > \alpha^{-1}$. By Lemma 3.6 the operator $\hat{K}_{\rho_0}$ is completely continuous in $L_2\Omega$, and for its s-numbers we have the bound

$$s_{2n-1}(\hat{K}_{\rho_0}) \leqslant \lambda_n(M_{p,\rho_0}) + \lambda_n(\hat{M}_{p,\rho_0}) \leqslant 2C_{10}\,\|a^{-1}\|_\alpha n^{-2(d-\rho_0)/m}.$$

Here we used the fact that $\|\hat{a}^{-1}\|_\alpha \leqq \|a^{-1}\|_\alpha$. An asymptotic bound for the singular numbers of the operator $\hat{K}_0 = \mathbf{A}_{(p)}^{-1} - \hat{\mathbf{A}}_{(p)}^{-1}$ is given by (2.19). If we apply Lemma 1.5 with $J_1 = J_2 = \mathbf{F}_d^{\rho_0}$, $T = \hat{K}_0$ and $\kappa = \rho/\rho_0$, we obtain (3.13). The lemma is proved.

5. In this subsection we will consider operators in $E_d^\rho\Omega$ generated by forms $B_{(p)}$ of the type (2.3).

LEMMA 3.9. *If $b \in C_0^\infty\Omega$ and $d \geqq r + 2p$, the form $B_{(p)}$ is bounded in $E_d^\rho\Omega$.*

PROOF. If r is an integer the assertion is clear. We consider the case of a half-integer r.

We let Π denote an operator which extends functions in Ω onto the whole space $\mathbf{R}^m$ and which is bounded in the classes H^ρ for all $\rho \in [0, d]$. We extend the matrix b outside Ω by letting it be zero there, and we denote the corresponding form in $\mathbf{R}^m$ by $\widetilde{B}_{(p)}$. If we recall that functions in $C_0^\infty\mathbf{R}^m$ are multiplicators in $H^\rho\mathbf{R}^m$, then by means of Fourier transforms we can easily obtain a bound for $u \in C_0^\infty\Omega$:

$$|B_{(p)}[u, u]| = |\widetilde{B}_{(p)}[\Pi u, \Pi u]| \leqslant cN^2 (\Pi u \,|\, H^{r+2p}\mathbf{R}^m)$$
$$\leqslant c_1 N^2 (u \,|\, H^{r+2p}\Omega) \leqslant c_2 \, \| \mathbf{F}_d^{r+2p} u \|^2.$$

The last inequality is justified by (3.1). The lemma is proved.

(We note that we cannot extend functions u outside Ω by letting them equal zero, since for half-integers r such an operator is unbounded in $\overset{\circ}{H}{}^r$; see [7].)

From Lemma 3.9 it follows that the form $B_{(p)}$ generates an unbounded operator in $E_d^\rho\Omega$ for arbitrary $\rho \geqq r + 2p$ and $d \geqq \rho$. We denote this operator by $\mathbf{B}_{(p,\rho)} = \mathbf{B}_{(p,\rho,d)}$. For $\rho > r + 2p$ the operator $\mathbf{B}_{(p,\rho)}$ is completely continuous.

LEMMA 3.10. *Suppose $b \in C_0^\infty\Omega$, and let $\rho > r + 2p$ and $d \geqq \rho$. Then*

$$s_n(\mathbf{B}_{(p,\rho)}) = O(n^{-2(\rho - r - 2p)/m}). \tag{3.14}$$

PROOF. From Lemma 3.9 we have

$$\frac{|B_{(p)}[u, u]|}{\|\mathbf{F}_d^\rho u\|^2} \leqslant c \frac{\|\mathbf{F}_d^{r+2p} u\|^2}{\|\mathbf{F}_d^\rho u\|^2},$$

and if we let $\mathbf{F}_d^\rho u = v$, we get (3.3).

§4. Proof of Theorem 2

The proof of Theorem 2 will be carried out in several steps. In the arguments in subsections 1–3 we will assume that the measure μ is absolutely continuous. $b(x) = d\mu/dx \in C_0^\infty\Omega$, $a \in L_1\Omega$ and $\partial\Omega \in C^\infty$. These assumptions will be removed in subsection 4.

1. We begin with the case of a nondegenerate matrix a; that is, we assume that $a^{-1} \in L_\infty\Omega$. It is clear that under suitable assumptions the hypothesis of Theorem 2 is satisfied. We choose $p \geqq 0$ and $\rho > 0$ such that for given l and r we have

$$2(l + 2p) > m, \tag{4.1}$$

(4.2) $$r+2p<\rho<l+2p,\quad 2(l+2p-\rho)<m.$$

With the idea of using Lemma 3.2 we compare the spectra of the ratios

(4.3) $$B_{(p)}[u,\ u]/A_{(p)}[u,\ u],\quad u\in \overset{\circ}{H}(u,\ p),$$

(4.4) $$B_{(p)}[u,\ u]/\widetilde{A}_{(p)}[u,\ u],\quad u\in \overset{\circ}{H}{}^{l+2p}\,\Omega.$$

Here $\widetilde{A}_{(p)}$ is the form constructed from the "smooth" matrix $\widetilde{a}$ (see Lemma 3.2). We note that the asymptotics of the spectrum of (4.4) were found in Lemma 2.2.

We set $d=l+2p$ and introduce the operators $\mathbf{A}_{(p,\rho)}$, $\widetilde{\mathbf{A}}_{(p,\rho)}$ and $\mathbf{B}_{(p,\rho)}$ which are generated by the forms $A_{(p)}$, $\widetilde{A}_{(p)}$ and $B_{(p)}$ in the space $E^{\rho}_{d}\Omega$. It is easy to see that the spectrum of (4.3) coincides with the spectrum of the following operator which is selfadjoint in $E^{\rho}_{d}\Omega$:

(4.5) $$T_{p,\rho}=\mathbf{A}^{-1/2}_{(p,\rho)}\mathbf{B}_{(p,\rho)}\mathbf{A}^{-1/2}_{(p,\rho)}.$$

Analogously the spectrum of the following operator acting in $E^{\rho}_{d}\Omega$

$$\widetilde{T}_{p,\rho}=\widetilde{\mathbf{A}}^{-1/2}_{(p,\rho)}\mathbf{B}_{(p,\rho)}\widetilde{\mathbf{A}}^{-1/2}_{(p,\rho)}$$

coincides with the spectrum of (4.4). The singular numbers of the operator $T_{p,\rho}-\widetilde{T}_{p,\rho}$ can be bounded by means of inequalities (I.1.9) and (I.1.10):

$$\begin{aligned}
&s_{6n-5}(T_{p,\rho}-\widetilde{T}_{p,\rho})\\
&\leqslant s_{3n-2}[(\mathbf{A}^{-1/2}_{(p,\rho)}-\widetilde{\mathbf{A}}^{-1/2}_{(p,\rho)})\mathbf{B}_{(p,\rho)}\mathbf{A}^{-1/2}_{(p,\rho)}]\\
&\quad+s_{3n-2}[\widetilde{\mathbf{A}}^{-1/2}_{(p,\rho)}\mathbf{B}_{(p,\rho)}(\mathbf{A}^{-1/2}_{(p,\rho)}-\widetilde{\mathbf{A}}^{-1/2}_{(p,\rho)})]\\
&\leqslant s_n(\mathbf{A}^{-1/2}_{(p,\rho)}-\widetilde{\mathbf{A}}^{-1/2}_{(p,\rho)})\,s_n(\mathbf{B}_{(p,\rho)})\\
&\quad\times[s_n(\mathbf{A}^{-1/2}_{(p,\rho)})+s_n(\widetilde{\mathbf{A}}^{-1/2}_{(p,\rho)})].
\end{aligned}$$

If we bound the singular numbers of $\mathbf{A}^{-1/2}_{(p,\rho)}$ and $\widetilde{\mathbf{A}}^{-1/2}_{(p,\rho)}$ by means of (3.11), and of $\mathbf{B}_{(p,\rho)}$ by means of (3.14), we find that

(4.6) $$s_{6n-5}(T_{p,\rho}-\widetilde{T}_{p,\rho})\leqslant cn^{-(l+\rho-2p-2r)/m}\,s_n(\mathbf{A}^{-1/2}_{(p,\rho)}-\widetilde{\mathbf{A}}^{-1/2}_{(p,\rho)}).$$

In order to bound the last factor in (4.6) we compare Lemmas 3.2 and 1.2. The hypothesis of Lemma 3.2 is satisfied since (2.12) for $\alpha=\infty$ coincides with (4.1). Lemma 1.2 is applicable by virtue of (4.2). Thus from (3.6) we obtain

$$\limsup_{n\to\infty} n^{(l+2p-\rho)/m}\,s_n(\mathbf{A}^{-1/2}_{(p,\rho)}-\widetilde{\mathbf{A}}^{-1/2}_{(p,\rho)})\leqslant c\sqrt{\varepsilon}.$$

Together with (4.6) this leads to a relation of the form

(4.7) $$\limsup_{n\to\infty} n^{2(l-r)/m}\,s_n(T_{p,\rho}-\widetilde{T}_{p,\rho})=O\left(\sqrt{\varepsilon}\right).$$

Since $\epsilon>0$ in (3.6) can be chosen arbitrarily small for the appropriate choice of $\widetilde{a}=\widetilde{a}_{\epsilon}$, we can use (4.7) when applying Lemma I.1.5.

Since we know the asymptotics of the spectrum of the operator $\widetilde{T}_{p,\rho}$, we just have to establish our right to take the limit as $\epsilon \to 0$ in the asymptotic coefficient $\omega_{\pm}(\tilde{a}_{\epsilon}, b; \Omega)$. We let $\omega_{\pm}(x|a, b)$ denote the inner integral in (0.10):

$$\omega_{\pm}(a,\ b;\ \Omega) = \int_{\Omega} w_{\pm}(x \,|\, a,\ b)\, dx.$$

By virtue of (I.2.5) we have

$$|\mathbf{a}^{-1}(x,\xi)| \leqq |a^{-1}(x)| \qquad (|\xi| = 1),$$

and consequently

$$w_{\pm}(x|a,b) \leqq c|a^{-1}(x)|^{\theta}. \tag{4.8}$$

Then by (I.5.14)

$$w_{\pm}(x|\tilde{a}_{\epsilon}, b) \leqq c\|\tilde{a}_{\epsilon}^{-1}\|_{\infty} \leqq c(4\|a^{-1}\|_{\infty})^{\theta}. \tag{4.9}$$

Furthermore,

$$|\tilde{a}_{\varepsilon}^{-1} - a^{-1}| \leqslant |\tilde{a}_{\varepsilon}^{-1/2}|\, |a^{-1/2}|\, |\tilde{a}_{\varepsilon}^{1/2}(\tilde{a}_{\varepsilon}^{-1} - a^{-1})\, a^{1/2}|$$
$$\leqslant 2\|a^{-1}\|_{\infty} |\tilde{a}_{\varepsilon}^{1/2}(\tilde{a}_{\varepsilon}^{-1} - a^{-1})\, a^{1/2}|,$$

and according to (I.5.13) (with $\bar{\kappa} = 1$) we have

$$\|\tilde{a}_{\epsilon}^{-1} - a^{-1}\|_2 \leqq 2\epsilon\|a^{-1}\|_{\infty}.$$

Thus it follows that for some sequence $\epsilon_n \to 0$ for almost all $x \in \Omega$ we have

$$\tilde{a}_{\epsilon_n}^{-1}(x) \to a^{-1}(x).$$

It is then easy to see that $w_{\pm}(x|\tilde{a}_{\epsilon_n}, b) \to w_{\pm}(x|a, b)$ as $\epsilon_n \to 0$ almost everywhere in Ω. Together with (4.9) this enables us to take the limit under the integral sign:

$$\omega_{\pm}(\tilde{a}_{\epsilon_n}, b; \Omega) \to \omega_{\pm}(a, b; \Omega).$$

Thus we have the following result.

LEMMA 4.1. *Suppose* $\partial\Omega \in C_0^{\infty}$, $b \in C_0^{\infty}\Omega$, $a \in L_1\Omega$ *and* $a^{-1} \in L_{\infty}\Omega$. *If* $p \geqq 0$ *is an integer which satisfies* (4.1), *then the asymptotics of the spectrum of the ratio* (4.3) *are given by* (0.8) – (0.10).

2. We now consider the "degenerate" case; that is, we let $1 \leqq \alpha < \infty$. We choose $p \geqq 0$ such that (2.12) is satisfied. We compare the asymptotic behavior of the spectra of (4.3) and

$$B_{(p)}[u,u]/\hat{A}_{(p)}[u,u], \qquad u \in \mathring{H}(\hat{a}, p), \tag{4.10}$$

where $\hat{a}$ is a nondegenerate matrix and $\hat{A}_{(p)}$ is the corresponding form constructed in Lemma 3.8; thus we have (3.13). Since $\hat{a} \in L_{\infty}\Omega$, by Lemma 4.1 we know that the asymptotics of the spectrum of (4.10) are given by (0.8) – (0.10).

We set $\rho = r + 2p$ and $d = l + 2p$. From (0.5) it follows that (3.5) is satisfied.

And in the space $E_d^\rho\Omega$ the operators $\mathbf{A}_{(p,\rho)}$, $\hat{\mathbf{A}}_{(p,\rho)}$ and $\mathbf{B}_{(p,\rho)}$ are defined and $\mathbf{B}_{(p,\rho)}$ is bounded by virtue of Lemma 3.9. We consider the operator (4.5) in $E_d^\rho\Omega$ and also the operator

$$\hat{T}_{p,\rho} = \hat{\mathbf{A}}_{(p,\rho)}^{-1/2}\mathbf{B}_{(p,\rho)}\hat{\mathbf{A}}_{(p,\rho)}^{-1/2};$$

the spectrum of this operator coincides with the spectrum of (4.10). Along with the operator $\hat{T}_{p,\rho}$ we also consider a unitarily equivalent operator which acts in the same space $E_d^\rho\Omega$.

First of all we set $Q = Q_{p,\rho} = \mathbf{A}_{(p,\rho)}^{1/2}\hat{\mathbf{A}}_{(p,\rho)}^{-1/2}$ and $\Gamma = |Q^+|^2 = QQ^+$; the symbol "+" denotes the operation of conjugation for operators in $E_d^\rho\Omega$. Since

$$0 \leq A_p[u,u] \leq \hat{A}_p[u,u], \qquad u \in \mathring{H}(\hat{a},p),$$

we have $0 \leq \Gamma \leq I$. The ranges of the operators Q and Q^+ are dense in $E_d^\rho\Omega$. This follows from the fact that $\mathring{H}(\hat{a},p)$ is dense in $\mathring{H}(a,p)$ and the fact that $\mathbf{A}_{(p,\rho)}$ and $\hat{\mathbf{A}}_{(p,\rho)}$ are positive definite in $E_d^\rho\Omega$. From the properties of the operator modulus it follows that there exists a unitary operator U in $E_d^\rho\Omega$ such that $Q = \sqrt{\Gamma}\,U^+$. Then from the definition of Q it follows that

$$\hat{\mathbf{A}}_{(p,\rho)}^{-1/2}U = \mathbf{A}_{(p,\rho)}^{-1/2}\Gamma^{1/2}, \quad U^+\hat{\mathbf{A}}_{(p,\rho)}^{-1/2} = \Gamma^{1/2}\mathbf{A}_{(p,\rho)}^{-1/2}.$$

From these relations we know that the following operator which acts in $E_d^\rho\Omega$

$$S_{p,\rho} = \sqrt{\Gamma}\,T_{p,\rho}\sqrt{\Gamma}$$

is unitarily equivalent to $\hat{T}_{p,\rho}$:

$$S_{p,\rho} = U^+\hat{T}_{p,\rho}U.$$

We will compare $T_{p,\rho}$ not with $\hat{T}_{p,\rho}$ but with $S_{p,\rho}$, whose spectrum coincides with the spectrum of $\hat{T}_{p,\rho}$ and therefore with the spectrum of (4.10).

Further calculations can be carried out by analogy with step 1. We have

$$\begin{aligned} s_{4n-3}(T_{p,\rho} - S_{p,\rho}) &\leqslant s_{2n-1}[(I - \sqrt{\Gamma})\mathbf{A}_{(p,\rho)}^{-1/2}\mathbf{B}_{(p,\rho)}\mathbf{A}_{(p,\rho)}^{-1/2}] \\ &+ s_{2n-1}\left[\sqrt{\Gamma}\,\mathbf{A}_{(p,\rho)}^{-1/2}\mathbf{B}_{(p,\rho)}\mathbf{A}_{(p,\rho)}^{-1/2}\left(I - \sqrt{\Gamma}\right)\right] \\ &\leqslant 2s_n\left[\left(I - \sqrt{\Gamma}\right)\mathbf{A}_{(p,\rho)}^{-1/2}\right]s_n(\mathbf{A}_{(p,\rho)}^{-1/2})\,\|\mathbf{B}_{(p,\rho)}\|. \end{aligned}$$

The operator $\mathbf{B}_{(p,\rho)}$ is bounded by virtue of Lemma 3.9. A bound for $s_n(\mathbf{A}_{(p,\rho)}^{-1/2})$ can be obtained from (3.11), and it remains to find a bound for $s_n[(I - \sqrt{\Gamma})\mathbf{A}_{(p,\rho)}^{-1/2}]$. We use Lemma 1.3, according to which

$$\begin{aligned} s_n^2\left[\left(I - \sqrt{\Gamma}\right)\mathbf{A}_{(p,\rho)}^{-1/2}\right] &\leqslant s_n^2\left[\sqrt{I - \Gamma}\,\mathbf{A}_{(p,\rho)}^{-1/2}\right] \\ &= \lambda_n[\mathbf{A}_{(p,\rho)}^{-1/2}(I - \Gamma)\mathbf{A}_{(p,\rho)}^{-1/2}] = \lambda_n(\mathbf{A}_{(p,\rho)}^{-1} - \hat{\mathbf{A}}_{(p,\rho)}^{-1}). \end{aligned}$$

If we recall the bound (3.13) and combine the above bounds, we obtain

$$\limsup_{n\to\infty} n^{-2(l-r)/m}s_n(T_{p,\rho} - S_{p,\rho}) = O\left(\sqrt{\varepsilon}\right). \tag{4.11}$$

For the appropriate choice of the matrix $\hat{a} = \hat{a}_\epsilon$ the upper limit in (4.11) can be made arbitrarily small; this enables us to apply Lemma I.1.5 for calculating the spectral asymptotics of $T_{p,\rho}$.

The asymptotics of the spectrum of $S_{p,\rho}$, as we already mentioned, are determined by (0.8) – (0.10), and we only have to justify taking the limit as $\epsilon \to 0$ in the asymptotic coefficient $\omega_\pm(\hat{a}_\epsilon, b; \Omega)$.

From (4.8) and the inequality $\hat{a}_\epsilon(x) \geqq a(x)$ we find that

$$w_\pm(x|\hat{a}_\epsilon, b) \leqq c|\hat{a}_\epsilon^{-1}(x)|^\theta \leqq c|a^{-1}(x)|^\theta.$$

By (0.5) we have $\theta < \alpha$, and thus $|a^{-1}x|^\theta \in L_1\Omega$. The convergence $w_\pm(x|\hat{a}_\epsilon, b) \to w_\pm(x|a, b)$ almost everywhere in Ω is clear. By Lebesgue's theorem we know that as $\epsilon \to 0$

$$\omega_\pm(\hat{a}_\epsilon, b; \Omega) \to \omega_\pm(a, b; \Omega).$$

We have proved the following assertion.

LEMMA 4.2. *Suppose* $\partial\Omega \in C^\infty$, $b \in C_0^\infty\Omega$, $a \in L_1\Omega$ *and* $a^{-1} \in L_\alpha\Omega$ *for* $\alpha > \theta$. *If* $p \geqq 0$ *is an integer for which* (2.14) *is satisfied, the asymptotics of the spectrum of* (4.3) *are given by* (0.8) – (0.10).

3. We now establish a result which will enable us to conclude the proof of Theorem 2.

LEMMA 4.3. *Suppose the hypothesis of Theorem* 2 *and also the following conditions are satisfied*:

a) $\partial\Omega \in C^\infty$.

b) $a \in L_1\Omega$.

c) $b \in C_0^\infty\Omega$.

d) *The measure* μ *does not have a singular component.*

Then for the spectrum of the problem $\mathscr{D}(A, B; \Omega)$ *we have the asymptotic formula* (0.8) – (0.10).

PROOF. We supplement the conditions in the lemma by the following assumption: *Suppose there exists an open set* Ω_0 *such that* $\overline{\Omega}_0 \subset \Omega$ *and* $a, a^{-1} \in L_\infty(\Omega \backslash \overline{\Omega}_0)$. Then the assertion in the lemma follows from Lemmas 4.2 and 2.7.

First we eliminate the assumption that $a \in L_\infty(\Omega\backslash\overline{\Omega}_0)$, still supposing that $a^{-1} \in L_\infty(\Omega\backslash\overline{\Omega}_0)$. Let $a_N(x)$ be the matrix defined by the formula $a_N = \varphi_N(a)$, where $\varphi_N(\lambda) = \min(\lambda, N)$, $N > 0$. We denote the corresponding quadratic form by A_N. For the problem $\mathscr{D}(A_N, B; \Omega)$ the formula (0.8) – (0.10) is valid, and $A_N[u, u] \leqq A[u, u]$ for $u \in \mathring{H}(a)$. Thus from (2.8) we find that

$$n_\pm(\lambda;\ A_{(p)},\ B_{(p)},\ \Omega;\ \mathscr{D}) \leqslant n_\pm(\lambda;\ A,\ B,\ \Omega;\ \mathscr{D}) \leqslant n_\pm(\lambda;\ A_N,\ B,\ \Omega;\ \mathscr{D}).$$

If we multiply by λ^θ and take the limit as $\lambda \to 0$ and use Lemma 4.2, we obtain

$$\omega_\pm(a,\ b;\ \Omega) \leqslant \delta_\pm(a,\ b,\ \Omega;\ \mathscr{D}) \leqslant \Delta_\pm(a,\ b,\ \Omega;\ \mathscr{D}) \leqslant \omega_\pm(a_N,\ b;\ \Omega).$$

Since $\omega_\pm(a_N, b; \Omega) \to \omega_\pm(a, b; \Omega)$ as $N \to \infty$, formula (0.8)–(0.10) is valid for the problem $\mathscr{D}(A, B; \Omega)$.

Now we will eliminate the assumption $a^{-1} \in L_\infty(\Omega \backslash \overline{\Omega}_0)$. Let K be a sphere containing the set Ω. We extend the matrix a outside Ω by letting it be the identity there, and we extend b by letting it be zero. The extended matrices will be denoted by a^* and b^*, and the corresponding forms in the sphere K will be denoted by A^* and B^*. For the problem $\mathscr{D}(A^*, B^*; K)$ the lemma is true by what we have already shown. From (2.8) and the minimax principle we obtain

$$n_\pm(\lambda;\ A_{(p)},\ B_{(p)};\ \Omega) \leqslant n_\pm(\lambda;\ A,\ B;\ \Omega) \leqslant n_\pm(\lambda;\ A^*,\ B^*;\ K).$$

Then after we multiply by λ^θ and take the limit as $\lambda \to 0$ we find that

$$\omega_\pm(a,\ b;\ \Omega) \leqslant \delta_\pm(a,\ b,\ \Omega;\ \mathscr{D}) \leqslant \Delta_\pm(a,\ b,\ \Omega;\ \mathscr{D}) \leqslant \omega_\pm(a^*,\ b^*;\ K)$$

(the limit on the left side is calculated by means of Lemma 4.2). We now note that $\omega_\pm(a, b; \Omega) = \omega_\pm(a^*, b^*; K)$. The lemma is proved.

4. Now we will complete the proof of Theorem 2. We just have to remove the assumptions a)–d) in the hypothesis of Lemma 4.3.

a) Let $\Omega \subset \mathbf{R}^m$ be an arbitrary bounded open set, and suppose, as in the proof of Lemma 4.3, that K is a sphere containing $\overline{\Omega}$. Suppose that Ω' is an open set such that $\partial\Omega' \in C^\infty$, $\overline{\Omega}' \subset \Omega$ and $b \in C_0^\infty \Omega'$. According to the minimax principle,

$$n_\pm(\lambda;\ A,\ B,\ \Omega';\ \mathscr{D}) \leqslant n_\pm(\lambda;\ A,\ B,\ \Omega;\ \mathscr{D}) \leqslant n_\pm(\lambda;\ A^*,\ B^*,\ K;\ \mathscr{D}).$$

The asymptotic behavior of the first and last terms of this inequality is known from Lemma 4.3. Since

$$\omega_\pm(a,\ b;\ \Omega') = \omega_\pm(a^*,\ b^*;\ K) = \omega_\pm(a,\ b;\ \Omega),$$

formula (0.8)–(0.10) is valid for the problem $\mathscr{D}(A, B; \Omega)$.

b) Suppose $a \in L_{1,\mathrm{loc}}\Omega$. Let a_N be the matrices introduced in the proof of Lemma 4.3, and let Ω' be an open set such that $\overline{\Omega}' \subset \Omega$ and $b \in C_0^\infty \Omega'$. Then

$$n_\pm(A,\ B,\ \Omega';\ \mathscr{D}) \leqslant n_\pm(A,\ B,\ \Omega;\ \mathscr{D}) \leqslant n_\pm(A_N,\ B,\ \Omega;\ \mathscr{D}).$$

Since $a \in L_1\Omega'$ and $a_N \in L_\infty\Omega$, the asymptotic behavior of the first and last terms of the inequality is known, and thus

$$\omega_\pm(a,\ b;\ \Omega') \leqslant \delta_\pm(a,\ b,\ \Omega;\ \mathscr{D}) \leqslant \Delta_\pm(a,\ b,\ \Omega;\ \mathscr{D}) \leqslant \omega_\pm(a_N,\ b;\ \Omega).$$

Clearly $\omega_\pm(a, b; \Omega') = \omega_\pm(a, b; \Omega)$ and $\omega_\pm(a_N, b; \Omega) \to \omega_\pm(a, b; \Omega)$ as $N \to \infty$. Thus formula (0.8)–(0.10) has been established for the case $a \in L_{1,\mathrm{loc}}\Omega$.

c) Without loss of generality we can assume that $\beta_{\tau_1\tau_2} < \infty$ for all admissible pairs τ_1, τ_2. Then the condition $b \in C_0^\infty \Omega$ can be removed by quoting Lemma I.5.2.

d) The assumption that the measure μ be absolutely continuous can be eliminated by means of Lemma I.5.1.

The proof of Theorem 2 is complete.

§5. Some corollaries of the fundamental result

To conclude this work we present some corollaries and generalizations of the above results. We will discuss the possibility of applying our techniques to the Neumann problem, the Dirichlet problem in unbounded domains, and problems containing a small parameter.

1. As in [1], we will consider the Neumann problem ($\mathscr{N}_t$) under more restrictive conditions than those imposed on the problem $\mathscr{D}$. We will assume that the set Ω belongs to the class $\mathscr{K}$ (see §2.4 of [1]), that $a, a^{-1} \in L_\infty \Omega$, and that the denominator of the ratio (0.6) contains a lower term which removes the degeneracy of the seminorm determined by the form A. The problem $\mathscr{N}_t$, $t > 0$, consists in finding the spectrum (successive extrema) of the ratio

$$\frac{B[u, u]}{A[u, u] + t(u, u)}, \qquad t > 0, \quad u \in H^l\Omega, \tag{5.1}$$

on the whole class $H^l\Omega$ (which corresponds to the natural boundary condition).

A refinement of Theorem I.3.5 is given by the following result.

THEOREM 5.1. *Suppose $\Omega \in \mathscr{K}$, $a, a^{-1} \in L_\infty\Omega$ and Condition* II *is satisfied for $\alpha = \infty$. Then for the spectrum of the ratio* (5.1) *the asymptotic formula* (0.8) – (0.10) *is valid.*

The assertion in Theorem 5.1 is a simple corollary of Theorem 2 by virtue of Lemma I.5.4. We note that the possibility of including (for $2(l-r) > m$) a singular component in the measure μ is presented in Lemma I.5.1.

It is clear that if the ratio (5.1) is considered on some subspace of $H^l\Omega$ which contains $\overset{\circ}{H}{}^l\Omega$, Theorem 5.1 remains valid. Finally, the result in Theorem 5.1 applies directly to the problem $\mathscr{N}$ for $t = 0$ if this problem can be well-posed (see [9] for an analysis of the formulation of problem $\mathscr{N}$).

We note that the assumptions on the matrix a can be weakened. The formulation of the problem $\mathscr{N}_t$ and the result in Theorem 5.1 can be extended to include the case where the hypothesis of Theorem 2 is satisfied in $\Omega \in \mathscr{K}$ and there exists an open subset $\Omega_0, \overline{\Omega}_0 \subset \Omega$, such that $a, a^{-1} \in L_\infty(\Omega \setminus \overline{\Omega}_0)$.

2. We consider the question of extending Theorems 1 and 2 to the case of unbounded domains. The first steps in this direction were achieved in [10] and [11] (for a description of these results see §11.3 in [1]). Further important results were obtained in [12]. After refining the techniques in [9] and [13], G. B. Rozenbljum proved the following result. Suppose that under the conditions of Theorem 1 the form B is of the type (0.4) (condition (0.5a) in this case follows from (0.5)) and $2(l-r) < m$. Then the assertion in Theorem 1 remains valid if

(5.2) $$\beta_{\tau_1\tau_2} = \beta, \quad \alpha^{-1} + \beta^{-1} = 2(l - r)m^{-1} < 1,$$

where the constant C_1 in (0.7) does not depend on Ω. This fact enables us to extend the result in Theorem 1 to an arbitrary (unbounded) set if (5.2) is satisfied.[6)]

Rozenbljum's considerations carry over to the general case of a form of the type (0.3). As a result we obtain the following assertion.

THEOREM 5.2. *Let $\Omega \subset \mathbf{R}^m$ be an arbitrary open set and suppose that Condition I is satisfied. Suppose that the form B of the type* (0.3) *contains only terms for which*

$$2m^{-1}(l - |\tau_i|) < \alpha^{-1} + 1, \qquad i = 1, 2,$$

and suppose (5.2) *is satisfied. Then for the spectrum of the problem $\mathscr{D}(A, B; \Omega)$ the bound* (0.7) *is valid with a constant C_1 which does not depend on Ω.*

In [12] Rozenbljum obtained some interesting results concerning the possibility of extending the asymptotics (0.8)–(0.10) to the problem $\mathscr{D}$ in an unbounded domain. These results concern forms of the type (0.4) and are not free from additional restrictions of the type (0.11). And the case $2(l - r) \geqq m$ was not considered (there are results in [10] concerning the case $m = 1$, $r = 0$). Thus the case of unbounded domains has not yet been treated systematically.

We introduce one example, not included in the results of Rozenbljum, in which (0.8)–(0.10) has been established for $\Omega \subseteq \mathbf{R}^m$. This example enables us to demonstrate certain elementary methods, and it is also interesting from the point of view of applications (see Theorem 5.5 below).

We consider the problem of finding the spectrum of the ratio

(5.3) $$\frac{\int_\Omega [u\langle \mathfrak{b}, \nabla u\rangle + \bar{u}\langle \nabla u, \mathfrak{b}\rangle + b_0|u|^2]\,dx}{\int_\Omega |\nabla u|^2\,dx}, \quad u \in \overset{\circ}{H}{}^1\Omega.$$

Here[7)] $\Omega \subseteq \mathbf{R}^m$, $m \geqq 3$; the functions b_0 and u are scalars; $\mathfrak{b} = (b_1, \cdots, b_m)$ is a vector, and $\operatorname{Im} b_0 = 0$, $b_0 \in L_{m/2}\Omega$ and $\mathfrak{b} \in L_m\Omega$. We assume at first that $\Omega = \mathbf{R}^m$. By virtue of Theorem 5.2 the spectrum of the ratio

$$\int b_0|u|^2\,dx \Big/ \int |\nabla u|^2\,dx, \qquad u \in \overset{\circ}{H}{}^1\mathbf{R}^m$$

admits the bound

---

6) Certain difficulties in formulating the problem $\mathscr{D}$ in $\mathbf{R}^m$ arise when $2l > m$. With condition (5.2) these difficulties are easily overcome; we will not treat this topic here.

7) By $\overset{\circ}{H}{}^1\Omega$ we mean the completion of the set $C_0^\infty\Omega$ in the metric of the Dirichlet integral. We recall that in accordance with [1] the notation ∇u differs from the ordinary usage by a factor of $-i$.

(5.4) $$\lambda_n^{\pm}(b_0, \mathbf{R}^m; \mathscr{D}) \leq c\| b_0 \|_{m/2} n^{-2/m}.$$

Theorem 5.2 also gives us the bound

(5.5) $$\lambda_n^{\pm}(\mathfrak{b}, \mathbf{R}^m; \mathscr{D}) \leq c\| \mathfrak{b} \|_m n^{-1/m}$$

for the spectrum of

(5.6) $$2\operatorname{Re}\int u\langle \mathfrak{b}, \nabla u\rangle\, dx \Big/ \int |\nabla u|^2\, dx, \qquad u \in \mathring{H}^1\mathbf{R}^m.$$

By comparing (5.4) and (5.5) we can see that the asymptotics of the spectra of (5.3) and (5.6) coincide (see Lemma I.1.4).

We now calculate the asymptotics of the spectrum of (5.6); we will use the method given in the proofs of Lemmas 2.7 and I.5.4. The bound (5.5) allows us to assume that $\mathfrak{b} \in C_0^\infty \mathbf{R}^m$. Suppose $\zeta \in C_0^\infty \mathbf{R}^m$ and $\zeta \mathfrak{b} = \mathfrak{b}$. Then for each $\epsilon > 0$ we have

$$\int |\nabla u|^2\, dx \geqslant \varepsilon \int |\nabla u|^2\, dx + (1-\varepsilon)\int |\nabla(\zeta u)|^2\, dx$$
$$+ (1-\varepsilon)\int [\zeta |\nabla u|^2 - |\nabla(\zeta u)|^2]\, dx \succ (1-\varepsilon)\int |\nabla(\zeta u)|^2\, dx.$$

Since the numerator of (5.6) does not change if we replace u by ζu, we are justified in making the substitution $\zeta u = v \in \mathring{H}^1 K$, where K is a sphere of sufficiently large radius. By virtue of Remark I.2.4 and Lemma I.1.2 we now have

(5.7) $$\limsup_{\lambda\to 0} \lambda^m n_{\pm}(\lambda;\, \mathfrak{b},\, \mathbf{R}^m;\, \mathscr{D}) \leqslant (1-\varepsilon)^{-m} \lim_{\lambda\to 0} \lambda^m n_{\pm}(\lambda;\, \mathfrak{b},\, K;\, \mathscr{D}).$$

And clearly

(5.8) $$n_{\pm}(\lambda; \mathfrak{b}, K; \mathscr{D}) \leq n_{\pm}(\lambda; \mathfrak{b}, \mathbf{R}^m; \mathscr{D}).$$

From Theorem 2 it follows that the limit on the right side of (5.7) exists and is determined by (0.8) – (0.10). If we compare (5.7) and (5.8) and take the limit as $\epsilon \to 0$ we can see that the asymptotics of the spectrum of (5.6) (and consequently of (5.3)) are given by (0.8) – (0.10).

This result immediately carries over to the case of an arbitrary open set $\Omega \subset \mathbf{R}^m$. It suffices to restrict our attention to finite b_0 and $\mathfrak{b}$. Suppose $\Omega_0 \subset \Omega$ is a bounded open set outside of which $b_0 - |\mathfrak{b}| = 0$. Then clearly

$$n_{\pm}(\lambda;\, \Omega_0;\, \mathscr{D}) \leqslant n_{\pm}(\lambda;\, \Omega;\, \mathscr{D}) \leqslant n_{\pm}(\lambda;\, \mathbf{R}^m;\, \mathscr{D}).$$

The first and last terms of this inequality have the same asymptotic behavior. If we explicitly calculate the asymptotic coefficient (0.10) we arrive at the following result.

THEOREM 5.3. *Let $\Omega \subseteq \mathbf{R}^m$, $m \geq 3$, be an open set; let $b_0 \in L_{m/2}\Omega$ and $\mathfrak{b} \in L_m\Omega$. Then the asymptotics of the spectrum of* (5.3) *are given by the formula*

$$\lim_{\lambda\to 0} \lambda^m n_{\pm}(\lambda;\, \Omega;\, \mathscr{D}) = \frac{(2\sqrt{\pi})^{-m}}{\Gamma(1+m/2)} \int_\Omega |\mathfrak{b}|^m\, dx.$$

3. Theorem 2 enables us to refine the results in §10 of [1] on spectral asymptotics for problems containing a small parameter. For the sake of brevity we will not repeat the statement of the problem and the notation introduced there. We note that Theorem I.10.1 can now be replaced by the following more precise theorem.

THEOREM 5.4. *Let the forms A, B and G be the forms in §10.1 of [1]. Then the following assertions are true:*

1. *If $r_b > r_g$, or $r_b \leqq r_g$ and $\Lambda = 0$, then*

$$\lim_{h\to 0} h^\theta \mathbf{N}(\Lambda;\ hA + B,\ G;\ \mathscr{D}) = \omega_-(a,\ b;\ \Omega), \qquad 2\theta = m(l - r_b)^{-1}.$$

2. *If $r_g > r_b$ and $\Lambda \neq 0$, then*

$$\lim_{h\to 0} h^\theta \mathbf{N}(\Lambda;\ hA + B,\ G;\ \mathscr{D}) = |\Lambda|^\theta \omega_\varepsilon(a,\ g;\ \Omega),$$

$$2\theta = m(l - r_g)^{-1}, \quad \varepsilon = \operatorname{sign}\Lambda.$$

3. *If $r_b = r_g$, then*

$$\lim_{h\to 0} h^\theta \mathbf{N}(\Lambda;\ hA + B,\ G;\ \mathscr{D}) = \omega_+(a,\ \Lambda g - b;\ \Omega).$$

We note that the conditions on A, B and G formulated in §10.1 of [1] essentially reduce to Conditions I and II for each of the pairs A, B and A, G together with the requirement that the form G be strictly positive. In contrast to Theorem I.10.1, in Theorem 5.4 there is no requirement that the conditions in one of the Theorems I.3.2–I.3.4 be satisfied. This requirement drops out since Theorem 2 assumes only that Conditions I and II are satisfied. A similar result is true for the problem $\mathscr{N}_t$; it is a corollary of Theorem 5.1.

Theorem 5.4 is concerned with bounded sets. On the basis of Theorem 5.3 we will consider one example involving the case $\Omega = \mathbf{R}^m$, $m \geqq 3$. Let b_0 and $\mathfrak{b}$ be the same as in Theorem 5.3. In $L_2\mathbf{R}^m$ we consider the problem of finding the spectrum of the quadratic form

$$h\int |\nabla u|^2\,dx - 2\operatorname{Re}\int u\langle\mathfrak{b},\ \nabla u\rangle\,dx - \int b_0|u|^2\,dx \qquad (u \in H^1\mathbf{R}^m). \tag{5.9}$$

For each $h > 0$ this quadratic form is semibounded from below and is closed in $L_2\mathbf{R}^m$. In the case of a differentiable vector $\mathfrak{b}$ the form (5.9) corresponds to the Schrödinger operator

$$-h\Delta u + i\sum_{j=1}^{m}\left[\frac{\partial}{\partial x_j}(b_j u) + \bar{b}_j\frac{\partial u}{\partial x_j}\right] - b_0 u. \tag{5.10}$$

We let $\mathbf{N}(h)$ denote the number of negative eigenvalues of (5.9). If we use arguments like those in §10 of [1] we can show that $\mathbf{N}(h) = n_+(h)$, where $n_+(\lambda)$ is the distribution function of the spectrum of (5.3) (for $\Omega = \mathbf{R}^m$). From Theorem 5.3 we now obtain the following result.

THEOREM 5.5. *Let* $\operatorname{Im} b_0 = 0$, $b_0 \in L_{m/2}\mathbf{R}^m$, $\mathfrak{b} \in L_m\mathbf{R}^m$ *and* $m \geqq 3$. *Then the number* $\mathbf{N}(h)$ *of negative eigenvalues of* (5.9) (*of* (5.10)) *is finite, and as* $h \to 0$ *we have the asymptotic relation*

$$\lim_{h \to 0} h^m \mathbf{N}(h) = \frac{(2\sqrt{\pi})^{-m}}{\Gamma(1+m/2)} \int |\mathfrak{b}|^m \, dx.$$

An analogous result is true for a form of the type (5.9) under the Dirichlet hypothesis in $L_2\Omega$, $\Omega \subset \mathbf{R}^m$, $m \geqq 3$.

Received 15/JUNE/71

Bibliography

[1] M. Š. Birman and M. Z. Solomjak, *Spectral asymptotics of nonsmooth elliptic operators*. I, Trudy Moskov. Mat. Obšč. **27** (1972), 3-52 = Trans. Moscow Math. Soc. **27** (1972), 1-52.

[2] ———, *Spectral asymptotics of non-smooth elliptic operators*, Dokl. Akad. Nauk SSSR **205** (1972), 267-270 = Soviet Math. Dokl. **13** (1972), 906-910. MR **46** #4278.

[3] M. A. Krasnosel'skiĭ, *The operator of translation along the trajectories of differential equations*, "Nauka", Moscow, 1966; English transl., Transl. Math. Monographs, vol. 19, Amer. Math. Soc., Providence, R. I., 1968. MR **34** #3012; **36** #6688.

[4] F. R. Gantmaher and M. G. Kreĭn, *Oscillation matrices and kernels and small oscillations of mechanical systems*, 2nd ed., GITTL, Moscow, 1950; German transl., Akademie-Verlag, Berlin, 1960. MR **14**, 178; **22** #5161.

[5] H. Weyl, *Inequalities between the two kinds of eigenvalues of a linear transformation*, Proc. Nat. Acad. Sci. U.S.A. **35** (1949), 408-411. MR **11**, 37.

[6] I. C. Gohberg and M. G. Kreĭn, *Introduction to the theory of linear nonselfadjoint operators in Hilbert space*, "Nauka", Moscow, 1965; English transl., Transl. Math. Monographs, vol. 18, Amer. Math. Soc., Providence, R. I., 1969. MR **36** #3137; **39** #7447.

[7] J. L. Lions and E. Magenes, *Problèmes aux limites non homogènes et applications*. Vol. 1, Travaux et Recherches Mathématiques, no. 17, Dunod, Paris, 1968; Russian transl., "Nauka", Moscow, 1970. MR **40** #512.

[8] M. Š. Birman and M. Z. Solomjak, *On equivalent renormalizations of the classes* W_p^r *under partitionings of the domain*, Izv. Vysš. Učebn. Zaved. Matematika **1973**, no. 3 (130), 19-27. (Russian)

[9] ———, *The principal term of the spectral asymptotics for "nonsmooth" elliptic problems*, Funkcional. Anal. i Priložen. **4** (1970), no. 4, 1-13 = Functional Anal. Appl. **4** (1970), 265-275. MR **43** #3857.

[10] M. Š. Birman and V. V. Borzov, *On the asymptotics of the discrete spectrum of certain singular differential operators*, Problemy Mat. Fiz., vyp. 5, Izdat. Leningrad. Gos. Univ., Leningrad, 1971, pp. 24-38 = Topics in Math. Phys., no. 5, Plenum Press, New York, 1972, pp. 19-30. MR **46** #726.

[11] G. V. Rozenbljum, *On the distribution of eigenvalues of the first boundary value problem in unbounded regions*, Dokl. Akad. Nauk SSSR **200** (1971), 1034-1036 = Soviet Math. Dokl. **12** (1971), 1539-1542. MR **44** #4412.

[12] ———, *The distribution of the discrete spectrum for singular differential operators*, Dokl. Akad. Nauk SSSR **202** (1972), 1012-1015 = Soviet Math. Dokl. **13** (1972), 245-249. MR **45** #4216.

[13] M. Š. Birman and M. Z. Solomjak, *The asymptotics of the spectrum of "nonsmooth" elliptic equations*, Funkcional. Anal. i Priložen. **5** (1971), no. 1, 69-70 = Functional Anal. Appl. **5** (1971), 56-57. MR **44** #659.

Translated by
F. A. CEZUS

Trudy Moskov Mat. Obšč.
Tom 28 (1973)

Trans. Moscow Math. Soc.
Vol. 28 (1973)

ON THE PLANE PROBLEM OF THE MOTION OF A BODY IMMERSED IN A FLUID

B. R. VAĬNBERG and V. G. MAZ'JA

Contents

Introduction

In this paper we study the question of the unique solvability of the plane problem of finding the velocity potential $u(x,y)$ due to a body moving uniformly under the free surface of a fluid. The problem is as follows.

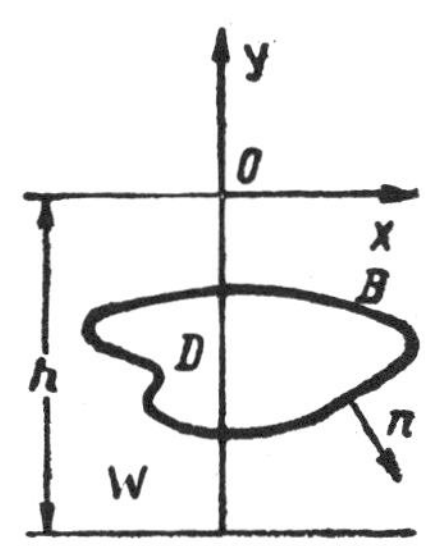

We denote by R^2_- the half-plane $\{(x,y): y<0\}$ and by S the strip $\{(x,y): -h<y<0\}$; let D be a bounded simply connected domain with boundary B of class C^3 such that $\overline{D}\subset S$. Let $W=S\setminus\overline{D}$ (see the figure).

We are required to find a function $u(x,y)$ that is harmonic in W and satisfies the boundary conditions

$$\frac{\partial u}{\partial n}=v\cos(n,\,x)\ \text{on}\ B,\qquad \frac{\partial^2 u}{\partial x^2}+gv^{-2}\frac{\partial u}{\partial y}=0\quad \text{for}\ y=0,$$

$$\frac{\partial u}{\partial y}=0\ \text{for}\ y=-h.$$

Here n is the normal to B directed into the exterior of D, g the acceleration of free fall, and v the velocity of the body. In addition we require that the gradient of the solution is bounded everywhere and tends to zero as $x\to+\infty$.

When the fluid has infinite depth ($h=\infty$) the boundary condition for $y=-h$ is omitted.

This problem, its modifications and its three-dimensional variant has

AMS (MOS) subject classifications (1970). Primary 31A05, 31A35, 35A05, 35B40, 35J25, 76D99; Secondary 45B05.

been the subject of numerous investigations (see the monographs [1]–[3] and the papers [4]–[7] and [15], which contain an exhaustive bibliography). We mention especially Kočin's paper [4], in which he investigates, in particular, the equation of the potential theory of this problem and proves it is uniquely solvable for large and small values of v. The results of the present paper refer to any value of v.

We say that a function u is a *solution of problem* I if it is twice continuously differentiable in $\overline{W}$ and

$$\Delta u = 0, \quad (x, y) \in W, \tag{1.0}$$

$$\frac{\partial^2 u}{\partial x^2} + \nu \frac{\partial u}{\partial y} = 0, \quad y = 0, \quad \nu = \text{const} > 0, \tag{2.0}$$

$$\frac{\partial u}{\partial n} = f(x, y), \quad (x, y) \in B, \quad f \in C^{1,\alpha}(B), \tag{3.0}$$

$$\frac{\partial u}{\partial y} = 0, \quad y = -h, \tag{4.0}$$

$$\sup_W |\nabla u| < \infty, \quad \lim_{x \to +\infty} |\nabla u| = 0. \tag{5.0}$$

As we noted above, the condition (4.0) is not required when $h = \infty$.

In this paper, under the assumption that $x \cos(n, x) \geqq 0$ on B[1)] we obtain the following results:

1°. We show that for $\nu h < 1$ (the case of a supercritical velocity $v > (gh)^{1/2}$) problem I has one and only one solution (up to a constant term) for all f.

2°. We show that for $\nu h > 1$ the homogeneous problem I with the additional condition

$$\lim_{x \to -\infty} |\nabla u| = 0 \tag{6.0}$$

has only the trivial solution $u \equiv \text{const}$.

3°. We prove that for $\nu h > 1$ the homogeneous problem I has not more than one nontrivial solution.

4°. We find a necessary and sufficient condition on the domain D for the nontrivial solution in 3° to be absent, so that problem I is uniquely solvable for any function f.

For example, for $h = \infty$ a possible formulation of this condition is

$$i \int_B G e^{\nu(y - ix)}\, ds \neq -1, \tag{7.0}$$

where G is the solution of a special (certainly uniquely solvable) Fredholm integral equation on the contour B.

---

1) This condition is satisfied if, in particular, B is split by the axis $x = 0$ into two parts that can be specified explicitly: $x = g_{\pm}(y)$.

Unfortunately, we do not know whether there exist domains for which the condition (7.0) does not hold.

In the first four sections we consider in detail the case $h = \infty$, for which many of the formulas are simpler. The proofs when $h < \infty$ are not significantly different from those in the limiting case when $h = \infty$, and are therefore merely sketched (§5). In §1 we prove the uniqueness theorem for the solution with a finite Dirichlet integral. In the next section we pose a certain auxiliary problem II with "radiation" conditions at infinity, and give asymptotic formulas at infinity for the solutions of problems I and II. In §3 we prove the unique solvability of problem II, and in §4 we prove assertions 3° and 4° for $h = \infty$.

The results of this paper were announced in the note [**16**].

§1. The uniqueness of a solution with a finite Dirichlet integral

Throughout §§1–4 we confine ourselves to the case $h = \infty$.

We begin with the following simple assertion of a technical nature.

LEMMA 1.1. *Suppose that a function u, which is continuously differentiable in $\overline{W}$, satisfies the conditions*

$$\int_{y=0} |u_x|^2 dx + \iint_W |\nabla u|^2 dxdy < \infty, \tag{1.1}$$

$$\int_{-\infty}^{0} |u(a, y)|^2 dy < \infty \tag{2.1}$$

for all a. Then

$$\lim_{x\to\infty} x^{-1} |u(x, 0)|^2 = 0, \tag{3.1}$$

$$\lim_{x\to\infty} x^{-1} \int_{-\infty}^{0} |u(x, y)|^2 dy = 0, \tag{4.1}$$

$$\lim_{y\to-\infty} y^{-1} \int_{-\alpha}^{\alpha} |u(x, y)|^2 dx = 0, \quad 0 < \alpha < \infty, \tag{5.1}$$

$$\varliminf_{x\to\infty} |x| \Big[|u_x(x, 0)|^2 + \int_{-\infty}^{0} |\nabla u(x, y)|^2 dy \Big] = 0, \tag{6.1}$$

$$\varliminf_{y\to-\infty} |y| \int_{-\infty}^{\infty} |\nabla u(x, y)|^2 dx = 0. \tag{7.1}$$

PROOF. Let a be any real number, $y \leq 0$ and $x > a$. Obviously

$$|u(x, y)|^2 \leqslant 2|u(a, y)|^2 + 2(x-a) \int_a^x |u_x(x, y)|^2 dx, \tag{8.1}$$

and therefore

$$\overline{\lim_{x\to+\infty}}\, x^{-1}\,|\,u(x,\ 0)\,|^2 \leqslant 2\int_a^\infty |\,u_x(x,\ 0)\,|^2\,dx.$$

Hence $x^{-1}|u(x,0)|^2 \to 0$ as $x \to +\infty$. The case $x \to -\infty$ is considered similarly. The equality (3.1) is proved.

When we divide both sides of the inequality (8.1) by x and integrate with respect to y, we obtain

$$\overline{\lim_{x\to+\infty}}\, x^{-1}\int_{-\infty}^{0} |\,u(x,\ y)\,|^2\,dy \leqslant 2\int_{-\infty}^{0}\int_a^\infty |\,\nabla u\,|^2\,dxdy.$$

To prove (4.1) it remains to note that the right-hand side tends to zero as $a \to \infty$. We obtain (5.1) in exactly the same way.

The equalities (6.1) and (7.1) are obvious. The lemma is proved.

The following lemma contains an integral identity from which the uniqueness theorem follows directly.

LEMMA 2.1. *Let u be twice continuously differentiable in $\overline{W}$ and satisfy the conditions* (1.0)–(3.0) *and* (1.1), (2.1). *Then*

$$2\iint_W |\,u_y\,|^2\,dxdy + \int_B x\cos(n,\ x)\,|\,\nabla u\,|^2\,ds = \operatorname{Re}\int_B (2x\bar{u}_x - \bar{u})\,f\,ds. \tag{9.1}$$

PROOF. The following identity can be verified without difficulty:

$$\operatorname{Re}[(2x\bar{u}_x - \bar{u})\,\Delta u] = 2\,|\,u_y\,|^2 - (x\,|\,\nabla u\,|^2)_x + \operatorname{div}\operatorname{Re}[(2xu_x - \bar{u})\,\nabla u\ ]. \tag{10.1}$$

Let $R_{\alpha,\beta}$ be the rectangle $\{|x| < \alpha,\ -\beta < y < 0\}$, which contains $\overline{D}$, let $W_{\alpha,\beta} = R_{\alpha,\beta}\backslash\overline{D}$, and let $p_{\alpha,0}$, $p_{\alpha,\beta}$ and $q_{\pm\alpha,\beta}$ be the top, bottom, right and left sides of the rectangle $R_{\alpha,\beta}$.

When we integrate (10.1) over $W_{\alpha,\beta}$ we obtain

$$\begin{aligned} &2\iint_{W_{\alpha,\beta}} |\,u_y\,|^2\,dxdy + \int_B x\cos(n,\ x)\,|\,\nabla u\,|^2\,ds + \int_{q_{-\alpha,\beta}} x\,|\,\nabla u\,|^2\,dy \\ &- \int_{q_{\alpha,\beta}} x\,|\,\nabla u\,|^2\,dy - \operatorname{Re}\int_{q_{-\alpha,\beta}} (2x\bar{u}_x - \bar{u})\,u_x\,dy - \operatorname{Re}\int_{p_{\alpha,\beta}} (2x\bar{u}_x - \bar{u})\,u_y\,dx \\ &+ \operatorname{Re}\int_{q_{\alpha,\beta}} (2x\bar{u}_x - \bar{u})\,u_x\,dy + \operatorname{Re}\int_{p_{\alpha,0}} (2x\bar{u}_x - \bar{u})\,u_y\,dx = \operatorname{Re}\int_B (2x\bar{u}_x - \bar{u})\,f ds. \end{aligned} \tag{11.1}$$

We transform the last term on the left-hand side by using the boundary condition (2.0):

$$\begin{aligned} \operatorname{Re}\int_{p_{\alpha,0}} (2x\bar{u}_x - \bar{u})\,u_y\,dx &= -\nu^{-1}\operatorname{Re}\int_{p_{\alpha,0}} (2x\bar{u}_x - \bar{u})\,u_{xx}\,dx \\ &= -\nu^{-1}\,(x\,|\,u_x\,|^2 - \operatorname{Re}\bar{u}\,u_x)\,\Big|_{-\alpha}^{\alpha}. \end{aligned} \tag{12.1}$$

By (5.1) and (7.1), for any fixed $\alpha < \infty$ and some sequence $\beta_k \to +\infty$ we have

$$\lim_{k\to\infty} \operatorname{Re} \int_{p_{\alpha,\beta_k}} (2x\bar{u}_x - \bar{u})\, u_y\, dx = 0.$$

By (3.1), (4.1) and (6.1) the quantity (12.1) and the integrals over $q_{\pm\alpha,\beta}$ in (11.1), where $\beta = \infty$, tend to zero for some choice of the sequence $\alpha_k \to \infty$. When we take the limit in (11.1), first as $\beta_k \to \infty$ and then as $\alpha_k \to \infty$, we obtain (9.1). The lemma is proved.

THEOREM 1.1. *Let the contour B satisfy the condition $x\cos(n,x) \geqq 0$. If u is twice continuously differentiable in $\overline{W}$ and satisfies the conditions* (1.0), (2.0) *and* (3.0), *where $f \equiv 0$, and also the conditions* (1.1) *and* (2.1), *then $u \equiv 0$.*

PROOF. By the identity (9.1) we have $u(x,y) = u(x)$. It remains to use the inequality (2.1).

REMARK 1.1. The assertion of Theorem 1.1 and its proof can be carried over without difficulty to the m-dimensional case when the problem is posed as follows. Let D be an m-dimensional domain with boundary B of class C^3 and compact closure in $R^m_- = \{x = (x_1, \cdots, x_m);\ x_m < 0\}$. We are required to find a function that is harmonic in $R^m_- \setminus \overline{D}$ and satisfies the boundary conditions

$$\frac{\partial^2 u}{\partial x_1^2} + \nu \frac{\partial u}{\partial x_m} = 0 \quad \text{for} \quad x_m = 0, \quad \nu > 0,$$

$$\frac{\partial u}{\partial n} = f \quad \text{on} \quad B.$$

The role of the condition (1.1) is played by the inequality

$$\int_{x_m=0} \left| \frac{\partial u}{\partial x_1} \right|^2 dx_1 \ldots dx_{m-1} + \int_{R^m_- \setminus D} |\nabla u|^2\, dx < \infty;$$

instead of (2.1) we require that u is square-summable on any $(m-1)$-dimensional plane parallel to the coordinate planes $x_i = 0$, $1 \leqq i < m$; and the fundamental identity (9.1) is rewritten in the form

$$\operatorname{Re}\left[(2x_1\bar{u}_{x_1} - \bar{u})\, \Delta u\right] = 2\sum_{j=2}^{m} \left| \frac{\partial u}{\partial x_j} \right|^2 - (x_1 |\nabla u|^2)_{x_1}$$
$$+ \operatorname{div} \operatorname{Re}\left[(2x_1\bar{u}_{x_1} - \bar{u})\, \nabla u\right].$$

It is assumed that $x_1 \cos(n, x_1) \geqq 0$ on B.

The following two lemmas and Theorem 2.1 are not used to prove the main results mentioned in the Introduction but are required only for a more detailed study of the spectrum of the integral equations of problems I and II.

LEMMA 3.1. *Let u be a function that is harmonic in W and D, continuous in*

R^2_-, and whose first derivatives have limiting values from both sides on B. Suppose also that u satisfies the conditions (2.0), (1.1) and (2.1) and the boundary condition

$$\left(\frac{\partial u}{\partial n}\right)_e = \mu\left(\frac{\partial u}{\partial n}\right)_i, \tag{13.1}$$

where $(v)_e$ and $(v)_i$ denote the limiting values of the function v on B from the side of W and D, respectively. If $\operatorname{Im}\mu = 0$, then

$$\begin{aligned} 2\iint_W |u_y|^2\,dxdy + 2\mu\iint_D |u_y|^2\,dxdy + \mu(1-\mu)\int_B x\cos(n,x)\left|\left(\frac{\partial u}{\partial n}\right)_i\right|^2 ds \\ + (1-\mu)\int_B x\cos(n,x)\left|\frac{\partial u}{\partial t}\right|^2 ds = 0, \end{aligned} \tag{14.1}$$

where t is the direction of the tangent to B.

Proof. Because

$$\left(\frac{\partial u}{\partial x}\right)_e = \left(\frac{\partial u}{\partial n}\right)_e \cos(n,x) + \frac{\partial u}{\partial t}\cos(t,x),$$

$$\left(\frac{\partial u}{\partial x}\right)_i = \left(\frac{\partial u}{\partial n}\right)_i \cos(n,x) + \frac{\partial u}{\partial t}\cos(t,x),$$

we have

$$\left(\frac{\partial u}{\partial x}\right)_e = \left(\frac{\partial u}{\partial x}\right)_i + (\mu-1)\left(\frac{\partial u}{\partial n}\right)_i \cos(n,x).$$

Hence the right-hand side of (9.1) is equal to

$$\mu\operatorname{Re}\int_B \left[2x\left(\frac{\partial \bar u}{\partial x}\right)_i - \bar u\right]\left(\frac{\partial u}{\partial n}\right)_i ds + 2\mu(\mu-1)\int_B x\cos(n,x)\left|\left(\frac{\partial u}{\partial n}\right)_i\right|^2 ds. \tag{15.1}$$

We transform the first term:

$$\begin{aligned} \mu\operatorname{Re}\int_B \left[2x\left(\frac{\partial \bar u}{\partial x}\right)_i - \bar u\right]\left(\frac{\partial u}{\partial n}\right)_i ds &= \mu\operatorname{Re}\iint_D \nabla\left(2x\frac{\partial \bar u}{\partial x} - \bar u\right)\nabla u\,dxdy \\ &= 2\mu\iint_D \left|\frac{\partial u}{\partial x}\right|^2 dxdy - \mu\iint_D |\nabla u|^2\,dxdy + 2\mu\operatorname{Re}\iint_D x\nabla\frac{\partial \bar u}{\partial x}\nabla u\,dxdy, \end{aligned}$$

which is obviously equal to

$$\begin{aligned} 2\mu\iint_D \left|\frac{\partial u}{\partial x}\right|^2 dxdy - 2\mu\iint_D |\nabla u|^2\,dxdy + \mu\int_B x\cos(n,x)\,|(\nabla u)_i|^2\,ds \\ = -2\mu\iint_D \left|\frac{\partial u}{\partial y}\right|^2 dxdy + \mu\int_B x\cos(n,x)\,|(\nabla u)_i|^2\,ds. \end{aligned}$$

Hence the expression (15.1) can be represented in the form

$$-2\mu\iint_D \left|\frac{\partial u}{\partial y}\right|^2 dxdy + \mu\int_B x\cos(n,x)\left[|(\nabla u)_i|^2 - 2(1-\mu)\left|\left(\frac{\partial u}{\partial n}\right)_i\right|^2\right] ds.$$

On the other hand, by (9.1) it is equal to the sum

$$2\iint_{W}\left|\frac{\partial u}{\partial y}\right|^2 dxdy+\int_{B} x\cos(n,\ x)\,|(\nabla u)_e|^2\,ds.$$

When we compare the last two expressions and use (13.1) we obtain the identity (14.1).

LEMMA 4.1. *Suppose that the hypotheses of Lemma* 3.1 *hold and the constant* μ *in* (13.1) *is an arbitrary complex number. Then*

$$\iint_{W}|\nabla u|^2\,dxdy+\mu\iint_{D}|\nabla u|^2\,dxdy-\nu^{-1}\int_{y=0}\left|\frac{\partial u}{\partial x}\right|^2 dx=0. \tag{16.1}$$

PROOF. We use the notation introduced in the proof of Lemma 2.1. By Green's theorem and the boundary condition (2.0) we have

$$\iint_{W_{\alpha,\beta}}|\nabla u|^2\,dxdy=-\int_{B}\bar{u}\left(\frac{\partial u}{\partial n}\right)_e ds+\nu^{-1}\int_{P_{\alpha,0}}\left|\frac{\partial u}{\partial x}\right|^2 dx+\dots.$$

Here the dots denote quantities that tend to zero for some choice of the sequences $\beta_k\to\infty$ and $\alpha_k\to\infty$ (compare this with the proof of Lemma 2.1). Hence

$$\iint_{W}|\nabla u|^2\,dxdy-\nu^{-1}\int_{y=0}\left|\frac{\partial u}{\partial x}\right|^2 dx=-\int_{B}\bar{u}\left(\frac{\partial u}{\partial n}\right)_e ds$$

$$=-\mu\int_{B}\bar{u}\left(\frac{\partial u}{\partial n}\right)_i ds=-\mu\iint_{D}|\nabla u|^2\,dxdy.$$

The lemma is proved.

THEOREM 2.1. *Suppose that* $x\cos(n,x)\geqq 0$ *on* B *and the conditions of Lemma* 3.1 *hold with either* $\operatorname{Im}\mu\neq 0$ *or* $1\geqq\mu\geqq 0$. *Then* $u=0$ *in* R^2_-.

This assertion is an obvious consequence of the identities (16.1) and (14.1) and the condition (2.1).

The observation in Remark 1.1 applies equally to Theorem 2.1.

§2. The behavior of solutions at infinity. The case $h=\infty$

Let $z=x+iy$, $r=|z|$, $\zeta=\xi+i\eta$, $\rho=|\zeta|$. Explicit formulas for the solutions of problem I in a half-plane and a strip with point singularities are given in [4] and [8]–[11]. In particular (see [8] or [4]), the fundamental solution of problem I in R^2_- is

$$E_1(z,\ \zeta)=\frac{-1}{2\pi}\operatorname{Re}\left\{\log(z-\zeta)+\log(z-\bar{\zeta})+2\int_0^\infty\frac{e^{it}\,dt}{t+\nu(z-\bar{\zeta})}\right\}, \tag{1.2}$$

where $y\leqq 0$ and $\eta<0$. In other words,

(2.2) $$-\Delta_{x,y}E_1=\delta(x-\xi,\ y-\eta),\quad y\leqslant 0,\ \eta<0,$$

(3.2) $$\frac{\partial^2 E_1}{\partial x^2}+\nu\frac{\partial E_1}{\partial y}=0,\quad y=0,\quad \eta<0,$$

(4.2) $$\overline{\lim_{r\to\infty}}|\nabla_{x,y}E_1|<\infty,\quad \lim_{x\to+\infty}|\nabla_{x,y}E_1|=0.$$

The next lemma follows easily from (1.2).

LEMMA 1.2. *The following asymptotic representation holds for* $\rho<R=\text{const}$ *and as* $r\to\infty$:

$$E_1=\begin{cases}-\dfrac{1}{\pi}\log r+\varphi_+, & x>0,\\[2mm] -\dfrac{1}{\pi}\log r-2e^{\nu(y+\eta)}\sin[\nu(x-\xi)]+\varphi_-, & x<0,\end{cases}$$

where $\varphi_\pm=O(r^{-1})$ *and* $|\nabla\varphi_\pm|=O(r^{-2})$.

PROOF. It is clear that

$$\int_0^\infty\frac{e^{it}\,dt}{t+\nu(z-\bar\zeta)}=\frac{i}{\nu(z-\bar\zeta)}+i\int_0^\infty\frac{e^{it}\,dt}{[t+\nu(z-\bar\zeta)]^2}.$$

For $x>0$ we have

$$\begin{aligned}2|t+\nu(z-\bar\zeta)|&\geqslant|t+\nu x-\nu\xi|+\nu|y+\eta|\\&\geqslant t+\nu x-\nu|\xi|+\nu|y|\geqslant t+\nu(r-R),\end{aligned}$$

which gives the asymptotic behavior of E_1 and ∇E_1 for $x>0$ as $r\to\infty$.

The asymptotic behavior for $x<0$ is obtained similarly by noting that

$$\int_0^\infty\frac{e^{it}\,dt}{t+\nu(z-\bar\zeta)}=2\pi ie^{-i\nu(z-\bar\zeta)}-\int_{-\infty}^0\frac{e^{it}\,dt}{t+\nu(z-\bar\zeta)}.$$

To study problem I, we must first consider another boundary-value problem. This is problem II, which is as follows: Find a solution of (1.0) that is twice continuously differentiable in $\overline{W}$ and satisfies the conditions (2.0) and (3.0), and the radiation conditions at infinity:

(5.2) $$u(x,y)=\gamma\log r+v(x,y),$$

where γ is some constant, and

(6.2) $$\sup_W|\nabla u|<\infty,\qquad \lim_{r\to\infty}\left(\frac{\partial v}{\partial|x|}-i\nu v\right)=0.$$

It follows from Lemma 1.2 that the function

(7.2) $$E_2(z,\zeta)=E_1(z,\zeta)-ie^{\nu(y+\eta-i(x-\xi))}$$

is a fundamental solution of problem II, and for $\rho\leqq R=\text{const}$ and as $r\to\infty$ we have

(8.2) $$E_2(z, \zeta) = -\frac{1}{\pi} \log r - ie^{\nu(y+\eta+i(x-\xi))} + \varphi,$$

where $\varphi = O(r^{-1})$ and $|\nabla\varphi| = O(r^{-2})$.

LEMMA 2.2. *If the function u is twice continuously differentiable in R^2_- and such that*

$$\Delta u = 0 \quad in \quad R^2_-,$$

$$\frac{\partial^2 u}{\partial x^2} + \nu \frac{\partial u}{\partial y} = 0 \quad for \quad y = 0,$$

$$\sup_{R^2_-} |\nabla u| < \infty,$$

then

(9.2) $$u = k_1 e^{\nu(y+ix)} + k_2 e^{\nu(y-ix)} + k_3 x + k_4.$$

PROOF. We put $U = \partial^2 u/\partial x^2 + \nu \partial u/\partial y$ for $y \leq 0$ and $U(x,y) = -U(x,-y)$ for $y > 0$ (compare with [7], pp. 416-417). Because $U = 0$ for $y = 0$, we have $\Delta U = 0$ in R^2. In addition,

$$\pi |U(z)| = \Big| \iint_{|\zeta - z| < 1} U(\zeta)\, d\xi d\eta \Big|$$

$$\leqslant \Big| \iint_{\substack{|\zeta - z| < 1 \\ \eta < 0}} (u_{\xi\xi} + \nu u_\eta)\, d\xi d\eta \Big| + \Big| \iint_{\substack{|\zeta - \bar{z}| < 1 \\ \eta < 0}} (u_{\xi\xi} + \nu u_\eta)\, d\xi d\eta \Big|.$$

When we integrate the function $u_{\xi\xi}$ with respect to ξ we obtain that U is uniformly bounded. Hence $U \equiv 0$; that is, $\partial^2 u/\partial x^2 + \nu \partial u/\partial y = 0$ in R^2_-. Since $\Delta U = 0$, we have $\partial^2 u/\partial y^2 + \nu \partial u/\partial y = 0$ in R^2_-, and hence $u(x,y) = C_1(x) + C_2(x) e^{\nu y}$. When we use once more the assumption that u is harmonic, we obtain (9.2).

LEMMA 3.2. *Let u_1 and u_2 be solutions of problems* I *and* II, *respectively. Then*

(10.2) $$u_i(z) = \int_B \left[u_i(\zeta) \frac{\partial E_i(z, \zeta)}{\partial n_\zeta} - \frac{\partial u_i(\zeta)}{\partial n_\zeta} E_i(z, \zeta) \right] ds_\zeta + c_i,$$

where $c_1 =$ const and $c_2 = 0$.

PROOF. Let g_i $(i = 1, 2)$ be a twice continuously differentiable function in R^2_- obtained by extending the function u_i to D in an arbitrary manner. Then

$$-\Delta g_i = f_i \quad \text{in } R^2_-, \ \operatorname{supp} f_i \subset \bar{D},$$

$$\frac{\partial^2 g_i}{\partial x^2} + \nu \frac{\partial g_i}{\partial y} = 0 \quad \text{for } y = 0,$$

and the functions g_i, $i = 1, 2$, satisfy the conditions (5.0) or (5.2), (6.2), respectively. The following functions have the same properties:

$$h_i(z) = \iint_D E_i(z, \zeta) f_i(\zeta)\, d\xi d\eta. \tag{11.2}$$

But because both the condition (5.0) and the conditions (5.2), (6.2) hold for the functions (9.2) only if $k_1 = k_2 = k_3 = 0$ in the case of problem I, and if $k_1 = k_2 = k_3 = k_4 = 0$ in the case of problem II, by Lemma 2.2 we have $u_i(z) - h_i(z) = c_i$. When we apply Green's theorem to the integral (11.2) for $z \in W$ we obtain (10.2). The lemma is proved.

We obtain the following assertions directly from Lemmas 1.2 and 3.2 and equation (8.2):

Theorem 1.2. *Let u be a solution of problem* I. *Then*

$$u(x, y) = \begin{cases} \gamma \log r + c + \psi_+(x, y), & x > 0, \\ \gamma \log r + c + e^{\nu y}(\mathcal{A} \sin \nu x + \mathcal{B} \cos \nu x) + \psi_-(x, y), & x < 0, \end{cases} \tag{12.2}$$

where $\psi_\pm = O(r^{-1})$, $|\nabla \psi_\pm| = O(r^{-2})$ *as* $r \to \infty$ *and the constants* γ, $\mathcal{A}$ *and* $\mathcal{B}$ *are given by the formulas*

$$\gamma = \frac{1}{\pi} \int_B \frac{\partial u}{\partial n} ds, \quad \mathcal{A} = -2 \int_B \left\{ u \frac{\partial}{\partial n} [e^{\nu y} \cos \nu x] - e^{\nu y} \cos \nu x \frac{\partial u}{\partial n} \right\} ds,$$

$$\mathcal{B} = 2 \int_B \left\{ u \frac{\partial}{\partial n} [e^{\nu y} \sin \nu x] - e^{\nu y} \sin \nu x \frac{\partial u}{\partial n} \right\} ds.$$

Theorem 2.2. *Let u be a solution of problem* II. *Then*

$$u(x, y) = \begin{cases} \gamma \log r + \mathcal{D}_+ e^{\nu(y+ix)} + \psi_+(x, y), & x > 0, \\ \gamma \log r + \mathcal{D}_- e^{\nu(y-ix)} + \psi_-(x, y), & x < 0, \end{cases} \tag{13.2}$$

where $\psi_\pm = O(r^{-1})$, $|\nabla \psi_\pm| = O(r^{-2})$ *as* $r \to \infty$, *and the constants* γ, $\mathcal{D}_+$ *and* $\mathcal{D}_-$ *are given by the formulas*

$$\gamma = \frac{1}{\pi} \int_B \frac{\partial u}{\partial n} ds, \quad \mathcal{D}_\pm = i \int_B \left(e^{\nu(y \mp ix)} \frac{\partial u}{\partial n} - u \frac{\partial}{\partial n} e^{\nu(y \mp ix)} \right) ds. \tag{14.2}$$

§3. The solvability of problem II for $h = \infty$

From Theorem 1.1 and the asymptotic formulas proved in Theorem 1.2 we immediately obtain the following result, which is mentioned in the Introduction (for the time being only for $h = \infty$):

Theorem 1.3. *Suppose that* $x \cos(n, x) \geqq 0$ *on* B, *and the function* u *is twice continuously differentiable in* $\overline{W}$, *satisfies the conditions* (1.0), (2.0), *and* (3.0) *for* $f \equiv 0$ *and is such that*

$$\sup_W |\nabla u| < \infty, \quad \lim_{x \to \infty} |\nabla u| = 0.$$

Then $u \equiv$ const.

From the next lemma it follows, in particular, that the solution of the homogeneous problem II satisfies the conditions of Theorem 1.3.

Lemma 1.3. *Let u be the solution of problem* II *and* $\int_B (\partial u/\partial n)ds = 0$. *Then*

$$|\mathscr{D}_+|^2 + |\mathscr{D}_-|^2 = 2\operatorname{Im}\int_B u \frac{\partial \bar{u}}{\partial n}\, ds. \tag{1.3}$$

Proof. We use the notation of Lemma 2.1. By Green's theorem we obtain

$$\int_{\partial R_{\alpha,\beta}} \left(\bar{u}\frac{\partial u}{\partial n} - u\frac{\partial \bar{u}}{\partial n}\right) ds = -\int_B \left(\bar{u}\frac{\partial u}{\partial n} - u\frac{\partial \bar{u}}{\partial n}\right) ds = 2i\operatorname{Im}\int_B u\frac{\partial \bar{u}}{\partial n}\, ds, \tag{2.3}$$

where the normal n is directed into $W_{\alpha,\beta}$. We now let first β and then α tend to infinity. It follows immediately from Theorem 2.2 that

$$\lim_{\beta\to\infty}\int_{p_{\alpha,\beta}} \left(\bar{u}\frac{\partial u}{\partial n} - u\frac{\partial \bar{u}}{\partial n}\right) ds = 0. \tag{3.3}$$

Further, from the same theorem we obtain

$$\bar{u}\frac{\partial u}{\partial |x|} - u\frac{\partial \bar{u}}{\partial |x|} = \begin{cases} 2i\nu|\mathscr{D}_+|^2 e^{2\nu y} + h_+(x, y), & x > 0, \\ 2i\nu|\mathscr{D}_-|^2 e^{2\nu y} + h_-(x, y), & x < 0, \end{cases}$$

where $h_\pm = O(r^{-2} + r^{-1}e^{\nu y})$ as $r \to \infty$.

Hence we obtain the equalities

$$\begin{gathered} \lim_{\alpha\to\infty}\int_{q_{\pm\alpha,\infty}} \left(\bar{u}\frac{\partial u}{\partial n} - u\frac{\partial \bar{u}}{\partial n}\right) ds = -i|\mathscr{D}_\pm|^2, \\ \lim_{\alpha\to\infty}\left(\bar{u}\frac{\partial u}{\partial x}(x, 0) - u\frac{\partial \bar{u}}{\partial x}(x, 0)\right)\Big|_{-\alpha}^{\alpha} = 2i\nu(|\mathscr{D}_+|^2 + |\mathscr{D}_-|^2). \end{gathered} \tag{4.3}$$

It follows from the last equality and the boundary condition (2.0) that as $a \to \infty$

$$\begin{aligned} \int_{p_{\alpha,0}} \left(\bar{u}\frac{\partial u}{\partial n} - u\frac{\partial \bar{u}}{\partial n}\right) ds &= \int_{p_{\alpha,0}} \left(u\frac{\partial \bar{u}}{\partial y} - \bar{u}\frac{\partial u}{\partial y}\right) dx \\ &= \nu^{-1}\int_{p_{\alpha,0}} \left(\bar{u}\frac{\partial^2 u}{\partial x^2} - u\frac{\partial^2 \bar{u}}{\partial x^2}\right) dx \\ &= \nu^{-1}\left(\bar{u}\frac{\partial u}{\partial x}(x, 0) - u\frac{\partial \bar{u}}{\partial x}(x, 0)\right)\Big|_{-\alpha}^{\alpha} \to 2i(|\mathscr{D}_+|^2 + |\mathscr{D}_-|^2). \end{aligned}$$

This in conjunction with (2.3) and (4.3) proves Lemma 1.3.

Theorem 2.3. *If* $x\cos(n, x) \geqq 0$ *on* B, *then problem* II *has one and only one solution for any function* $f \in C^{1,\alpha}(B)$.

PROOF. The uniqueness of the solution follows from Theorems 1.3 and 2.2 and Lemma 1.3.

To prove the existence of the solution we seek it in the form of a potential

$$u(z)=\int_B E_2(z,\zeta)\,g(\zeta)\,ds_\zeta \tag{5.3}$$

with an unknown density $g\in C^{1,\alpha}(B)$. Then u is twice continuously differentiable in $\overline{W}$ and satisfies the conditions (1.0), (2.0), (5.2) and (6.2). Because on B

$$\frac{\partial u(z)}{\partial n}=-\frac{1}{2}g(z)+\int_B \frac{\partial E_2(z,\zeta)}{\partial n_z}g(\zeta)\,ds_\zeta, \tag{6.3}$$

when we require u to satisfy the condition (3.0) we obtain a Fredholm integral equation on B:

$$-g+Tg=2f,\quad (Tg)(z)=2\int_B \frac{\partial E_2(z,\zeta)}{\partial n_z}g(\zeta)\,ds_\zeta. \tag{7.3}$$

Let g_0 be a solution of the homogeneous equation (7.3). Then the potential u_0 with the density g_0 is zero in W because, as we have proved, the solution of problem II is unique. Because u_0 is continuous in R^2_- and harmonic in D, we have $u_0\equiv 0$ in R^2_-. Hence and from the equality

$$g_0=\left(\frac{\partial u_0}{\partial n}\right)_i-\left(\frac{\partial u_0}{\partial n}\right)_e$$

it follows that the homogeneous equation (7.3) has only the trivial solution. Hence (7.3) is solvable for any $f\in C^{1,\alpha}(B)$.

Theorem 2.3 is proved.

REMARK 1.3. In the proof of Theorem 2.3 we have shown that (7.3) is uniquely solvable for any right-hand side.

REMARK 2.3. We consider a more general problem than problem II: To find a function u that is harmonic in D and W and continuous in R^2_-, whose first derivatives have limiting values on B from both sides, and for which (2.0), (5.2), (6.2), and the following condition hold:

$$\left(\frac{\partial u}{\partial n}\right)_e-\mu\left(\frac{\partial u}{\partial n}\right)_i=f \text{ on } B, \tag{8.3}$$

where $\operatorname{Im}\mu>0$ or $0\leq\mu\leq 1$. If $x\cos(n,x)\geq 0$ on B, then this problem has one and only one solution for any $f\in C^{1,\alpha}(B)$.

PROOF. Let $f\equiv 0$. Then we have

$$\int_B\left(\frac{\partial u}{\partial n}\right)_e ds=\mu\int_B\left(\frac{\partial u}{\partial n}\right)_i ds=\mu\iint_D \Delta u\,dx\,dy=0.$$

Then by Lemma 1.3

$$|\mathscr{D}_+|^2+|\mathscr{D}_-|^2=2\operatorname{Im}\left[\bar{\mu}\int_B u\left(\frac{\partial \bar{u}}{\partial n}\right)_i ds\right]=2\operatorname{Im}\bar{\mu}\iint_D|\nabla u|^2\,dx\,dy.$$

Hence because $\operatorname{Im}\mu \geqq 0$ we obtain $\mathscr{D}_+ = \mathscr{D}_- = 0$, which in conjunction with Theorems 2.2 and 1.3 ensures that the solution of the problem is unique.

We seek a solution of the inhomogeneous problem in the form (5.3). From (6.3) and the equality

$$2(\partial u/\partial n)_i = g + Tg$$

we obtain the following equation to determine g:

$$-(1+\mu)g + (1-\mu)Tg = 2f. \tag{9.3}$$

The remaining arguments are as in Theorem 2.3.

REMARK 3.3. From solvability of (9.3), which we have just proved, it follows that the poles of the resolvent $R_\lambda = (I - \lambda T)^{-1}$, where I is the identity operator, lie in the half-plane $\operatorname{Im}\lambda > 0$ or on the real axis outside the segment $[0,1]$.

§4. The solvability of problem I for $h = \infty$

THEOREM 1.4. *If $x\cos(n,x) \geqq 0$ on B, then the homogeneous problem* I *has not more than one nontrivial solution (up to a constant factor and term).*

PROOF. Let u_1 and u_2 be two different (with the given accuracy) solutions of the homogeneous problem I. Then by Theorem 1.2

$$u_i(x,y) = \begin{cases} \psi_+(x,y), & x > 0 \\ e^{\nu y}(\mathscr{A}_i \sin \nu x + \mathscr{B}_i \cos \nu x) + \psi_-(x,y), & x < 0, \end{cases}$$

where

$$\psi_\pm = O(r^{-1}),\ |\nabla\psi_\pm| = O(r^{-2}) \quad \text{as } r \to \infty.$$

It follows from Theorem 1.3 that the vectors $(\mathscr{A}_1, \mathscr{B}_1)$ and $(\mathscr{A}_2, \mathscr{B}_2)$ are linearly independent. Then a certain linear combination of u_1 and u_2 has the same asymptotic behavior with the constants $\mathscr{A} = -i$ and $\mathscr{B} = 1$. Thus we have obtained a solution of the homogeneous problem I with the asymptotic form (13.2), where $\gamma = 0$, $\mathscr{D}_+ = 0$ and $\mathscr{D}_- = 1$; that is, a nontrivial solution of the homogeneous problem II. This contradicts Theorem 2.3, and so Theorem 1.4 is proved.

If we seek a solution of problem I in the form

$$u = \int_B E_1(z,\zeta)\, g(\zeta)\, ds_\zeta + c, \tag{1.4}$$

then by analogy with (7.3) we obtain for the function $g(z)$ the equation

$$-g + T_1 g = 2f, \quad (T_1 g)(z) = 2\int_B \frac{\partial E_1(z,\zeta)}{\partial n_z}\, g(\zeta)\, ds_\zeta. \tag{2.4}$$

THEOREM 2.4. *Suppose that $x\cos(n,x) \geqq 0$ on B; then problem* I *and equation* (2.4) *are equivalent in the sense that any solution of problem* I *can be represented, and moreover uniquely, as a potential* (1.4) *with a density g that satisfies* (2.4); *conversely, for any solution g of* (2.4), *formula* (1.4) *gives a solution of problem* I.

PROOF. Let u be a solution of problem I with the boundary function f. It then follows from Theorem 1.2 that the function $v = u + \frac{1}{2}(\mathcal{A}_i - \mathcal{B})e^{\nu(y+ix)} - c$ is a solution of problem II (with another boundary function). As we showed in the proof of Theorem 2.3, the solution of problem II can be represented as the potential

$$v(z) = \int_B E_2(z, \zeta)\, g(\zeta)\, ds_\zeta.$$

In conjunction with (7.2) this gives

$$u + \frac{1}{2}(\mathcal{A}i - \mathcal{B})\, e^{\nu(y+ix)} - c = \int_B E_1(z, \zeta)\, g(\zeta)\, ds_\zeta - i\int_B e^{\nu(y+\eta+i(x-\xi))}\, g(\zeta)\, ds_\zeta.$$

Therefore

$$u = \int_B E_1(z, \zeta)\, g(\zeta)\, ds_\zeta + c_1 e^{\nu(y+ix)} + c.$$

From the asymptotic formulas of Theorem 1.2 and from Lemma 1.2 (for $x > 0$) we find that $c_1 = 0$; that is, a solution of problem I can be represented as a potential (1.4).

Because

$$\int_B E_1(z, \zeta)\, g(\zeta)\, ds_\zeta \not\equiv \text{const}$$

in W for $g \not\equiv 0$ (see the proof of Theorem 2.3), this representation is unique. The first part of the theorem is proved. The converse is obvious.

THEOREM 3.4. *Suppose that $x\cos(n, x) \geqq 0$ on B. Then the following assertions are equivalent*:

1°. *Problem* I *has one and only one solution* (*up to a constant term*) *for any function $f \in C^{1,\alpha}(B)$.*

2°. *For the solution U of problem* II *with the boundary function*

$$f = \partial e^{\nu(y+ix)}/\partial n$$

the constant $\mathcal{D}_+$ in the asymptotic form (13.2) *is not equal to* 1.

3°. *The following inequality holds for the function U in* 2°:

$$\int_B U \frac{\partial}{\partial n} e^{\nu(y-ix)}\, ds \neq 2\nu^2 \iint_D e^{2\nu y}\, dx\, dy + i. \tag{3.4}$$

4°. *For the function U in* 2°, $|\mathcal{D}_-| \neq 1$.

5°. *The following inequality holds*:

$$\left| \int_B U \frac{\partial}{\partial n} e^{\nu(y+ix)}\, ds \right| \neq 1. \tag{4.4}$$

6°. *For the solution G of the equation*

$$-G+TG=2\frac{\partial}{\partial n}e^{\nu(y+ix)}, \tag{5.4}$$

where T is the operator defined in (7.3), *we have*

$$-i\int_B Ge^{\nu(y-ix)}\,ds\neq 1. \tag{6.4}$$

7°. *For the solution G of equation* (5.4) *we have*

$$\Big|\int_B Ge^{\nu(y+ix)}\Big|\neq 1. \tag{7.4}$$

PROOF. $1°\leftrightarrow 2°$. Let $\mathscr{D}_+=1$. Then it follows from the asymptotic behavior of the function U (Theorem 2.2) that the function $U-e^{\nu(y+ix)}$ is a nontrivial solution of the homogeneous problem I.

Conversely, let u_0 be a nontrivial solution of problem I. By Theorem 1.4 we can assume that it is real. According to Theorem 1.2, the principal term in the asymptotic form of u_0 for $x<0$ and $r\to\infty$ has the form

$$\frac{1}{2}(\mathscr{B}-i\mathscr{A})\,e^{\nu(y+ix)}+\frac{1}{2}(\mathscr{B}+i\mathscr{A})\,e^{\nu(y-ix)}+c,$$

where $\mathscr{A}$ and $\mathscr{B}$ are real constants, which by Theorem 1.3 are not simultaneously zero.

We consider the function

$$W=\frac{2}{i\mathscr{A}-\mathscr{B}}(u_0-c)+e^{\nu(y+ix)}.$$

It is clear that

$$\frac{\partial W}{\partial n}=\frac{\partial}{\partial n}e^{\nu(y+ix)}\quad\text{on } B.$$

It follows from Theorem 1.2 that W is a solution of problem II and that for it $\mathscr{D}_+=1$. But because problem II has a unique solution, $W=U$.

$2°\leftrightarrow 3°$. By Theorem 2.2 we have

$$\mathscr{D}_+=-i\int_B U\frac{\partial}{\partial n}e^{\nu(y-ix)}\,ds+i\int_B e^{\nu(y-ix)}\frac{\partial}{\partial n}e^{\nu(y+ix)}\,ds. \tag{8.4}$$

It remains to note that

$$\int_B e^{\nu(y-ix)}\frac{\partial}{\partial n}e^{\nu(y+ix)}\,ds=\iint_D|\nabla e^{\nu(y+ix)}|^2\,dxdy=2\nu^2\iint_D e^{2\nu y}\,dxdy. \tag{9.4}$$

$2°\leftrightarrow 4°$. It follows from (8.4) and (9.4) that

$$\operatorname{Re}\mathscr{D}_+=\operatorname{Im}\int_B U\frac{\partial}{\partial n}e^{\nu(y-ix)}\,ds;$$

in conjunction with (1.3), in which $u=U$, this gives for the function U

$$|\mathscr{D}_+|^2 + |\mathscr{D}_-|^2 = 2\,\mathrm{Re}\,\mathscr{D}_+. \tag{10.4}$$

Hence $\mathscr{D}_+ = 1$ if and only if $|\mathscr{D}_-| = 1$.

4° ↔ 5°. From the formula (14.2) for U and from the equality

$$\int_B e^{\nu(y+ix)} \frac{\partial}{\partial n} e^{\nu(y+ix)}\, ds = \frac{1}{2} \int_B \frac{\partial}{\partial n} e^{2\nu(y+ix)}\, ds = 0 \tag{11.4}$$

we obtain

$$\mathscr{D}_- = -i \int_B U \frac{\partial}{\partial n} e^{\nu(y+ix)}\, ds,$$

which proves the equivalence of assertions 4° and 5°.

1° ↔ 6°. By Theorem 2.4 the unique solvability of problem I is equivalent to the triviality of the solution of the equation

$$-g + T_1 g = 0.$$

By (7.2) this last equation can be rewritten as

$$-g + Tg = -2i \left[\frac{\partial}{\partial n} e^{\nu(y+ix)}\right] \int_B e^{\nu(\eta - i\xi)} g\, ds. \tag{12.4}$$

Let g be a nontrivial solution of (12.4). By Remark 1.3 the equation $-g + Tg = 0$ has only the trivial solution, and hence

$$k = -i \int_B e^{\nu(\eta - i\xi)} g\, ds \neq 0.$$

Therefore the function $k^{-1}g$ satisfies (5.4); that is, $k^{-1}g = G$. Hence

$$-i \int_B G e^{\nu(y-ix)}\, ds = -ik^{-1} \int_B g e^{\nu(\eta - i\xi)}\, ds = 1.$$

Conversely, let

$$-i \int_B G e^{\nu(y-ix)}\, ds = 1.$$

Then (5.4) can be rewritten as

$$-G + TG = -2i \left[\frac{\partial}{\partial n} e^{\nu(y+ix)}\right] \int_B G e^{\nu(\eta - i\xi)}\, ds,$$

or, equivalently, as $-G + T_1 G = 0$. We have obtained a nontrivial solution of equation (12.4). This proves the equivalence of assertions 1° and 6°.

5° ↔ 7°. Let G be a solution of (5.4). Then the potential

$$V(z) = \int_B E_2(z, \zeta)\, G(\zeta)\, ds_\zeta$$

satisfies the boundary condition (3.0) with the boundary function $f = \partial(e^{\nu(y+ix)})/\partial n$; hence by Theorem 2.3 we have $V(z) = U(z)$ in W. Therefore

$$G=\left(\frac{\partial U}{\partial n}\right)_i-\left(\frac{\partial V}{\partial n}\right)_e$$

and

$$\int_B Ge^{\nu(y+ix)}\,ds=\int_B e^{\nu(y+ix)}\left(\frac{\partial V}{\partial n}\right)_i ds-\int_B e^{\nu(y+ix)}\frac{\partial}{\partial n}e^{\nu(y+ix)}\,ds. \tag{13.4}$$

Because V satisfies Laplace's equation in D and $V=U$ on B, we have

$$\int_B e^{\nu(y+ix)}\left(\frac{\partial V}{\partial n}\right)_i ds=\int_B U\frac{\partial}{\partial n}e^{\nu(y+ix)}\,ds.$$

In conjunction with (11.4) and (13.4) this gives

$$\int_B Ge^{\nu(y+ix)}\,ds=\int_B U\frac{\partial}{\partial n}e^{\nu(y+ix)}\,ds.$$

The theorem is proved.

REMARK 1.4. As is clear from the proof of Theorem 3.4, the following equalities hold for the function U:

$$i\mathcal{D}_-=\int_B U\frac{\partial}{\partial n}e^{\nu(y+ix)}\,ds=\int_B Ge^{\nu(y+ix)}\,ds. \tag{14.4}$$

By (10.4) for any contour B the modulus of this number does not exceed 1. Thus, if for some curve B the solution of problem I is not unique, then the modulus of the quantity in (14.4) attains an absolute maximum.

REMARK 2.4. Problem I is uniquely solvable for all $\nu>0$ except possibly for a finite number of values of ν. For by Theorem 2.4 it is sufficient to have the corresponding assertion for equation (2.4). Because the function

$$\frac{1}{\pi}\operatorname{Re}\frac{\partial}{\partial n_z}\int_0^\infty\frac{e^{it}\,dt}{t+\nu(z-\bar{\zeta})},\qquad z\in B,\quad \zeta\in B,$$

as an element of the space $C^{1,\alpha}(B\times B)$ depends analytically on the parameter ν for $\nu>0$, it follows from (1.2) that the operator T_1 depends analytically on $\nu>0$. Therefore, if there is a single value of the parameter $\nu>0$ for which the operator $I-T_1$ is invertible, then it is invertible for almost all $\nu>0$ and the operator $(I-T_1)^{-1}$ is a meromorphic function of $\nu>0$ (see [**12**]). It remains to use Kočin's result [**4**] on the unique solvability of (2.4) for all sufficiently large and small values of $\nu>0$.

REMARK 3.4. We consider the boundary-value problem formulated in Remark 2.3 with the conditions (5.2) and (6.2) replaced by (5.0). When we use the arguments employed in the proof of Remark 2.3, we obtain for this problem a theorem that is completely similar to Theorem 3.4. In it the role of the function U is played by the solution of the problem in Remark 2.3

with the boundary function $f=(1-\mu)\partial(e^{\nu(y+ix)})/\partial n$, and the role of the integral equation (3.4) is played by the equation

$$-(1+\mu)G+(1-\mu)TG=2(1-\mu)\partial(e^{\nu(y+ix)})/\partial n.$$

§5. The case of a fluid of finite depth

Let S be the strip $\{(x,y):-h<y<0\}$, $h<\infty$, and let W denote the domain $S\backslash\overline{D}$. Here we consider the question of the unique solvability of problem I in W as formulated in the Introduction. Since the proofs in this section are similar to those used in the case $h=\infty$, we confine ourselves to a brief exposition.

The fundamental solution $E_1(z,\zeta)$ of problem I in the strip S has been constructed in [13] and [9]; its asymptotic form as $r\to\infty$ has also been obtained in these papers. Let Q be a function that is analytic in z (for $z\neq\zeta$) and such that $\operatorname{Re}Q=E_1$. Then

$$\frac{dQ}{dz}=\frac{\nu}{2(\nu h-1)}-\frac{1}{2\pi(z-\zeta)}-\frac{1}{2\pi(z-\overline{\zeta}+2ih)}$$

$$-\frac{1}{\pi}\int_0^\infty(\nu+\lambda)e^{-\lambda h}\frac{\cosh\lambda(\eta+h)\sin\lambda(z-\zeta+ih)}{\nu\sinh\lambda h-\lambda\cosh\lambda h}\,d\lambda \tag{1.5}$$

$$+\nu\frac{\cosh\lambda_0(\eta+h)}{\nu h-\cosh^2\lambda_0 h}\cos\lambda_0(z-\xi+ih),$$

where λ_0 is a positive root of the equation $\nu\tanh\lambda h=\lambda$. This equation has no positive solutions for $\nu h<1$ and only one such solution for $\nu h>1$. In the first case the last term in (1.5) must be omitted, and in the second case the integral in (1.5) is to be understood in the sense of the principal value. accordingly, (see [13] and [9]) the fundamental solution in the first case has as $r\to\infty$ the form

$$E_1=O(x^{-1}),\quad |\nabla E_1|=O(x^{-2})\quad\text{as }x\to+\infty, \tag{2.5}$$

$$E_1=\frac{\nu}{\nu h-1}(x-\xi)+\varphi,\quad \varphi=O(|x|^{-1}),\quad |\nabla\varphi|=O(x^{-2})\quad\text{as }x\to-\infty,$$

and in the second case the estimates as $x\to+\infty$ remain the same as for $x\to-\infty$:

$$E_1=\frac{\nu}{\nu h-1}(x-\xi)-2\frac{\nu\cosh\lambda_0(\eta+h)\cosh\lambda_0(y+h)}{\lambda_0(\cosh^2\lambda_0 h-\nu h)}\sin\lambda_0(x-\xi)+\psi,$$

$$\psi=O(|x|^{-1}),\qquad |\nabla\psi|=O(x^{-2}).^{2)}$$

---

2) By using the arguments in [14] one can show that ψ, $|\nabla\psi|$, and all the quantities in (2.5) that tend to 0 as $|x|\to\infty$ also tend to 0 as $|x|\to\infty$ exponentially (this result is not essential for us). Hence in all the asymptotic formulas given below the remainder terms decrease exponentially.

When we take the Fourier transformation with respect to x we obtain for the strip S an assertion analogous to Lemma 2.2. Here the solutions of the homogeneous problem have the form c_1x+c_2 for $\nu h<1$, and the form

$$c_1x+c_2+\cosh\lambda_0(y+h)\,[c_3e^{i\lambda_0x}+c_4e^{-i\lambda_0x}]$$

for $\nu h>1$. The subsequent arguments of the second section can be automatically carried over to the problem being considered here. As a result we obtain the following theorem.

Theorem 1.5. *Let u be a solution of problem* I. *Then for $\nu h<1$*

$$u(x,y)=\begin{cases} c_+ + \psi_+(x,y), & x>0,\\ \gamma x + c_- + \psi_-(x,y), & x<0,\end{cases}$$

where

$$\gamma=\frac{\nu}{1-\nu h}\int_B \frac{\partial u}{\partial n}\,ds,$$

the constants $c_\pm$ are connected by the relation

$$c_+-c_-=\frac{2\nu}{1-\nu h}\int_B\left[x\frac{\partial u}{\partial n}-u\cos(x,n)\right]ds$$

and $\psi_\pm=O(|x|^{-1})$ and $|\nabla\psi_\pm|=O(x^{-2})$ as $|x|\to\infty$. For $\nu h>1$ we have

$$u(x,y)=\begin{cases} c_+ + \psi_+(x,y), & x>0,\\ \gamma x+c_- + \cosh\lambda_0(y+h)(\mathcal{A}\sin\lambda_0x+\mathcal{B}\cos\lambda_0x)+\psi_-(x,y), & x<0,\end{cases}$$

where the constants γ and c_+-c_- and the estimates for $\psi_\pm$ and $\nabla\psi_\pm$ are the same as in the case $\nu h<1$, and the constants $\mathcal{A}$ and $\mathcal{B}$ are given by

$$\mathcal{A}=\frac{-2\nu}{\lambda_0(\cosh^2\lambda_0h-\nu h)}\int_B\left\{u\frac{\partial}{\partial n}\left[\cosh\lambda_0(y+h)\cos\lambda_0x\right]-\frac{\partial u}{\partial n}\cosh\lambda_0(y+h)\cos\lambda_0x\right\}ds,$$

$$\mathcal{B}=\frac{2\nu}{\lambda_0(\cosh^2\lambda_0h-\nu h)}\int_B\left\{u\frac{\partial}{\partial n}[\cosh\lambda_0(y+h)\sin\lambda_0x]-\frac{\partial u}{\partial n}\cosh\lambda_0(y+h)\sin\lambda_0x\right\}ds.$$

We now note that Lemma 2.1 remains true for the domain W considered in this section. Its proof, although essentially the same, is actually somewhat simpler. Therefore if $x\cos(n,x)\geqq 0$ on B, then the solution of the homogeneous problem I that satisfies the additional condition (1.1) is a constant.

By Theorem 1.5, for $\nu h<1$ any solution u of the homogeneous problem

I satisfies the condition (1.1), and hence $u \equiv$ const. To prove the existence theorem we seek a solution of problem I in the form of a potential with the kernel $E_1(z, \zeta)$ and an unknown density g. Then, in exactly the same way as in §3, we obtain a Fredholm equation on the contour B. When we repeat the proof of Theorem 2.3 we obtain the following result:

THEOREM 2.5. *Suppose that $x\cos(n,x) \geq 0$ on B. Then for $\nu h < 1$ problem* I *has one and only one solution (up to a constant term) for any $f \in C^{1,\alpha}(B)$.*

To analyze the case $\nu h > 1$ we consider the auxiliary problem II, which differs from problem I only by the conditions at infinity. The relations (5.0) are replaced by the "radiation conditions"

$$u(x,y) = \gamma|x| + c\,\mathrm{sign}\,x + v(x,y),$$

where γ and c are certain constants, and

$$\sup_W |\nabla u| < \infty, \qquad \lim_{|x|\to\infty}\left(\frac{\partial v}{\partial |x|} - i\lambda_0 v\right) = 0.$$

The fundamental solution of this problem is the function

$$E_2(z, \zeta) = E_1(z, \zeta) - i\frac{\nu\cosh\lambda_0(\eta + h)\cosh\lambda_0(y+h)}{\lambda_0(\cosh^2\lambda_0 h - \nu h)}e^{i\lambda_0(x-\xi)} + \frac{\nu}{2(1-\nu h)}(x-\xi). \tag{3.5}$$

As in the case of Theorem 1.5, we obtain the following assertion:

THEOREM 3.5. *Suppose that $\nu h > 1$ and u is a solution of problem* II. *Then*

$$u(x, y) = \begin{cases} \gamma x + c + \mathcal{D}_+\cosh\lambda_0(y+h)\,e^{i\lambda_0 x} + \psi_+(x, y), & x > 0, \\ -\gamma x - c + \mathcal{D}_-\cosh\lambda_0(y+h)\,e^{-i\lambda_0 x} + \psi_-(x, y), & x < 0, \end{cases} \tag{4.5}$$

where $\psi_\pm = O(|x|^{-1})$ and $|\nabla\psi_\pm| = O(x^{-2})$ as $x \to \infty$, and the constants γ, c, $\mathcal{D}_+$ and $\mathcal{D}_-$ are given by the formulas

$$\gamma = \frac{\nu}{2(\nu h - 1)}\int_B \frac{\partial u}{\partial n}\,ds, \qquad c = \frac{\nu}{2(1-\nu h)}\int_B\left[x\frac{\partial u}{\partial n} - \cos(x, n)\,u\right]ds,$$

$$\mathcal{D}_\pm = \frac{i\nu}{\lambda_0(\cosh^2\lambda_0 h - \nu h)}\int_B\left\{\frac{\partial u}{\partial n}\cosh\lambda_0(y+h)\,e^{\mp i\lambda_0 x} - u\frac{\partial}{\partial n}\left[\cosh\lambda_0(y+h)\,e^{\mp i\lambda_0 x}\right]\right\}ds.$$

When $\int_B(\partial u/\partial n)\,ds = 0$ the following formula, which corresponds to (1.3), holds for the solutions of problem II:

$$|\mathcal{D}_+|^2 + |\mathcal{D}_-|^2 = \frac{2\nu}{\lambda_0(\cosh^2\lambda_0 h - \nu h)}\,\mathrm{Im}\int_B u\frac{\partial\bar{u}}{\partial n}\,ds. \tag{5.5}$$

To prove this we must also write down Green's theorem for the domain $W \cap \{(x,y) : |x| < \rho\}$, which must now be averaged with respect to ρ with limits from R to $2R$. We then take the limit $R \to \infty$. The details of the proof are mainly the same as for Lemma 1.3.

It follows from Theorem 3.5 and equation (5.5) that the solution u of the homogeneous problem II satisfies the condition (1.1); hence if $x\cos(n,x) \geqq 0$ on B, then $u \equiv \text{const}$. When we apply Theorem 3.5 again we find that $u \equiv 0$. The solvability of the inhomogeneous problem II is also obtained in the same way as in the case $h = \infty$ (Theorem 2.3).

Then by complete analogy with Theorems 1.4 and 2.4 we find that the homogeneous problem I has not more than one nontrivial solution (up to a constant term and a factor) and that the formula

$$u = \int_B E_1(z, \zeta)\, g(\zeta)\, ds_\zeta + c$$

establishes a one-to-one correspondence between the solutions of problem I and the pairs (g,c), where $c = \text{const}$ and g is the solution of the Fredholm integral equation

$$-g + T_1 g = 2f, \quad (T_1 g)(z) = 2\int_B \frac{\partial E_1(z, \zeta)}{\partial n_z} g(\zeta)\, ds_\zeta. \tag{6.5}$$

Finally, the following theorem holds.

THEOREM 4.5. *Suppose that* $x\cos(n,x) \geqq 0$ *on* B *and* $\nu h > 1$, $h < \infty$. *Then the following assertions are equivalent*:

1°. *Problem* I *has one and only one solution* (*up to a constant term*) *for any* $f \in C^{1,\alpha}(B)$.

2°. *For the solution* U *of problem* II *with the boundary function*

$$f = \frac{\partial}{\partial n}\left[\cosh \lambda_0 (y+h)\, e^{i\lambda_0 x}\right]$$

the constant $\mathscr{D}_+$ *in the asymptotic form* (4.5) *is not equal to* 1.

3°. *The following inequality holds for the function* U *in* 2°:

$$\int_B U \frac{\partial}{\partial n} \cosh \lambda_0 (y+h)\, e^{-i\lambda_0 x}\, ds \neq \lambda_0^2 \iint_D \cosh 2\lambda_0 (y+h)\, dxdy$$
$$+ \lambda_0 \nu^{-1} (\cosh^2 \lambda_0 h - \nu h)\, i.$$

4°. *For the function* U *in* 2°, $|\mathscr{D}_-| \neq 1$.

5°. *The following inequality holds*:

$$\frac{\nu}{\lambda_0 (\cosh^2 \lambda_0 h - \nu h)} \left| \int_B U \frac{\partial}{\partial n} \left[\cosh \lambda_0 (y+h)\, e^{i\lambda_0 x}\right] ds + \lambda_0^2 \iint_D e^{2i\lambda_0 x}\, dxdy \right| \neq 1.$$

6°. *For the solution* G *of the equation*

(7.5) $$-G+TG=2\frac{\partial}{\partial n}[\cosh\lambda_0(y+h)e^{i\lambda_0 x}],$$

where

$$(TG)(z)=2\int_B\frac{\partial E_2(z,\zeta)}{\partial n_z}g(\zeta)\,ds_\zeta,$$

we have the inequality

$$-i\int_B G\cosh\lambda_0(y+h)e^{-i\lambda_0 x}\,ds\neq\lambda_0\nu^{-1}(\cosh^2\lambda_0 h-\nu h).$$

7°. *For the solution* G *of* (7.5) *we have*

$$\left|\int_B G\cosh\lambda_0(y+h)e^{i\lambda_0 x}\,ds\right|\neq\lambda_0\nu^{-1}(\cosh^2\lambda_0 h-\nu h).$$

This theorem is proved in exactly the same way as Theorem 3.4; one must make only a small addition in the proof of the equivalence of 1° and 6°. When we use (3.5) instead of (7.2) we find that the equation

(8.5) $$-g+T_1g=0,$$

where the operator T_1 is given by (6.5), is equivalent to the equation

(9.5) $$-g+Tg=-2i\left[\frac{\partial}{\partial n}\frac{\nu\cosh\lambda_0(y+h)e^{i\lambda_0 x}}{\lambda_0(\cosh^2\lambda_0 h-\nu h)}\right]\int_B\cosh\lambda_0(\eta+h)e^{-i\lambda_0\xi}g(\zeta)\,ds_\zeta - \frac{\nu\cos(n,x)}{1-\nu h}\int_B g(\zeta)\,ds_\zeta.$$

Because $E_1(z,\zeta)$ differs from $-\log|z-\zeta|/2\pi$ by a harmonic function, for $\zeta\in B$ we have

$$\int_B\frac{\partial E_1(z,\zeta)}{\partial n_z}ds_z=\frac{-1}{2\pi}\int_B\frac{\partial}{\partial n_z}\log|z-\zeta|\,ds_z=-\frac{1}{2}.$$

Therefore when we integrate (8.5) over B we find that $\int_B g(\zeta)ds_\zeta=0$. But then we can discard the last term in (9.5), after which we obtain an equation similar to (12.4). The subsequent arguments are similar to those used in the proof of Theorem 3.4. This proves Theorem 4.5.

Received 12/NOV/71

Bibliography

[1] J. J. Stoker, *Water waves: the mathematical theory with applications*, Pure and Appl. Math., vol. 4, Interscience, New York, 1957; Russian transl., IL, Moscow, 1959. MR **21** #2438.

[2] A. A. Kostjukov, *Water resistance to the motion of ships*, "Sudostroenie", Leningrad, 1966. (Russian) RŽMeh **1967** #8B504.

[3] H. Lamb, *Hydrodynamics*, 6th ed., Cambridge Univ. Press, Cambridge, 1932; Russian transl., OGIZ, Moscow, 1947.

[4] N. E. Kočin, *On the wave-making resistance and lift of bodies submerged in water,* Proc. Conf. on Theory of Wave Resistance, Central Aero-Hydrodynamical Inst., Moscow, 1937, pp. 65-134; reprinted in *Collected works.* Vol. 2, Izdat. Akad. Nauk SSSR, Moscow, 1949, pp. 105-182; English transl., Soc. Naval Architects Marine Engrs. Tech. Res. Bull. **1**–**8** (1951). MR **11**, 573, 812; **13**, 997.

[5] ———, *A conference on wave resistance,* Uspehi Mat. Nauk **3** (1937), 194-221; reprinted in *Collected works.* Vol. 2, Izdat. Akad. Nauk SSSR, Moscow, 1949, pp. 183-214. (Russian) MR **11**, 573, 872.

[6] N. N. Moiseev, Introduction to the book *Theory of surface waves,* IL, Moscow, 1959, pp. 5-27. (Russian)

[7] A. S. Peters and J. J. Stoker, *The motion of a ship, as a floating rigid body, in a seaway,* Comm. Pure Appl. Math. **10** (1957), 399-490; Russian transl. in *Theory of surface waves,* IL, Moscow, 1959, pp. 43-144. MR **19**, 703.

[8] M. V. Keldyš, *Remarks on some movements of a heavy fluid,* Tehn. Zametki Central. Aero-Gidrodinam. Inst. **52** (1935), 5-9. (Russian)

[9] A. I. Tihonov, *The two-dimensional problem on the motion of a wing under the surface of a heavy fluid of finite depth,* Izv. Akad. Nauk SSSR Otd. Tehn. Nauk **1940**, no. 4, 57-78. (Russian)

[10] T. H. Havelock, *Wave resistance,* Proc. Roy. Soc. London Ser. A **118** (1928), 24-33.

[11] L. N. Sretenskiĭ, *Motion of a cylinder under the surface of a heavy fluid,* Trudy Central. Aero-Gidrodinam. Inst. Vyp. 346 (1938); English transl., NACA Tech. Memo. no. 1335, 1952. MR **14**, 508.

[12] I. C. Gohberg and M. G. Kreĭn, *Introduction to the theory of linear nonselfadjoint operators in Hilbert space,* "Nauka", Moscow, 1965; English transl., Transl. Math. Monographs, vol. 18, Amer. Math. Soc., Providence, R. I., 1969. MR **36** #3137; **39** #7447.

[13] M. D. Haskind, *Translation of bodies under the free surface of a heavy fluid of finite depth,* Prikl. Mat. Meh. **9** (1945), 67-78. (Russian) MR **8**, 235.

[14] G. Kreisel, *Surface waves,* Quart. Appl. Math. **7** (1949), 21-44. MR **11**, 227.

[15] M. V. Keldyš and M. A. Lavrent'ev, *On the motion of a wing under the surface of a heavy fluid,* Proc. Conf. on Theory of Wave Resistance, Central Aero-Hydrodynamical Inst., Moscow, 1937, pp. 31-64. (Russian)

[16] B. R. Vaĭnberg and V. G. Maz'ja, *On some stationary problems in the linear theory of surface waves,* Dokl. Akad. Nauk SSSR **205** (1972), 310-313 = Soviet Physics Dokl. **17** (1972/73), 640-643.

Translated by
J. B. BARBOUR

Trudy Moskov. Mat. Obšč.
Tom 28 (1973)

Trans. Moscow Math. Soc.
Vol. 28 (1973)

ON THE PROBLEM OF THE STEADY-STATE OSCILLATIONS OF A FLUID LAYER OF VARIABLE DEPTH

B. R. VAĬNBERG and V. G. MAZ'JA

Contents

Introduction

In this paper we find conditions for the unique solvability of the problem of finding the velocity potential $u(x,y,z,t) = \mathrm{Re}\{e^{-i\omega t}u(x,y,z)\}$ of a layer of heavy incompressible fluid of variable depth that oscillates under the influence of a periodic surface pressure or a periodic force.

It is assumed that the fluid occupies a domain V in the half-space $R^3_- = \{(x,y,z) : z < 0\}$, whose boundary consists of two nonintersecting components: the plane $z = 0$ and a continuously differentiable surface B at a finite distance from the plane $z = 0$. Additional assumptions about V will be formulated below.

The problem of the steady-state oscillations of a fluid under the influence of a surface pressure $p(x,y,t) = p(x,y)\sin \omega t$ reduces to the following stationary problem: To find a function $u(x,y,z)$ that is harmonic in V and satisfies the boundary conditions

$$\frac{\partial u}{\partial z} - \nu u = -\frac{\omega}{\rho g} p(x,y) \text{ for } z = 0, \quad \frac{\partial u}{\partial n} = 0 \text{ on } B, \tag{1.0}$$

where $\nu = \omega^2/g$, g is the acceleration of free fall, and ρ is the density of the fluid. A solution must also satisfy a radiation condition, which, for example, in the case when V coincides at infinity with the layer $0 > z > -h$ has the form

$$u = O(r^{-1/2}), \quad \frac{\partial u}{\partial r} - i\lambda_0 u = o(r^{-1/2}) \text{ for } r = (x^2 + y^2)^{1/2} \to \infty;$$

here λ_0 is a positive root of the equation $\lambda \tanh \lambda h = \nu$.

AMS (MOS) subject classifications (1970). Primary 31A05, 31A25, 35A05, 35B40, 35D05, 35J25, 76D99; Secondary 30A24, 46E30, 46E35.

Similar problems are obtained when at the point (x_0, y_0, z_0) of the fluid, or its surface or bottom, there is a source whose intensity oscillates harmonically; in this case, either Laplace's equation in V is replaced by the equation $\Delta u = c\delta(x - x_0, y - y_0, z - z_0)$, or there is a δ-function on the right-hand side of one of the boundary conditions.

Below we also consider the two-dimensional analog of this problem.

We briefly dwell on the results known so far that have in one way or another a bearing on our subject. Much attention has been devoted to cases that allow an exact solution (see, for example, [1]–[11]); in particular, an exhaustive study has been made of the problem of the steady-state oscillations of a fluid in a two-dimensional strip, in a layer between two planes, in a half-plane, and in a half-space.

Oscillations of a fluid of infinite depth in the presence of an immersed body of arbitrary form have been studied by Kočin [8], [9], who proved that the problem is solvable for large and small values of ν. Haskind [10] extended Kočin's method to the case of a fluid of finite depth. In a series of papers, Ursell ([12]–[16] and elsewhere) has made a detailed study of the same problem for an immersed or semi-immersed circular cylinder. For example, he has proved in [12] that the plane problem of an immersed circular cylinder is uniquely solvable for all values of ν. He has given in [13] two examples when the homogeneous three-dimensional problem has a nontrivial solution with finite energy; one of these refers to the case of a circular cylinder of fairly small radius immersed in a fluid of finite extent.

Subsequently, this restriction on the radius was lifted (by means of a different method) by Jones [17], who also showed that nontrivial solutions can arise in the case of a channel of finite width whose bottom has a projection. The uniqueness of the solution of the problem of the oscillations of a fluid of constant depth due to a floating body has been proved by John [18] under the assumption that any vertical straight line intersecting the body does not intersect the free surface. The existence of a solution is proved in [18] under the condition that the immersed surface is the lower part of a smooth convex surface that is symmetric about the plane $z = 0$. Similar results are obtained in [18] for two-dimensional motion in the presence of cylindrical obstacles.

Kreisel [6] has discussed the plane problem of the oscillations of a fluid of variable depth. He proved in [6] that this problem is uniquely solvable under the assumption that the domain occupied by the fluid can be mapped into a strip by a conformal mapping that is near (in some known sense) to the identity.

In the theorems of the present paper there are no restrictions on smallness. The number ν can take all positive values, and the surface B must satisfy one of the two basic requirements dictated by the method of proving the uniqueness theorems for a solution with finite energy. The first of these conditions (the inequality (2.1)) means that the intersection of $\overline{V}$ with any

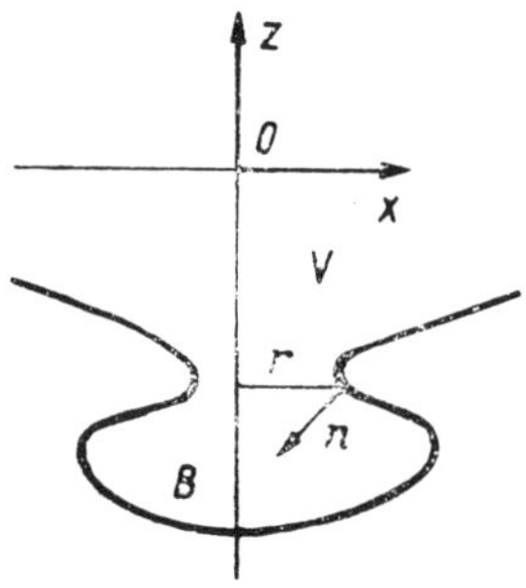

FIGURE 1

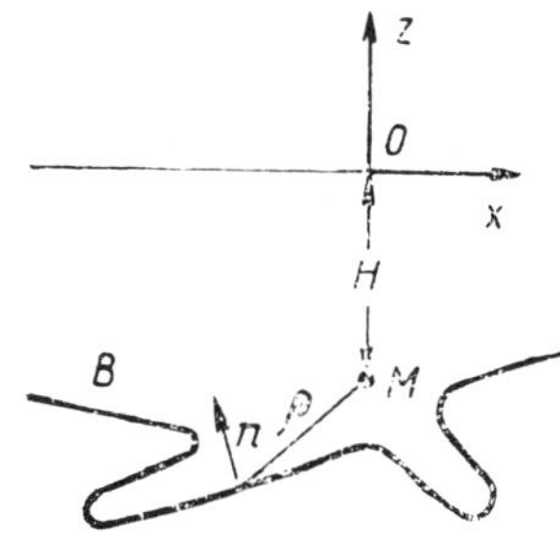

FIGURE 2

horizontal plane $z = -H$, $H > 0$, is starlike relative to the point $(0,0,-H)$ (see Figure 1). The alternative second condition (the inequality (10.1)) is that $\overline{V}$ is starlike relative to some point at a depth H, $H\nu \leqq 1$ (see Figure 2). The restrictions on the curve B in the two-dimensional case are reformulated accordingly.

This volume also contains our paper [**19**], which is devoted to the plane problem of the motion of a body immersed in a fluid. Conceptually the two papers are to some degree close, but they can be read independently.

The results of this paper were published without proofs in the note [**22**].

§1. The uniqueness theorems for a solution with finite energy

Let (r,φ,z) be cylindrical coordinates in R^3. We introduce the following notation: K_R is the disc $r < R$ on the plane $z = 0$; V_R is the intersection of the cylinder $r < R$ with the domain V specified in the Introduction; Γ_R is the lateral surface of V_R; $B_R = B \cap V_R$; and n is the normal to B directed into V.

We begin with the following simple assertion of a technical nature.

Lemma 1.1. *Suppose that the function u is continuously differentiable in $\overline{V}$ and satisfies the condition*

$$\int_{z=0} |u|^2\,dxdy + \int_V |\nabla u|^2\,dx\,dy\,dz < \infty \tag{1.1}$$

and suppose that on the surface B

$$\cos(n,r) \leqq 0. \tag{2.1}$$

Then

$$\lim_{\overline{R\to\infty}} R^2 \ln R\Big(\int_{\partial K_R} |u|^2\,d\varphi + \int_{\Gamma_R} |\nabla u|^2\,d\varphi\,dz\Big) = 0, \tag{3.1}$$

$$\lim_{R\to\infty} (\ln R)^{-1} \int_{\Gamma_R} |u|^2\,d\varphi\,dz = 0. \tag{4.1}$$

Proof. The equality (3.1) is obvious. We prove (4.1). For $R > \rho > 0$ we have

$$\int_{\Gamma_R} |u|^2 \, d\varphi \, dz \leqslant 2 \int_{\Gamma_\rho} |u|^2 \, d\varphi \, dz + 2 \ln \frac{R}{\rho} \int_{V_R \setminus V_\rho} \left| \frac{\partial u}{\partial \rho} \right|^2 r \, dr \, d\varphi \, dz. \tag{5.1}$$

(Here we have used the fact that by (2.1) the intersection of any ray $\{\varphi = \varphi_0, z = z_0\}$ with $\overline{V}$ is connected.) From (5.1) we obtain

$$\overline{\lim_{R\to\infty}} (\ln R)^{-1} \int_{\Gamma_R} |u|^2 \, d\varphi \, dz \leqslant \int_{V \setminus V_\rho} |\nabla u|^2 \, dx \, dy \, dz.$$

It remains to note that the right-hand side tends to zero as $\rho \to \infty$. The lemma is proved.

Lemma 2.1. *Let the function u be continuously differentiable in* $\overline{V}$, *harmonic in V, satisfy the condition* (1.1), *and be such that*

$$\partial u / \partial z - \nu u = 0 \quad \text{for } z = 0 \tag{6.1}$$

and $\partial u / \partial n$ *has compact support on B.*

Suppose that the surface B satisfies (2.1). *Then*

$$\begin{aligned} &2 \int_V (|u_x|^2 + |u_y|^2) \, dx \, dy \, dz - \int_B r \cos(r, n) |\nabla u|^2 \, ds \\ &= -2 \operatorname{Re} \int_B \left(r \frac{\partial \bar{u}}{\partial r} + \bar{u} \right) \frac{\partial u}{\partial n} \, ds \end{aligned} \tag{7.1}$$

and hence if $\partial u / \partial n \equiv 0$ *on B, then* $u \equiv 0$.

Proof. It is straightforward to verify that the following identity holds:

$$\begin{aligned} \operatorname{Re} [(r\bar{u}_r + \bar{u}) \Delta u] &= \operatorname{Re} \operatorname{div} [(r\bar{u}_r + \bar{u}) \nabla u] \\ &- |u_x|^2 - |u_y|^2 - \frac{1}{2} (x |\nabla u|^2)_x - \frac{1}{2} (y |\nabla u|^2)_y. \end{aligned} \tag{8.1}$$

When we integrate over V_R we obtain

$$\begin{aligned} &\operatorname{Re} \int_{K_R} (r\bar{u}_r + \bar{u}) u_z \, dx \, dy - \int_{V_R} (|u_x|^2 + |u_y|^2) \, dx \, dy \, dz \\ &+ \frac{1}{2} \int_{B_R} [x \cos(n, x) + y \cos(n, y)] |\nabla u|^2 \, ds \\ &= \operatorname{Re} \int_{B_R} (r\bar{u}_r + \bar{u}) u_n \, ds - \operatorname{Re} \int_{\Gamma_R} (r\bar{u}_r + \bar{u}) u_r \, ds \\ &+ \frac{1}{2} \int_{\Gamma_R} (x \cos(r, x) + y \cos(r, y)) |\nabla u|^2 \, ds. \end{aligned} \tag{9.1}$$

We transform the first term on the left-hand side by using the boundary condition (6.1):

$$\operatorname{Re}\int_{K_R}(r\bar{u}_r+\bar{u})\,u_z\,dx\,dy=\nu\operatorname{Re}\int_{K_R}\bar{u}_r\,ur^2\,dr\,d\varphi$$

$$+\nu\int_{K_R}|u|^2\,dx\,dy=\frac{\nu}{2}R^2\int_{\partial K_R}|u|^2\,d\varphi.$$

By Lemma 1.1 there exists a sequence $\{R_k\}$, $R_k\to\infty$ as $k\to\infty$, such that

$$\frac{\nu}{2}R_k^2\int_{\partial K_{R_k}}|u|^2\,d\varphi+\operatorname{Re}\int_{\Gamma_{R_k}}(r\bar{u}_r+\bar{u})\,u_r\,ds-\frac{R_k}{2}\int_{\Gamma_{R_k}}|\nabla u|^2\,ds\to 0.$$

Therefore when we take the limit in (9.1), where $R=R_k$, we obtain the lemma.

Lemma 3.1. *Suppose that the domain $\bar{V}$ is starlike relative to some point M at a depth H, $0\leq H\leq\nu^{-1}$; that is,*

$$\cos(n,\rho)\leq 0 \tag{10.1}$$

on B, where ρ is the radius vector of a point with origin at M. Let u be a function that is continuously differentiable in $\bar{V}$, harmonic in V, and satisfies the condition

$$\int_{z=0}|u|^2\,dx\,dy+\int_V(|\nabla u|^2+|\rho|^{-2}|u|^2)\,dx\,dy\,dz<\infty. \tag{11.1}$$

Suppose that u also satisfies the boundary condition (6.1) and $\partial u/\partial n$ has compact support on B. Then

$$\nu(1-H\nu)\int_{z=0}|u|^2\,dx\,dy+H\int_{z=0}(|u_x|^2+|u_y|^2)\,dx\,dy-\int_B|\rho|\cos(\rho,n)\,|\nabla u|^2\,ds$$

$$=-\operatorname{Re}\int_B\left(2|\rho|\frac{\partial\bar{u}}{\partial\rho}+\bar{u}\right)\frac{\partial u}{\partial n}\,ds \tag{12.1}$$

and so if $\partial u/\partial n\equiv 0$ on B, then $u\equiv 0$.

Proof. We can assume that M has the Cartesian coordinates $(0,0,-H)$. The following identity can be verified directly:

$$\begin{aligned}\operatorname{Re}[(2|\rho|\bar{u}_\rho+\bar{u})\,\Delta u]&=\operatorname{Re}\operatorname{div}[(2|\rho|\bar{u}_\rho+\bar{u})\nabla u]\\&-(x|\nabla u|^2)_x-(y|\nabla u|^2)_y-[(z+H)|\nabla u|^2]_z.\end{aligned} \tag{13.1}$$

When we integrate (13.1) over V_R we obtain

$$\operatorname{Re}\int_{K_R}(2|\rho|\bar{u}_\rho+\bar{u})\,u_z\,dx\,dy-H\int_{K_R}|\nabla u|^2\,dxdy$$

$$+\int_{B_R}|\rho|\cos(n,\rho)\,|\nabla u|^2\,ds+\operatorname{Re}\int_{\Gamma_R}(2|\rho|\bar{u}_\rho+\bar{u})\,u_r\,ds- \tag{14.1}$$

$$-\int_{\Gamma_R}[x\cos(r,x)+y\cos(r,y)]\,|\nabla u|^2\,ds=\operatorname{Re}\int_{B_R}\left(2|\rho|\frac{\partial\bar u}{\partial\rho}+\bar u\right)\frac{\partial u}{\partial n}\,ds.$$

When we use (6.1) we find that the first two terms on the left are equal to

$$2\nu\operatorname{Re}\int_{K_R} r\bar u_r u\,dx\,dy+H\nu^2\int_{K_R}|u|^2\,dx\,dy+\nu\int_{K_R}|u|^2\,dx\,dy$$

$$-H\int_{K_R}(|u_x|^2+|u_y|^2)\,dx\,dy=\nu(H\nu-1)\int_{K_R}|u|^2\,dx\,dy$$

$$-H\int_{K_R}(|u_x|^2+|u_y|^2)\,dx\,dy+\int_{\partial K_R}\nu R^2|u|^2\,d\varphi.$$

By (11.1) the integrals over Γ_R in (14.1) and the integral over ∂K_R in the last equality tend to zero when R ranges over a certain sequence R_k that tends to infinity. When in (14.1) we take the limit as $R=R_k\to\infty$, we obtain (12.1). The lemma is proved.

We now consider the two-dimensional case. Here we denote by V a domain in $R^2_-=\{(x,y):y<0\}$ whose boundary consists of the straight line $y=0$ and a continuously differentiable curve B that does not intersect this straight line and is a finite distance from it.

Lemma 4.1. *Let the function u be continuously differentiable in $\bar V$, harmonic in V, satisfy the condition*

$$\int_{y=0}|u|^2\,dx+\int_V|\nabla u|^2\,dx\,dy<\infty, \tag{15.1}$$

and be such that

$$\partial u/\partial y-\nu u=0 \quad \textit{for } y=0, \tag{16.1}$$

and the support of $\partial u/\partial n$ on B is compact. Suppose that the curve B satisfies the condition

$$x\cos(n,x)\leqq 0. \tag{17.1}$$

Then

$$2\int_V|u_x|^2\,dx\,dy-\int_B x\cos(n,x)\,|\nabla u|^2\,ds=-\operatorname{Re}\int_B(2x\bar u_x+\bar u)\,u_n\,ds \tag{18.1}$$

and so if $u_n=0$ on B, then $u\equiv 0$.

The proof is similar to that of Lemma 2.1 and is based on the identity

$$\begin{aligned}\operatorname{Re}[(2x\bar u_x+\bar u)\,\Delta u]&=\operatorname{Re}\operatorname{div}[(2x\bar u_x+\bar u)\nabla u]\\&\quad-2|u_x|^2-(x|\nabla u|^2)_x.\end{aligned} \tag{19.1}$$

Lemma 5.1. *Let the function u be continuously differentiable in $\bar V$, harmonic in V, satisfy the conditions* (15.1) *and* (16.1), *and be such that the support of*

$\partial u/\partial n$ on B is compact. Suppose that $\overline{V}$ is starlike relative to some point M at a depth H, $0 \leqq H \leqq \nu^{-1}$. Then

$$\nu(1 - H\nu) \int\limits_{y=0} |u|^2 dx + H \int\limits_{y=0} |u_x|^2 dx$$

$$- \int\limits_B |\rho| \cos(\rho, n) |\nabla u|^2 ds = -2\mathrm{Re} \int\limits_B |\rho| \frac{\partial \bar{u}}{\partial \rho} \frac{\partial u}{\partial n} ds,$$

where ρ is the radius vector of a variable point of the surface B with origin at the point $M = (0, -H)$. Hence if $\partial u/\partial n = 0$ on B, then $u \equiv 0$.

The proof is carried out in exactly the same way as that of Lemma 2.1. It is based on the identity

$$2\mathrm{Re}(|\rho| \bar{u}_\rho \Delta u) = 2\mathrm{Re}[\mathrm{div}(|\rho| \bar{u}_\rho \nabla u)]$$
$$- (x|\nabla u|^2)_x - [(y + H)|\nabla u|^2]_y.$$

REMARK. To conclude this section we consider the three-dimensional problem in the channel $K = V \times \{y : 0 < y < l\}$, where l is a positive constant, and V is a domain in the (x, z) plane of the type considered above, and satisfies the condition (17.1).

PROPOSITION. *Let the function u be continuously differentiable in $\overline{K}$, harmonic in K, satisfy the condition* (6.1) *and the condition $\partial u/\partial n = 0$ on $\partial K \cap \{z < 0\}$, and be such that*

$$\int\limits_{z=0} |u|^2 dx\, dy + \int\limits_K |\nabla u|^2 dx\, dy\, dz < \infty.$$

Then $u \equiv 0$.

For the proof it is sufficient to integrate the identity (19.1) over K; this identity holds for functions of any number of variables.

Thus the condition (17.1) ensures that the solution of the problem in the channel is unique. This condition is essential for the validity of the proposition; for, as D. S. Jones has shown in [**17**], a projection on the bottom of the channel generates nontrivial solutions of the homogeneous problem that decrease rapidly at infinity (trapping modes, in Ursell's terminology [**14**]). These solutions have the form $u(x, z)\cos(s\pi y/l)$, where s is a natural number and the function $u(x, z)$ is the solution of the plane problem

$$\Delta u - \lambda^2 u = 0 \text{ in } V, \quad \lambda = s\pi l^{-1}; \qquad u_z - \nu u = 0 \text{ for } z = 0, \quad \nu > 0; \tag{20.1}$$
$$u_n = 0 \text{ on } B.$$

The condition (17.1) then eliminates a raising of the bottom and allows a deepening. If this condition is fulfilled, it follows from the proposition formulated above that the problem (20.1) has not more than one solution (with finite energy) for all $\lambda \geqq 0$ (that is, the assertion of Lemma 4.1 about uniqueness is generalized to solutions of the equation $\Delta u - \lambda^2 u = 0$). In this connection, we note that in Lemma 5.1 uniqueness is also obtained for domains with projections on the bottom. However, in contrast to the work of Jones, here $\lambda = 0$.

§2. The solvability of the boundary-value problem in the three-dimensional case

Let S denote the layer $0 > z > -h$ $(0 < h < \infty)$ in the space R^3. We suppose that the domain $V \subset R^3$ is obtained from S by means of a one-to-one thrice continuously differentiable mapping $\psi : \bar{S} \to \bar{V}$ that is the identity for $r = (x^2 + y^2)^{1/2} > a$, $a = \text{const}$, and leaves the upper boundary of the layer fixed. We denote the lower component of ∂V by B. As in §1 we let $V_R = V \cap \{(x, y, z) : r < R\}$. We define S_R similarly.

Let $H^s(\Omega)$ (s is a nonnegative integer) denote the Sobolev space of functions defined in the domain $\Omega \subset R^3$ whose derivatives of order up to s inclusive are square-summable. We denote by $L_{2,R}(V)$, $R > 0$, the subspace of $L_2(V)$ consisting of functions that vanish for $r > R$. The space $L_{2,R}(S)$ is defined similarly.

Suppose that the function u belongs to $H^2(V_R)$ for every $R < \infty$ and satisfies the conditions

$$\Delta u = f \text{ in } V, \qquad f \in L_{2,b}(V) \tag{1.2}$$

for some $b < \infty$,

$$\partial u / \partial z - \nu u = 0 \quad \text{for} \quad z = 0, \quad \nu > 0, \tag{2.2}$$

$$\partial u / \partial n = 0 \quad \text{on } B, \tag{3.2}$$

$$u = O(r^{-1/2}), \ \partial u / \partial r - i\lambda_0 u = o(r^{-1/2}) \quad \text{as } r \to \infty, \tag{4.2}$$

where λ_0 is a positive root of the equation

$$\lambda \sinh \lambda h - \nu \cosh \lambda h = 0. \tag{5.2}$$

In this case we say that u is a solution of the problem (1.2)–(4.2).

The problem (1.2)–(4.2) in the layer S has been studied in great detail (see, for example, [**9**], [**11**] and [**17**]). It is well known that in this case the problem has one and only one solution; formulas for the solution of the problem and its fundamental solution $(f = \delta(x - x_0, y - y_0, z - z_0))$ and their asymptotic forms as $r \to \infty$ have been obtained. The fundamental solution of the problem in S has the form

$$\begin{aligned} E(x, y, z, \xi, \eta, \zeta) &= \frac{1}{4\pi}[(x-\xi)^2 + (y-\eta)^2 + (z-\zeta)^2]^{-\frac{1}{2}} \\ &+ \frac{1}{4\pi}[(x-\xi)^2 + (y-\eta)^2 + (2h + z + \zeta)^2]^{-\frac{1}{2}} \\ &+ \frac{1}{2\pi}\int_l \frac{(\nu + \mu)e^{-h\mu}\cosh\mu(\zeta + h)\cosh\mu(z + h)}{\mu\sinh\mu h - \nu\cosh\mu h} \\ &\times I_0\left(\mu\sqrt{(x-\xi)^2 + (y-\eta)^2}\right) d\mu, \end{aligned} \tag{6.2}$$

where I_0 is a Bessel function, and l is a contour in the complex μ-plane obtained from the real positive semiaxis by replacing the interval $|\mathrm{Re}\,\mu - \lambda_0| < \epsilon$ by the semicircle $|\mu - \lambda_0| = \epsilon$, $\mathrm{Im}\,\mu < 0$. Here $\epsilon > 0$ is sufficiently small that the disc $|\mu - \lambda_0| \leqq \epsilon$ contains no other zeros of the function $\mu \sinh \mu h - \nu \cosh \mu h$ except $\mu = \lambda_0$.

From this explicit formula for the fundamental solution one can obtain its asymptotic form for $\xi^2 + \eta^2 \leqq b^2$ and $r \to \infty$ (see [11] and [17]):

$$E = \frac{\nu \cosh \lambda_0(\zeta + h) \cosh \lambda_0(z + h)}{(\nu h + \sinh^2 \lambda_0 h)(2\pi\lambda_0 r)^{1/2}} e^{i\left(\lambda_0 \sqrt{(x-\xi)^2+(y-\eta)^2} + \frac{\pi}{4}\right)} + g(x, y, z, \xi, \eta, \zeta),$$

where $|g| + |\nabla_{x,y,z} g| < Cr^{-3/2}$ for $\xi^2 + \eta^2 \leqq b^2$ and $r \to \infty$.[1)] It follows immediately from this that any solution u of the problem (1.2) – (4.2) in the layer S has the following form as $r \to \infty$:

$$u = F(\theta, z) r^{-1/2} e^{i\lambda_0 r} + g(x, y, z), \tag{7.2}$$

where $F(\theta, z)$ is a certain smooth function of z and the polar angle θ in the (x, y) plane, and $|g| + |\nabla g| = O(r^{-3/2})$ as $r \to \infty$.

We now turn to a study of the problem in the domain V.

LEMMA 1.2. *Every solution u of the problem* (1.2) – (4.2) *has as $r \to \infty$ the form* (7.2), *where*

$$\lambda_0 \int_{-h}^{0} \int_{0}^{2\pi} |F(\theta, z)|^2 \, d\theta \, dz = \mathrm{Im} \int_V f \bar{u} \, dV. \tag{8.2}$$

PROOF. Let $\chi(r)$ be an infinitely differentiable function equal to 1 for $r > \max(a, b) + 1$ and to 0 for $r < \max(a, b)$. Then the function $u\chi(r)$ is a solution of the problem (1.2) – (4.2) in the layer S with some smooth function f on the right-hand side of (1.2) whose support is contained in the domain $\max(a, b) < r < \max(a, b) + 1$. But then (7.2) holds for $u\chi(r)$ and hence for u.

Further,

$$\int_{V_R} (\bar{u}\,\Delta u - u \Delta \bar{u})\, dV = \int_{\partial V_R} \left(u \frac{\partial \bar{u}}{\partial n} - \bar{u} \frac{\partial u}{\partial n} \right) ds,$$

where n is the inward drawn normal to V_R. We write the integral over ∂V_R as a sum of integrals over B_R, K_R and Γ_R (see §1 for the notation). It follows from (2.2) and (3.2) that the first two of these integrals are zero. Thus

1) If in this formula we replace

$$\left(\frac{1}{2} \pi \lambda_0 r\right)^{-\frac{1}{2}} e^{i\left(\lambda_0 \sqrt{(x-\xi)^2+(y-\eta)^2} - \frac{\pi}{4}\right)}$$

by the Hankel function $H_0^{(1)}(\lambda_0 \{(x-\xi)^2 + (y-\eta)^2\}^{1/2})$, the remainder term decreases exponentially.

$$\int_{\Gamma_R} \left(u \frac{\partial \bar{u}}{\partial n} - \bar{u} \frac{\partial u}{\partial n} \right) r \, d\theta \, dz = 2i \operatorname{Im} \int_{V_R} f \bar{u} \, dV.$$

It follows from (7.2) that the left-hand side of this equality is equal to

$$2i\lambda_0 \int_{-h}^{0} \int_{0}^{2\pi} | F(\theta, z) |^2 \, d\theta \, dz + O(R^{-1}) \quad \text{as} \quad R \to \infty.$$

Therefore when we take the limit as $R \to \infty$ we obtain (8.2). Lemma 1.2 is proved.

THEOREM 1.2. *Suppose that the domain* V *satisfies one of the conditions* (2.1) *or* (10.1). *Then the problem* (1.2) – (4.2) *has one and only one solution for any function* $f \in L_{2,b}(V)$.

PROOF. The uniqueness of the solution is an obvious consequence of Lemmas 2.1, 3.1 and 1.2. We now prove its existence.

Without loss of generality we may assume that the mapping $\psi : \bar{S} \to \bar{V}$ has the following properties: 1) ψ is the identity for $z \geqq -\epsilon$ for some $\epsilon > 0$; 2) ψ carries the normals to the plane $z = -h$ into the normals to B. This follows readily from the postulated properties of ψ and the fact that every nondegenerate mapping carries the field of normals to the plane $z = -h$ into a vector field on B that does not touch B at any point. Then without loss of generality we can assume that $b = a + 1$, because we can always replace a and b by $\max(a, b)$ and $\max(a, b) + 1$, respectively.

In the problem (1.2) – (4.2) we now make the change of variables $(x, y, z) \to \psi^{-1}(x, y, z)$. From the point of view of the unique solvability we obtain the equivalent problem:

(9.2) $$(\Delta + L) v = g \text{ in } S, \quad g \in L_{2,a+1}(S),$$

(10.2) $$\partial u / \partial z - \nu v = 0 \text{ for } z = 0,$$

(11.2) $$\partial v / \partial z = 0 \text{ for } z = -h,$$

(12.2) $$v = O(r^{-1/2}), \quad \partial v / \partial r - i\lambda_0 v = o(r^{-1/2}) \text{ for } r \to \infty,$$

where the coefficients of the differential operator

$$L = L(x, y, z, \partial/\partial x, \partial/\partial y, \partial/\partial z)$$

are zero outside $\bar{S}_{a,\epsilon}$. Here

$$S_{a,\epsilon} = S_a \cap \{ (x, y, z) : z < -\epsilon \},$$

and the constant $\epsilon > 0$ has been specified above.

Let I_ψ be the modulus of the Jacobian of ψ. Obviously the operator $\Delta + L_1 = I_\psi(\Delta + L)$ with the boundary conditions (10.2) and (11.2) is formally selfadjoint; that is, for any twice continuously differentiable functions φ_i,

$i = 1, 2$, satisfying (10.2) and (11.2) and vanishing in a neighborhood of infinity we have

$$\langle (\Delta + L_1)\varphi_1, \varphi_2 \rangle = \langle \varphi_1, (\Delta + L_1)\varphi_2 \rangle,$$

where $\langle u, v\rangle = \int_S u\bar{v}dS$. Here, the coefficients of the operator L_1 are also zero outside $\bar{S}_{a,\epsilon}$. When we replace the operator L in (9.2) by L_1, we obtain a problem that is equivalent in the sense of the unique solvability to the problem (9.2) – (12.2). This is obvious because one problem is obtained from the other by multiplying both sides of (9.2) by an everywhere nonzero function I_ψ. Thus, retaining the old notation, we may assume that in the problem (9.2) – (12.2) the operator $\Delta + L$ with the boundary conditions (10.2) and (11.2) is formally selfadjoint:

$$\langle (\Delta + L)\varphi_1, \varphi_2 \rangle = \langle \varphi_1, (\Delta + L)\varphi_2 \rangle. \tag{13.2}$$

We denote by F the set of functions f defined on S, twice continuously differentiable on $\bar{S}_{a,\epsilon}$, and satisfying the condition $\partial f/\partial z = 0$ for $z = -h$, $r \leqq a$. Since (13.2) holds when $L \equiv 0$, it follows from (13.2) that $\langle L\varphi_1, \varphi_2\rangle = \langle \varphi_1, L\varphi_2\rangle$. But because the coefficients of L are zero outside $\bar{S}_{a,\epsilon}$, we have

$$\langle Lf_1, f_2\rangle = \langle f_1, Lf_2\rangle, \quad f_i \in F. \tag{14.2}$$

We denote by R the operator that associates with each function $f \in L_{2,a+1}(S)$ a solution of the problem (1.2) – (4.2) in the layer S. Because $Rf \in H^2(S_b)$ for every $b < \infty$, the operator

$$LR: L_{2,a+1}(S) \to L_{2,a+1}(S)$$

is defined on the whole of the space $L_{2,a+1}(S)$. Hence, by the Hahn-Banach Theorem, the operator LR is bounded.

We seek a solution of the problem (9.2) – (12.2) in the form $v = Rq$ with an unknown function $q \in L_{2,a+1}(S)$. Then the function v is a solution of the problem if

$$q + LRq = g, \quad q, g \in L_{2,a+1}(S). \tag{15.2}$$

The converse is also true: if v is a solution of the problem (9.2) – (12.2), then $v = Rq$, where $q = -Lu + g$ and q satisfies (15.2).

Thus to prove Theorem 1.2 it is sufficient to show that equation (15.2) can be solved. We first prove that the range of the operator $I + LR$ (I is the identity operator) is closed.

Let v_n be a solution of the problem (9.2) – (12.2) with a function g_n on the right-hand side of (9.2). By well-known local a priori estimates for the solutions of elliptic problems (see, for example, [**20**]) we have

$$\| v_n \|_{H^2(S_a)} \leqslant C[\| v_n \|_{L_2(S_{a+1})} + \| g_n \|_{L_{2,a+1}(S)}]. \tag{16.2}$$

Because the operator $R: L_{2,a+1}(S) \to H^2(S_b)$ is bounded for every $b < \infty$ and

$$v_n = R(-Lv_n + g_n), \tag{17.2}$$

it follows from (16.2) that

$$\|v_n\|_{H^2(S_{a+1})} \leqslant C[\|v_n\|_{L_2(S_{a+1})} + \|g_n\|_{L_{2,a+1}(S)}].$$

Suppose that $g_n \to g_0$ in the space $L_{2,a+1}(S)$ as $n \to \infty$. Then by standard arguments we obtain from the last estimate the existence of a subsequence v_{n_k} that is convergent in $H^2(S_{a+1})$ to some function v as $k \to \infty$. This in conjunction with (17.2) entails the convergence of v_{n_k} to v in the space $H^2(S_b)$ with any $b < \infty$, and it follows from (17.2) that the limit function is a solution of the problem (9.2)–(12.2) with g_0 on the right-hand side of (9.2). Thus the set of functions $g \in L_{2,a+1}(S)$ for which the problem (9.2)–(12.2) is solvable is closed. By the equivalence (from the point of view of the solvability) of the problem (9.2)–(12.2) and the equation (15.2), we obtain from here that the range of the operator $I + LR$ is closed.

Now to prove Theorem 1.2 it is sufficient to show that the equation

$$w + (LR)^* w = 0, \qquad w \in L_{2,a+1}(S), \tag{18.2}$$

has only the trivial solution.

We prove that

$$(LR)^* f = \begin{cases} \overline{RL\bar{f}} & \text{for } r < a+1 \\ 0 & \text{for } r > a+1, \quad f \in L_{2,a+1}(S) \cap F, \end{cases} \tag{19.2}$$

and that every solution of (18.2) belongs to F.

We recall that the operator R is an integral operator with the kernel E defined in (6.2). It follows from (6.2) that $E(x,y,z,\xi,\eta,\zeta) = E(\xi,\eta,\zeta,x,y,z)$ and that the function

$$E(x, y, z, \xi, \eta, \zeta) - \frac{1}{4\pi}[(x-\xi)^2 + (y-\eta)^2 + (z-\zeta)^2]^{-\frac{1}{2}}$$

is a smooth function for $(x,y,z) \in \bar{S}$, $(\xi,\eta,\zeta) \in S$. Hence

$$\langle Rh, g\rangle = \langle h, \overline{R\bar{g}}\rangle, \quad h \in D(S_{a+1}), \quad g \in L_{2,a+1}(S), \tag{20.2}$$

where $D(S_{a+1})$ denotes the space of infinitely differentiable functions on S with a compact support belonging to S_{a+1}. Because the solution Rh of the problem (1.2)–(4.2) in the layer S with $f = h \in D(S_{a+1})$ on the right-hand side of (1.2) is infinitely differentiable in $\bar{S}$ (see, for example, [20]) and the coefficients of L are real, it follows from (14.2) and (20.2) that

$$\langle LRh, f\rangle = \langle h, \overline{RL\bar{f}}\rangle, \quad h \in D(S_{a+1}), \quad f \in F.$$

In this equality we can obviously take the limit with respect to h in the space $L_{2,a+1}(S)$. This proves (19.2).

We now show that any solution w of (18.2) belongs to F. We denote by Φ the space of twice continuously differentiable functions φ on S that vanish outside $\bar{S}_{a+1,\epsilon/2}$ and satisfy the condition $\partial\varphi/\partial z = 0$ for $z = -h$. Obviously the function $\varphi \in \Phi$ is a solution of the problem (1.2)–(4.2) in the layer S with the function $f = \Delta\varphi$ on the right-hand side of (1.2); that is, $R\Delta\varphi = \varphi$

for $\varphi \in \Phi$. Together with (18.2) this gives

$$0 = \langle w + (LR)^* w, \Delta\varphi \rangle = \langle w, \Delta\varphi + LR\Delta\varphi \rangle = \langle w, \Delta\varphi + L\varphi \rangle$$

for every $\varphi \in \Phi$. This means that the function w is a generalized (in the sense of the integral identity) solution of the problem

$$(\Delta + L) w = 0 \quad \text{in} \quad S_{a+1,\epsilon/2},$$
$$\partial w / \partial z = 0 \quad \text{for} \quad z = -h, \qquad r < a + 1.$$

It is well known (see [**21**], Chapter III) that the function w in this case is twice continuously differentiable in $\overline{S}_{a,\epsilon}$. Thus $w \in F$. But then by using (19.2) we can rewrite (18.2) as

(21.2) $$w + RLw = 0 \quad \text{for} \quad r < a + 1, \qquad w \in L_{2,a+1}(S) \cap F.$$

We now suppose that w is a nontrivial solution of equation (21.2). We define the function u on the complete layer S by the formula

(22.2) $$u = -RLw.$$

Because $Lw \in L_{2,a+1}(S)$ it follows from the properties of the operator R that u belongs to $H^2(S_b)$ for any $b < \infty$, and it satisfies the conditions (10.2) – (12.2). Because it follows from (21.2) and (22.2) that $u = w$ for $r < a + 1$, and the coefficients of L are zero outside $\overline{S}_{a,\epsilon}$, we can rewrite (22.2) as $u + RLu = 0$. When we apply the operator Δ to this equality we find that the function u also satisfies (9.2) with $g = 0$. Thus u is a solution of the homogeneous problem (9.2) – (12.2). Here $u \not\equiv 0$ if $w \not\equiv 0$. Thus it follows from the existence of a nontrivial solution of (18.2) that there exists a nontrivial solution of the problem (9.2) – (12.2); this contradicts the proved uniqueness of the solution of the problem (1.2) – (1.4). Hence equation (18.2) has only the trivial solution, and Theorem 1.2 is proved.

Let $q = (\xi, \eta, \zeta)$ be a point of $\overline{V}$, let ρ be the distance between the point (x, y, z) and q, and let K_ϵ be the ball $\rho \leqq \epsilon$. We denote by e_q^1 and e_q^2 the functions

$$e_q^1 = \frac{-1}{4\pi\rho}, \quad q \in V; \quad e_q^2 = \frac{1}{2\pi\rho} - \frac{\nu}{2\pi} \ln(\rho - z), \qquad \zeta = 0.$$

Let (α, β, γ) be Euclidean coordinates in R^3 such that the origin is at the point $q \in B$, the positive direction of the γ-axis coincides with the inward drawn normal n to B through the point q, and in these coordinates the surface B in a sufficiently small neighborhood of the point q can be written in the form

$$\gamma = C_1\alpha^2 + C_2\beta^2 + O\left(\sqrt{(\alpha^2 + \beta^2)^3}\right).$$

We denote by e_q^3 the function

$$e_q^3 = -\frac{\chi(\rho)}{2\pi} \left[\frac{1}{\rho} + \frac{(C_1 - C_2)(\rho - \gamma)}{4(\rho + \gamma)} \cos 2\varphi + \frac{C_1 + C_2}{2} \ln(\rho + \gamma) \right], \quad q \in B.$$

Here φ is the polar angle in the (α, β)-plane and $\chi(\rho)$ is an infinitely differentiable function that is equal to 1 for $\rho < \delta$ and 0 for $\rho > 2\delta$. The constant $\delta > 0$ is chosen sufficiently small that the sphere $K_{2\delta}$ does not intersect the plane $z = 0$, and the negative part of the γ-axis does not intersect $\overline{V}$ within $K_{2\delta}$. It obviously follows from the last condition that the function e_q^3 is twice continuously differentiable in $\overline{V}$ outside the point q.

We call the function E_q^i $(i = 1, 2, 3)$ *the velocity potential due to a periodic source* at the point q within the fluid $(i = 1)$, on the free surface of the fluid $(i = 2)$, or on the bottom of the fluid $(i = 3)$ if it satisfies the following conditions:

1°. The point q belongs to V, the plane $z = 0$, or the surface B, respectively.

2°. The function $E_q^i - e_q^i$ $(i = 1, 2, 3)$ belongs to $H^2(V_b)$ for every $b < \infty$.

3°. For any $\epsilon > 0$

(23.2) $$\Delta E_q^i = 0 \quad \text{in} \quad V \backslash K_\epsilon,$$

(24.2) $$\partial E_q^i/\partial z - \nu E_q^i = 0 \quad \text{for} \quad z = 0, \qquad \rho > \epsilon,$$

(25.2) $$\partial E_q^i/\partial n = 0 \quad \text{on} \quad B \backslash K_\epsilon,$$

(26.2) $$E_q^i = O(r^{-1/2}), \quad \partial E_q^i/\partial r - i\lambda_0 E_q^i = o(r^{-1/2}) \quad \text{as} \quad r \to \infty.$$

Equations (23.2) – (25.2) could be written down for the entire domain V, the plane $z = 0$, and the surface B, respectively. In this case one must replace by a δ-function the right-hand side of the first of these equations for $i = 1$, the second for $i = 2$, and the third for $i = 3$.

THEOREM 2.2. *Suppose that the domain V satisfies one of the conditions* (2.1) *or* (10.1). *For each of the three cases when the periodic source is within the fluid, on the free surface, or on the bottom, there exists one and only one velocity potential.*

PROOF. We consider first the case of a source within the fluid. Let $\chi(\rho)$ be an infinitely differentiable function that is equal to 1 for $\rho < \delta$ and 0 for $\rho > 2\delta$, where $\delta > 0$ is sufficiently small that $K_\delta \subset V$. We seek the function E_q^1 in the form $\chi(\rho) e_q^1 - u$. Then the function u must belong to $H^2(V_b)$ for every $b < \infty$, and it must satisfy the equation

(27.2) $$\Delta u = f \quad \text{in} \quad V \backslash K_\epsilon$$

for any $\epsilon > 0$, and also the conditions (2.2) – (4.2). Here $f = \Delta \chi(\rho) e_q^1$. Because $\Delta e_q^1 = 0$ in $V \backslash K_\epsilon$, we have

$$f = e_q^1 \Delta \chi(\rho) + 2\langle \nabla \chi(\rho), \nabla e_q^1 \rangle.$$

This function belongs to $L_{2,c}(V)$ for some $c < \infty$. Because $u \in H^2(V_b)$, it follows that $\Delta u \in L_2(V_b)$ for every $b < \infty$. But a function belonging to $L_2(V_b)$ that vanishes in $V_b \backslash K_\epsilon$ for any $\epsilon > 0$ also vanishes in V_b (as an element of the space $L_2(V_b)$). Hence it follows from (27.2) that

$$\Delta u = f \in L_{2,c}(V) \quad \text{in} \quad V.$$

It follows from Theorem 2.1 that such a function exists, and there is only one. Thus there exists one and only one velocity potential produced by a periodic source within the fluid.

Now suppose that the source is on the free surface. We seek the potential E_q^2 in the form $E_q^2 = \chi(\rho)e_q^2 - v$, where $\chi(\rho)$ is the same function as above but we only require $\delta > 0$ to be such that $K_\delta \cap B$ is empty. Because $\Delta e_q^2 = 0$ in V, and $(\partial/\partial z - \nu)e_q^2 = (\nu^2/2\pi)\ln\rho$ for $z = 0$ and $\rho > \epsilon$, by arguing as in the foregoing case we find that the function v must belong to $H^2(V_b)$ for every $b < \infty$, and it must satisfy the conditions

$$\begin{aligned} &\Delta v - e_q^2\,\Delta\chi(\rho) + 2\langle\nabla\chi(\rho), \nabla e_q^2\rangle \in L_{2,c}(V) \quad \text{in } V, \\ &\frac{\partial v}{\partial z} - \nu v = e_q^2\,\frac{\partial\chi(\rho)}{\partial z} + \frac{\nu^2}{2\pi}\chi(\rho)\ln\rho \quad \text{for } z = 0 \end{aligned} \tag{28.2}$$

and the conditions (3.2) and (4.2).

It is easily verified that the function

$$v_1 = ze_q^2\,\frac{\partial\chi(\rho)}{\partial z} + z\,\frac{\nu^2}{2\pi}\,\chi(\rho)\ln(\rho - z)$$

belongs to $H^2(V_b)$ with any $b < \infty$, is of compact support and satisfies the conditions (28.2), (3.2) and (4.2). Therefore, if we seek the function v in the form $v_1 + u$, then for the determination of u we have a problem for which the existence and the uniqueness of the solution follows from Theorem 1.2. Thus there exists one and only one potential E_q^2.

The case of a source on the bottom of the fluid is treated in exactly the same way as the foregoing case. Theorem 2.2 is proved.

It is obvious that Theorem 2.2 implies the unique solvability of the problem of the steady-state oscillations of a fluid due to a perturbation that is periodic in time and has compact support, without any assumptions about the smoothness of the perturbation.

§3. The plane problem

In this section S denotes the strip $0 > y > -h$ $(0 < h < \infty)$ on the plane R^2, and V denotes the domain obtained from S by a one-to-one thrice continuously differentiable mapping $\psi: \overline{S} \to \overline{V}$ that is the identity for $|y| > a$, $a = \text{const} < \infty$, and leaves the straight line $y = 0$ fixed. We denote the lower component of ∂V by B. The domain V_R and the spaces $H^S(V_R)$ and $L_{2,b}(V)$ are defined as in §2.

The plane problem of the steady-state oscillations of a heavy incompressible fluid occupying the domain V is as follows: to find a function u (the velocity potential) that belongs to $H^2(V_R)$ for every $R < \infty$ and satisfies the conditions

$$\Delta u = f \quad \text{in} \quad V, \qquad f \in L_{2,b}(V) \tag{1.3}$$

for some $b < \infty$,

$$\partial u/\partial y - \nu u = 0, \quad \text{for} \quad y = 0, \qquad \nu > 0, \tag{2.3}$$

$$\partial u/\partial n = 0 \quad \text{on} \quad B, \tag{3.3}$$

$$u = O(1), \quad \partial u/\partial |x| - i\lambda_0 u = o(1) \quad \text{as} \quad r \to \infty, \tag{4.3}$$

where λ_0 is a positive root of the equation (5.2).

THEOREM 1.3. *Suppose that the domain V satisfies one of the conditions $x\cos(n,x) \leqq 0$ on B or $\cos(n,\rho) \leqq 0$ on B (see §1). Then the problem (1.3)–(4.3) has one and only one solution for any function $f \in L_2(V_b)$.*

PROOF. This theorem is proved in exactly the same way as Theorem 1.2; therefore we confine ourselves to a short account of the proof. We note first that the corresponding problem in the strip S has been studied in detail (see, for example, [1], [4]–[6], or [17]). In particular, the existence and uniqueness of the solution of the problem in S have been proved; explicit formulas have been obtained for the fundamental solution $E(x,y,\xi,\eta)$ of the problem, and it follows from them, in particular, that $E(x,y,\xi,\eta) = E(\xi,\eta,x,y)$; the asymptotic form of $E(x,y,\xi,\eta)$ for $\xi^2+\eta^2 \leqq b^2$ and $r = (x^2+y^2)^{1/2} \to \infty$ has been obtained:

$$E = \frac{-i\nu\cosh\lambda_0(\eta+h)\cosh\lambda_0(y+h)}{\lambda_0(\nu h + \sinh^2\lambda_0 h)} e^{i\lambda_0|x-\xi|} + g(x, y, \xi, \eta), \tag{5.3}$$

where $|g| + |\nabla_{x,y} g| = O(r^{-1})$ for $\xi^2+\eta^2 \leqq b^2$ and $r \to \infty$.[2)]

It obviously follows from (5.3) that as $r \to \infty$ any solution of the problem (1.3)–(4.3) in S has the form

$$u = \begin{cases} c_+(y)\, e^{i\lambda_0 x} + g_+(x, y), & x \to +\infty, \\ c_-(y)\, e^{-i\lambda_0 x} + g_-(x, y), & x \to -\infty, \end{cases} \tag{6.3}$$

where $|g_\pm| + |\nabla g_\pm| = O(r^{-1})$ as $r \to \infty$. This asymptotic formula is similar to (7.2), which was obtained for the solution of the corresponding problem in the layer $0 > z > -h$ in three-dimensional space. In exactly the same way as the assertions of Lemma 1.2 followed from (7.2), we now obtain from (6.3) that any solution of the problem (1.3)–(4.3) in V has the form (6.3) as $r \to \infty$, and

$$\lambda_0 \int_{-h}^{0} [\,|c_+(y)|^2 + |c_-(y)|^2]\, dy = \operatorname{Im} \int_V f\bar{u}\, dV. \tag{7.3}$$

The uniqueness of the solution of the problem (1.3)–(4.3) follows from the formulas (6.3) and (7.3) and Lemmas 4.4 and 5.1. The existence of a solution is proved in exactly the same way as the corresponding assertion in Theorem 1.2. Theorem 1.3 is proved.

Suppose that $q = (\xi,\eta) \in \overline{V}$ and ρ is the distance between the points (x,y) and q. We denote by e_q^i $(i = 1,2,3)$ the functions

2) The remainder term in (5.3) decreases exponentially (see [6] and [7]), but for what follows it is sufficient to have a power estimate.

$$e_q^1 = \frac{1}{2\pi} \ln \rho, \quad q \in V;$$

$$e_q^2 = -\frac{1}{\pi} \ln \rho - \frac{1}{\pi}\left[y \ln \rho + (x-\xi) \arctan \frac{y}{x-\xi} + \frac{\pi}{2} |x-\xi| \right], \qquad \eta = 0;$$

$$e_q^3 = \frac{1}{\pi} \ln \rho, \quad q \in B.$$

The function E_q^i $(i = 1, 2, 3)$ is called the velocity potential due to a periodic source within the fluid $(i = 1)$, on the free surface of the fluid $(i = 2)$, or on the bottom of the fluid $(i = 3)$ if it satisfies the following conditions:

1°. The point q belongs to the domain V, the straight line $y = 0$, or the curve B, respectively.

2°. The function $E_q^i - e_q^i$ $(i = 1, 2, 3)$ belongs to the space $H^2(V_b)$ for every $b < \infty$.

3°. The following conditions hold for any $\epsilon > 0$:

$$\Delta E_q^i = 0 \quad \text{in} \quad V \quad \text{for} \quad \rho > \epsilon,$$

$$\partial E_q^i / \partial y - \nu E_q^i = 0 \quad \text{for} \quad z = 0, \qquad \rho > \epsilon,$$

$$\partial E_q^i / \partial n = 0 \quad \text{on} \quad B \quad \text{for} \quad \rho > \epsilon,$$

$$E_q^i = O(1), \quad \partial E_q^i / \partial |x| - i\lambda_0 E_q^i = o(1) \quad \text{as} \quad r \to \infty.$$

The assertion of Theorem 2.2 and its proof can be carried over automatically, in effect, to the case considered in this section.

Received 3/DEC/71

Bibliography

[1] H. Lamb, *Hydrodynamics*, 6th ed., Cambridge Univ. Press, Cambridge, 1932; Russian transl., OGIZ, Moscow, 1947.

[2] J. J. Stoker, *Water waves: the mathematical theory with applications*, Pure and Appl. Math., vol. 4, Interscience, New York, 1957; Russian transl., IL, Moscow, 1959. MR **21** # 2438.

[3] A. Weinstein, *Sur un problème aux limites dans une bande indéfinie*, C. R. Acad. Sci. Paris **184** (1927), 497-499.

[4] E. Isaacson, *Water waves over a sloping bottom*, Comm. Pure Appl. Math. **3** (1950), 11-31. MR **12**, 137; 1002.

[5] A. E. Heins, *Water waves over a channel of finite depth with a dock*, Amer. J. Math. **70** (1948), 730-748. MR **10**, 490.

[6] G. Kreisel, *Surface waves*, Quart. Appl. Math. **7** (1949), 21-44. MR **11**, 227.

[7] J. J. Stoker, *Surface waves in water of variable depth*, Quart. Appl. Math. **5** (1947), 1-54. MR **9**, 163.

[8] N. E. Kočin, *The plane problem of steady oscillations of bodies beneath the free surface of a heavy incompressible fluid*, Izv. Akad. Nauk SSSR Otd. Tehn. Nauk **1939**, no. 4, 37-62; reprinted in *Collected works*. Vol. 2, Izdat. Akad. Nauk SSSR, Moscow, 1949, pp. 244-276; English transl., Soc. Naval Architects and Marine Engrs. Tech. Res. Bull. no. **1 – 9** (1952). MR **11**, 573; 872; **13**, 997.

[9] ———, *Theory of waves generated by oscillations of a body beneath the free surface of a heavy incompressible fluid*, Učen. Zap. Moskov. Gos. Univ. Meh. **46** (1940), 85-106; reprinted in *Collected works*. Vol. 2, Izdat. Akad. Nauk SSSR, Moscow, 1949, pp. 277-304; English transl., Soc. Naval Architects and Marine Engrs. Tech. Res. Bull. **1 – 10** (1952). MR **11**, 573; 872; **12**, 59; **13**, 997.

[10] M. D. Haskind, *The oscillation of a body immersed in heavy fluid,* Prikl. Mat. Meh. **8** (1944), 287-300. (Russian) MR **6**, 192.
[11] ———, *Waves arising from oscillation of bodies in shallow water,* Prikl. Mat. Meh. **10** (1946), 475-480. (Russian) MR **8**, 236.
[12] F. Ursell, *Surface waves on deep water in the presence of a submerged circular cylinder.* I, II, Proc. Cambridge Philos. Soc. **46** (1950), 141-152, 153-158. MR **11**, 436; 480.
[13] ———, *Discrete and continuous spectra in the theory of gravity waves,* National Bureau of Standards Circular 521, U. S. Government Printing Office, Washington, D. C., 1952. MR **14**, 811.
[14] ———, *Trapping modes in the history of surface waves,* Proc. Cambridge Philos. Soc. **47** (1951), 347-358. MR **12**, 870.
[15] ———, *Water waves generated by oscillating bodies,* Quart. J. Mech. Appl. Math. **7** (1954), 427-437. MR **17**, 423.
[16] ———, *The expansion of water-wave potentials at great distances,* Proc. Cambridge Philos. Soc. **64** (1968), 811-826.
[17] D. S. Jones, *The eigenvalues of* $\nabla^2 u + \lambda u = 0$ *when the boundary conditions are given on semi-infinite domains,* Proc. Cambridge Philos. Soc. **49** (1953), 668-684. MR **15**, 319.
[18] F. John, *On the motion of floating bodies.* II. *Simple harmonic motions,* Comm. Pure Appl. Math. **3** (1950), 45-101. MR **12**, 214.
[19] B. R. Vaĭnberg and V. G. Maz′ja, *On the plane problem of the motion of a body immersed in a fluid,* Trudy Moskov. Mat. Obšč. **28** (1973), 35-56 = Trans. Moscow Math. Soc. **28** (1973), 33-55.
[20] S. Agmon, A. Douglis and L. Nirenberg, *Estimates near the boundary for solutions of elliptic partial differential equations satisfying general boundary conditions.* I, Comm. Pure Appl. Math. **12** (1959), 623-727; Russian transl., IL, Moscow, 1962. MR **23** # A2610.
[21] O. A. Ladyženskaja and N. N. Ural′ceva, *Linear and quasilinear equations of elliptic type,* "Nauka", Moscow, 1964; English transl., Academic Press, New York, 1968. MR **35** # 1955; **39** # 5941.
[22] B. R. Vaĭnberg and V. G. Maz′ja, *On some stationary problems in the linear theory of surface waves,* Dokl. Akad. Nauk SSSR **205** (1972), 310-313 = Soviet Physics Dokl. **17** (1972/73), 640-643.

Translated by
J. A. BARBOUR

Trudy Moskov. Mat. Obšč.
Tom 28 (1973)

Trans. Moscow Math. Soc.
Vol. 28 (1973)

BOUNDARY VALUE PROBLEMS AND THE PARAMETRIX FOR SYSTEMS OF ELLIPTIC PSEUDODIFFERENTIAL EQUATIONS

G. I. ÈSKIN

Contents

Introduction

Boundary value problems for systems of elliptic pseudodifferential equations have been considered by Višik and the author [**4**], [**5**], Shamir [**16**], Dynkin [**10**], Šubin [**17**] and Boutet de Monvel [**1,2**]. In §§1 and 3 of the present article we give a simpler proof of the theorem on the normal solvability of boundary value problems. In §2 we construct an algebra of operators on the half-line that contains the pseudodifferential operators (p.d.o.). This gives a new approach to the study of normal solvability and the calculation of the index of a p.d.o. on the half-line, and also allows us to study mixed boundary value problems for elliptic differential equations in regions containing an edge (see [**15**], §4). The present article, however, does not deal with the latter problem. The contents of §2 closely border on the author's articles [**18**], [**19**]. Cordes and Hermann [**11**], and Gohberg and Krupnik [**9**] have obtained results close to these.

In §§4, 5 and 6 we construct the parametrix of an elliptic boundary value problem. Boutet de Monvel [**1**], [**2**] has constructed a parametrix for elliptic systems satisfying an additional condition (the pseudodifferential operators are of class $D_\alpha^{(0)}$ (see [**4**])). The parametrix proposed in the present article

AMS (MOS) subject classifications (1970). Primary 35S15, 47G05.

is new even in the case of a p.d.o. of class $D_\alpha^{(0)}$, because it is constructed in a narrower class of operators than in [1] and [2].

We remark that the constructions in §6 allow us to study, for example, nonelliptic boundary value problems for elliptic pseudodifferential operators. The exposition in this article is rather condensed; detailed proofs are given in the author's book [22].

§1. Conditions for the normal solvability of a pseudodifferential operator on a half-line

1. Let $A(\xi) = \|a_{ij}(\xi)\|_{i,j=1}^{p}$ be a matrix consisting of symbols of the class $O_{\alpha+i\beta}^{\infty}$; that is, homogeneous symbols of order $\alpha + i\beta$ that are of class C^∞ for $\xi \neq 0$. The matrix $A(\xi)$ is called *elliptic* if $\det A(\xi) \neq 0$ for $\xi \neq 0$. We put $\hat{A}(\xi', \xi_n) = A((1+|\xi'|)\omega, \xi_n)$ and $\omega = \xi'/|\xi'|$, and consider a system of pseudodifferential equations in $\overline{\mathbf{R}}_+^n$ with the symbol $\hat{A}(\xi', \xi_n)$:

$$p\hat{A}u_+ = f, \tag{1.1}$$

where $f(x) = (f_1(x), \cdots, f_p(x))$, $u_+ = (u_1^+(x), \cdots, u_p^+(x))$, and $f_k(x) \in H_{s-\alpha}(\mathbf{R}_+^n)$ (the Sobolev-Slobodeckiĭ space in $\mathbf{R}_+^n$), $u_k^+(x) \in H_s^+(\mathbf{R}^n)$ (the subspace of $H_s(\mathbf{R}^n)$ consisting of functions with supports in $\overline{\mathbf{R}}_+^n$), and p is the restriction operator to $\mathbf{R}_+^n$. In what follows, for brevity we also denote the spaces of vector-valued functions $f(x)$ and $u_+(x)$ by $H_{s-\alpha}(\mathbf{R}_+^n)$ and $H_s^+(\mathbf{R}^n)$.

We show that the system (1.1) is equivalent to a system of pseudodifferential equations of the form

$$p\hat{A}_0 v_+ = g, \tag{1.2}$$

where

$$v_+ \in H_0^+(\mathbf{R}^n), \quad g \in H_0(\mathbf{R}_+^n), \quad \operatorname{ord}_\xi A_0(\xi', \xi_n) = 0.$$

Let $u_- = lf - \hat{A}u_+$, where lf is an arbitrary extension of f to $\mathbf{R}^n$ that belongs to $H_{s-\alpha}(\mathbf{R}^n)$. Then $u_- \in H_{s-\alpha}^-(\mathbf{R}^n)$ and the system (1.1) is equivalent to the system of equations

$$\hat{A}u_+ + u_- = lf \tag{1.3}$$

with the unknowns $u_+ \in H_s^+(\mathbf{R}^n)$ and $u_- \in H_{s-\alpha}^-(\mathbf{R}^n)$.

Let $\hat{\Lambda}_\pm^s$ be pseudodifferential operators with the symbols $\hat{\Lambda}_\pm^s(\xi) = (\xi_n \pm i|\xi'| \pm i)^s$. We put $v_+ = \Lambda_+^s u_+$ and $g = p\Lambda_-^{s-i-i\beta} lf$. Then $v_+ \in H_0^+(\mathbf{R}^n)$, $g \in H_0(\mathbf{R}_+^n)$, and g does not depend on the choice of the extension lf. We apply the p.d.o. $\hat{\Lambda}_-^{s-\alpha-i\beta}$ to (1.3) and put $u_+ = \hat{\Lambda}_+^{-s} v_+$. We obtain the following system of equations, which is equivalent to (1.3):

$$\hat{A}_0 V_+ + \hat{\Lambda}_-^{s-\alpha-i\beta} u_- = \hat{\Lambda}_-^{s-\alpha-i\beta} lf, \tag{1.4}$$

where $\hat{A}_0$ is the p.d.o. with the symbol $\hat{A}_0(\xi) = \hat{A}(\xi)\Lambda_-^{s-\alpha-i\beta}(\xi)\Lambda_+^{-s}(\xi)$, $\operatorname{ord}_\xi A_0(\xi) = 0$. Because $\hat{\Lambda}_-^{s-\alpha-i\beta} u_- \in H_0^-(\mathbf{R}^n)$ and so $p\hat{\Lambda}_-^{s-\alpha-i\beta} u_- = 0$, the system (1.4) is equivalent to (1.3), and $g = p\hat{\Lambda}_-^{s-\alpha-i\beta} lf$.

Let $g_+(x) \in H_0^+(\mathbf{R}^n)$ be a function equal to $g(x)$ for $x_n > 0$ and zero for $x_n < 0$. The system (1.2) in the half-space is obviously equivalent to the following system of pseudodifferential equations in $\mathbf{R}^n$:

$$\theta^+ \hat{A}_0 v_+ = g_+, \tag{1.5}$$

that is obtained from (1.2) by extending the left and right sides by zero for $x_n < 0$. In (1.5) θ^+ stands for the operation of multiplication by the function $\theta(x_n)$.

2. We fix $\omega = \xi'/|\xi'|$ and consider the following system of pseudodifferential equations on the half-line $\mathbf{R}_+^1$:

$$pA(\omega, D_n) u_+(x_n) = f(x_n), \tag{1.6}$$

where

$$A(\omega, D_n)\, u_+(x_n) = \frac{1}{2\pi} \int_{-\infty}^{\infty} A(\omega, \xi_n)\, \tilde{u}_+(\xi_n)\, e^{-ix_n\xi_n} d\xi_n$$

is the p.d.o. on $\mathbf{R}^1$ with the symbol $A(\omega, \xi_n)$, $f(x_n) \in H_{s-\alpha}(\mathbf{R}_+^1)$, and $u_+(x_n) \in H_s^+(\mathbf{R}^1)$.

We show in §3 that the study of a system of pseudodifferential equations of the form (1.1) in $\mathbf{R}_+^n$ is based on an investigation of the family of pseudodifferential operators $pA(\omega, D_n)u_+(x_n)$ in $\mathbf{R}_+^1$ that depend on the parameter $\omega = \xi'/|\xi'|$. Consequently the rest of this section will be devoted to a study of the system (1.6) or the equivalent system of pseudodifferential equations

$$\theta^+ A_0(\omega, D_n) v_+(x_n) = g_+(x_n), \tag{1.7}$$

where $g_+(x_n) \in H_0^+(\mathbf{R}^1)$, $v_+(x_n) \in H_0^+(\mathbf{R}^1)$, the parameter ω is fixed, and

$$A_0(\omega, \xi_n) = A(\omega, \xi_n)(\xi_n - i)^{s-\alpha-i\beta}(\xi_n + i)^{-s}. \tag{1.8}$$

In what follows we denote by O_0 the class of functions of form (1.8) with $A(\xi', \xi_n) \in O_{\alpha+i\beta}^{\infty}$.

Because $\operatorname{ord}_\xi A_0(\xi', \xi_n) = 0$, we have $A_0(\omega, \xi_n) = A(\omega/|\xi_n|, \xi_n/|\xi_n|)$. Let $\xi_n > 0$ and $\xi_n \to +\infty$. Then $A_0(\omega, \xi_n) = A_0(\omega/\xi_n, +1) \to A_0(0, +1) = a_+$. For $\xi_n < 0$ and $\xi_n \to -\infty$ we obtain $A_0(\omega, \xi_n) = A_0(\omega/|\xi_n|, -1) \to A_0(0, -1) = a_-$. We note that $\det a_\pm \neq 0$ by the ellipticity of $A_0(\xi', \xi_n)$.

3. We first consider the case

$$a_+ = a_-. \tag{1.9}$$

Lemma 1.1. *If condition* (1.9) *holds, then the elliptic system* (1.7) *defines a Fredholm* (*normally solvable*) *operator in* $H_0^+(\mathbf{R}^1)$.

Proof. We take the Fourier transformation with respect to x_n in (1.7) and obtain

$$\Pi^+ A_0(\omega, \xi_n) \tilde{v}_+(\xi_n) = \tilde{g}_+(\xi_n), \tag{1.10}$$

where the operator Π^+ is defined on the functions $\tilde{u}(\xi_n) \in S(\mathbf{R}^1)$ by

$$\Pi^+\tilde{u}(\xi_n)=\frac{i}{2\pi}\int_{-\infty}^{\infty}\frac{\tilde{u}(\eta_n)}{\xi_n-\eta_n+i0}\,d\eta_n,$$

and for any $\tilde{u}(\xi_n)\in\tilde{H}_0(\mathbf{R}^1)$ it is defined by continuity.

Let $\tilde{R}$ denote the operator defined by

$$\tilde{R}\tilde{g}_+=\Pi^+A_0^{-1}(\omega,\xi_n)\tilde{g}_+(\xi_n). \tag{1.11}$$

$\tilde{R}$ is bounded in $\tilde{H}_0^+(\mathbf{R}^1)$. We show that $\tilde{R}$ is a regularizer of the operator $\mathscr{A}$ defined by the system (1.10); that is,

$$\tilde{\mathscr{A}}\tilde{R}=I+T_1, \tag{1.12}$$

$$\tilde{R}\tilde{\mathscr{A}}=I+T_2, \tag{1.12'}$$

where the T_i are completely continuous operators in $H_0^+(\mathbf{R}^1)$, $i=1,2$. We have

$$\begin{aligned}\tilde{\mathscr{A}}\tilde{R}\tilde{g}_+&=\Pi^+A_0(\omega,\xi_n)\,\Pi^+A_0^{-1}(\omega,\eta_n)\,\tilde{g}_+(\eta_n)\\&=\Pi^+A_0(A_0^{-1}\Pi^+\tilde{g}_++C\tilde{g}_+)=\tilde{g}_++\Pi^+A_0C\tilde{g}_+,\end{aligned} \tag{1.13}$$

where

$$C\tilde{g}_+=\Pi^+A_0^{-1}\tilde{g}_+-A_0^{-1}\Pi^+\tilde{g}_+=\frac{i}{2\pi}\int_{-\infty}^{\infty}C(\omega,\xi_n,\eta_n)\,\tilde{g}_+(\eta_n)\,d\eta_n, \tag{1.14}$$

$$C(\omega,\xi_n,\eta_n)=\frac{A_0^{-1}(\omega,\eta_n)-A_0^{-1}(\omega,\xi_n)}{\xi_n-\eta_n}. \tag{1.14'}$$

We show that C is a Hilbert-Schmidt operator. Because $A(\xi',\xi_n)\in O^\infty_{\alpha+i\beta}$ the following estimates hold:

$$|A_0^{-1}(\omega,\xi_n)-a_+^{-1}|\leqslant C(1+|\xi_n|)^{-1}, \tag{1.15}$$

$$\left|\frac{\partial}{\partial\xi_n}A_0^{-1}(\omega,\xi_n)\right|\leqslant C(1+|\xi_n|)^{-2}. \tag{1.15'}$$

We prove that

$$|C(\omega,\xi_n,\eta_n)|\leqslant C(1+|\xi_n|)^{-1}(1+|\eta_n|)^{-1}. \tag{1.16}$$

For, let

$$|\xi_n-\eta_n|<\frac{1}{2}(1+|\xi_n|).$$

Then

$$1+|\eta_n|\leqslant 1+|\xi_n|+|\xi_n-\eta_n|<\frac{3}{2}(1+|\xi_n|).$$

By (1.15′) and the Lagrange formula we obtain

$$\begin{aligned}|C(\omega,\xi_n,\eta_n)|&\leqslant\left|\frac{\partial}{\partial\xi_n}A_0^{-1}(\omega,\xi_n+\Theta(\eta_n-\xi_n))\right|\\&\leqslant\frac{C_1}{(1+|\xi_n|)^2}\leqslant\frac{C_2}{(1+|\xi_n|)(1+|\eta_n|)}.\end{aligned}$$

Now let $|\xi_n - \eta_n| \geqq \frac{1}{2}(1+|\xi_n|)$. Then $1+|\eta_n| \leqq 3|\xi_n - \eta_n|$. Hence by (1.15)

$$|C(\omega, \xi_n, \eta_n)| = \frac{1}{|\xi_n - \eta_n|}(|A_0^{-1}(\omega, \eta_n) - a_+| + |A_0^{-1}(\omega, \xi_n) - a_+|$$
$$\ll \frac{C}{(1+|\xi_n|)(1+|\eta_n|)}.$$

Thus the estimate (1.16) holds for every ξ_n and η_n, so that C is a Hilbert-Schmidt operator. Hence $\Pi^+ A_0 \widetilde{C}$ is a completely continuous operator since it is the product of a bounded and a completely continuous operator.

The relation (1.12) is proved. The equality (1.12′) is proved similarly. We note that if only one equality (1.12) (or (1.12′)) is valid, then the operator R is called a left (or right) regularizer.

Let k be the dimension of the kernel of the operator $\mathscr{A}$, l the dimension of the kernel of $\mathscr{A}^*$, and $\kappa = k - l$ the index of $\mathscr{A}$.

LEMMA 1.2 (SEE, FOR EXAMPLE [7]). *Let the conditions of Lemma 1.1 hold. Then*

$$\varkappa = \frac{1}{2\pi} \Delta \arg \det A_0(\omega, \xi_n) \Big|_{\xi_n = +\infty}^{\xi_n = -\infty} \tag{1.17}$$

that is, the change in the argument of $\det A_0(\omega, \xi_n)$ *as* ξ_n *varies from* $+\infty$ *to* $-\infty$ *is* $2\pi x$.

4. We now consider the general case

$$a_+ \neq a_-. \tag{1.18}$$

We find conditions that ensure that the operator $\widetilde{\mathscr{A}} = \Pi_+ A_0$ is a Fredholm operator in $\widetilde{H}_0^+(\mathbf{R}^1)$.

Because $A_0(\xi', \xi_n)$ is elliptic the matrix $a_+^{-1}a_-$ is nondegenerate. Let $a_1, \cdots, a_p$ be the eigenvalues of the matrix $a_+^{-1}a_-$ with their multiplicity taken into account. We require that the matrix $a_+^{-1}a_-$ satisfies the following condition:

(1.19) *There are no real negative numbers amongst the eigenvalues* $a_j, \quad 1 \leqq j \leqq p.$

We define the logarithm of the matrix $a_+^{-1}a_-$ by the formula (see [6])

$$\ln a_+^{-1} a_- = \frac{i}{2\pi} \int_\Gamma \ln z\, (a_+^{-1}a_- - zI)^{-1} dz, \tag{1.20}$$

where I is the unit matrix, and Γ is a closed contour in the complex plane $z = \sigma + i\tau$, with a cut along the negative real axis, that encloses all the eigenvalues a_j, $1 \leqq j \leqq p$. The branch of $\ln z$ in (1.20) is taken to be real for $\tau = 0$, $\sigma > 0$. We write

$$b = \frac{1}{2\pi i} \ln a_+^{-1} a_-, \quad b_\pm(\xi_n) = \exp\{\ln(\xi_n \pm i) b\}. \tag{1.21}$$

In (1.21) we also choose the branch of the logarithm that is real on the positive real axis. Let b_j be the eigenvalues of the matrix $b = \ln(a_+^{-1}a_-)/2\pi i$. Then by (1.19) and (1.20) we have

$$-\tfrac{1}{2} < \operatorname{Re} b_j < \tfrac{1}{2}, \qquad 1 \leqq j \leqq p. \tag{1.22}$$

We note that if b is chosen arbitrarily, then the condition is equivalent to

$$\operatorname{Re} b_j \neq \tfrac{1}{2} (\operatorname{mod} k), \qquad 1 \leqq j \leqq p, \tag{1.23}$$

where k is any integer.

We show that the matrix $a_+ b_-^{-1}(\xi_n) b_+(\xi_n)$ has the same limits as the matrix $A_0(\omega, \xi_n)$ when $\xi_n \to \pm\infty$.

We have

$$a_+ b_-^{-1}(\xi_n)\, b_+(\xi_n) = a_+ \exp\left\{\ln \frac{\xi_n + i}{\xi_n - i}\, b\right\}.$$

As $\xi_n \to +\infty$,

$$\ln \frac{\xi_n + i}{\xi_n - i} \to 0.$$

Hence $a_+ b_-^{-1} b_+ \to a_+ = A_0(0, +1)$. As $\xi_n \to -\infty$,

$$\ln \frac{\xi_n + i}{\xi_n - i} = \arg(\xi_n + i) - \arg(\xi_n - i) \to 2\pi i.$$

Hence $a_+ b_-^{-1} b_+ \to a_+ e^{2\pi i b} = a_+ a_+^{-1} a_- = a_-$ as $\xi_n \to -\infty$.

We write

$$P_\pm \tilde{u}_+ = b_\pm^{-1} \Pi^+ b_\pm \tilde{u}_+. \tag{1.24}$$

Lemma 1.3. *If the condition* (1.22) *holds, then the operators* $P^\pm$ *are bounded in* $\widetilde{H}_0(\mathbf{R}^1) = L_2(\mathbf{R}^1)$.

Proof. Let b_0 be the Jordan normal form of the matrix b, and let N be a nondegenerate matrix that takes b into its Jordan normal form

$$b = N^{-1} b_0 N. \tag{1.25}$$

Then

$$b_\pm(\xi_n) = \exp\{\ln(\xi_n \pm i) N^{-1} b_0 N\} = N^{-1} b_0^\pm(\xi_n) N,$$

where

$$b_0^\pm(\xi_n) = \exp\{\ln(\xi_n \pm i)\, b_0\}.$$

When b_0 is a diagonal matrix the matrices $b_0^\pm(\xi_n)$ are also diagonal and have the form $b_0^\pm(\xi_n) = \| (\xi_n \pm i)^{b_j} \delta \|_{j,k=1}^p$. In the general case each r_k-dimensional Jordan block of the matrix corresponds to a triangular r_k-dimensional block $\| b_{pl}^{(i)}(\xi_n \pm i) \|_{p,l=1}^{r_k}$, where

$$\begin{aligned} b_{pl}^{(i)}(\xi_n \pm i) &= (\xi_n \pm i)^{b_i} \ln^{l-p}(\xi_n \pm i) \text{ when } 1 \leqq p \leqq l \leqq r_k, \\ b_{pl}^{(i)}(\xi_n \pm i) &= 0 \text{ when } p > l \end{aligned} \tag{1.26}$$

(see [**6**]). Because the matrix N is independent of ξ_n and so commutes with the operator Π^+, we have $P_\pm \tilde{u}_+ = b_\pm^{-1}\Pi^+ b_\pm \tilde{u}_+ = N^{-1} b_0^\pm \Pi^+ (b_0^\pm)^{-1} N \tilde{u}_+$. Hence to prove Lemma 1.3 it is sufficient to establish the estimate

$$\text{(1.27)} \qquad \| b_0^\pm \Pi^+ (b_0^\pm)^{-1} \tilde{v}_+ \|_0 \leqslant C \| \tilde{v}_+ \|_0, \qquad \forall_+ \tilde{v}_+ (\xi_n) \in \tilde{H}_0 (\mathbf{R}^1).$$

We have

$$b_0^\pm \Pi^+ (b_0^\pm)^{-1} \tilde{v}_+ = \frac{i}{2\pi} \int_{-\infty}^{\infty} \frac{\exp\left\{ \ln \frac{\xi_n \pm i}{\eta_n \pm i} b_0 \right\} \tilde{v}_+ (\eta_n)\, d\eta_n}{\xi_n - \eta_n + i0}.$$

Consequently by (1.26) the estimate (1.27) follows, for example, from [**20**], Lemma 5.3.

Lemma 1.4. *Let the matrix $B(\xi_n)$ satisfy the estimates (cf. (1.15) and (1.15′))*

$$\text{(1.28)} \qquad |B(\xi_n) - I| \leqslant C(1 + |\xi_n|)^{-1}, \qquad \left| \frac{\partial B(\xi_n)}{\partial \xi_n} \right| \leqslant C(1 + |\xi_n|)^{-2}.$$

Assume that condition (1.22) holds. Then the operator $C = P_+ B - B P_+$ (the commutator of P_+ and B) is a Hilbert-Schmidt operator.

Proof. We have

$$\text{(1.29)} \qquad C \tilde{u}_+ = \frac{i}{2\pi} \int_{-\infty}^{\infty} C(\xi_n, \eta_n) \tilde{u}_+ (\eta_n)\, d\eta_n,$$

where

$$\text{(1.29′)} \qquad C(\xi_n, \eta_n) = \frac{b_+^{-1}(\xi_n) b_+(\eta_n) B(\eta_n) - B(\xi_n) b_+^{-1}(\xi_n) b_+(\eta_n)}{\xi_n - \eta_n}.$$

We prove that

$$\text{(1.30)} \qquad |C(\xi_n, \eta_n)| \leqslant C(1 + |\xi_n|)^{-\frac{1}{2} - \delta_0} (1 + |\eta_n|)^{-\frac{1}{2} - \delta_0},$$

where

$$0 < \delta_0 < \frac{1}{2} - \max_{1 \leqslant j \leqslant p} |\operatorname{Re} b_j|.$$

Let $|\xi_n - \eta_n| < (1 + |\xi_n|)/2$. Then by (1.27) and Lagrange's formula we have

$$|B(\xi_n) - B(\eta_n)| \leqslant \frac{C |\xi_n - \eta_n|}{(1 + |\xi_n|)^2}.$$

Similarly

$$|b_+^{-1}(\xi_n) b_+(\eta_n) - I| = \left| \exp\left\{ \ln \left(1 - \frac{\xi_n - \eta_n}{\xi_n + i} \right) b \right\} - I \right| \leqslant C \frac{|\xi_n - \eta_n|}{1 + |\xi_n|}$$

for $|\xi_n - \eta_n| < (1 + |\xi_n|)/2$. Hence

$$
\begin{aligned}
|C(\xi_n, \eta_n)| &\leqslant \frac{|B(\xi_n) - B(\eta_n)|}{|\xi_n - \eta_n|} \\
&+ \frac{|b_+^{-1}(\xi_n) b_+(\eta_n) - I|\,(|B(\eta_n) - I| + |B(\xi_n) - I|)}{|\xi_n - \eta_n|} \qquad (1.31)\\
&\leqslant \frac{C}{(1+|\xi_n|)^2} + \frac{C}{(1+|\xi_n|)(1+|\eta_n|)} \leqslant \frac{C}{(1+|\xi_n|)(1+|\eta_n|)},
\end{aligned}
$$

since $1+|\eta_n| \leqq 3(1+|\xi_n|)/2$.

Now let $|\xi_n - \eta_n| \geqq (1+|\xi_n|)/2$. Then $1+|\eta_n| \leqq 3|\xi_n - \eta_n|$. By (1.26) we have

$$
|b_+^{-1}(\xi_n)| \leqslant C(1+|\xi_n|)^{\frac{1}{2}-\delta_0}, \quad |b_+(\eta_n)| \leqslant C(1+|\eta_n|)^{\frac{1}{2}-\delta_0}.
$$

Hence

$$
\begin{aligned}
|C(\xi_n, \eta_n)| &\leqslant \frac{|b_+^{-1}(\xi_n)|\,|b_+(\eta_n)|\,(|B(\xi_n) - I| + |B(\eta_n) - I|)}{|\xi_n - \eta_n|} \qquad (1.32)\\
&\leqslant C(1+|\xi_n|)^{-\frac{1}{2}-\delta_0}(1+|\eta_n|)^{-\frac{1}{2}-\delta_0}.
\end{aligned}
$$

Thus (1.30) is valid for every ξ_n and η_n, and so C is a Hilbert-Schmidt operator.

THEOREM 1.1. *Let $A_0(\omega, \xi_n)$ be an elliptic matrix, and let* (1.19) *or the equivalent condition* (1.23) *hold. Then $\Pi^+ A_0$ is a Fredholm operator in $\widetilde{H}_0^+(\mathbf{R}^1)$.*

PROOF. The operators P_+ and P_- satisfy the following relation:

$$
P_+ A_0^{-1} u_+ = A_0^{-1} a_+ P_- + C_1, \qquad (1.33)
$$

where C_1 is a Hilbert-Schmidt operator. For, since the matrices $a^+ b_-^{-1}(\xi_n) b_+(\xi_n)$ and $A_0(\omega, \xi_n)$ have the same limits as $\xi_n \to \pm\infty$, the matrix $B(\xi_n) = A_0^{-1} a_+ b_-^{-1}(\xi_n) b_+(\xi_n)$ satisfies the estimates (1.28). Hence by Lemma 1.4

$$
\begin{aligned}
P_+ A_0^{-1} a_+ &= P_+ \cdot B b_+^{-1} b_- = B P_+ b_+^{-1} b_- + C b_+^{-1} b_- \qquad (1.34)\\
&= A_0^{-1} a_+ b_-^{-1} b_+ P_+ b_+^{-1} b_- + C_1,
\end{aligned}
$$

where $C_1 = C b_- b_+^{-1}$. Because $b_+ P_+ b_+^{-1} = \Pi^+$ and $b_-(\xi_n) b_+^{-1}(\xi_n)$ are bounded matrices, the relation (1.33) follows from (1.34).

We show that the operator $\widetilde{R}$ defined by $\widetilde{R}\widetilde{g}_+ = P_+ A_0^{-1} a_+ P_- a_+^{-1} \widetilde{g}_+$ is a left and right regularizer of $\Pi^+ A_0$. We have $\Pi^+ A_0 \widetilde{R} \widetilde{g}_+ = \Pi^+ A_0 P_+ A_0^{-1} a_+ P_- a_+^{-1} \widetilde{g}_+$. By (1.33) and the obvious equality $P_-^2 = P_-$ we obtain

$$
\begin{aligned}
\Pi^+ A_0 \widetilde{R} \widetilde{g}_+ &= \Pi_+ A_0 (A_0^{-1} a_+ P_- + C_1) P_- a_+^{-1} \widetilde{g}_+ \qquad (1.35)\\
&= \Pi^+ a_+ P_- a_+^{-1} \widetilde{g}_+ + T_1 \widetilde{g}_+,
\end{aligned}
$$

where $T_1 = \Pi^+ A_0 C_1 P_- a_+^{-1}$ is a completely continuous operator in $\widetilde{H}_0^+(\mathbf{R}^1)$. Because a_+ is a constant matrix, $\Pi^+ a_+ = a_+ \Pi^+$. We show that $\Pi^+ P_- = \Pi^+$. In fact, $\Pi^+ P_- \widetilde{h} = \Pi^+ \widetilde{h}$ for every function $\widetilde{h} \in \widetilde{H}_0(\mathbf{R}^1)$; since $b_-^{-1}\Pi^- b_- \widetilde{h} \in \widetilde{H}_0^-(\mathbf{R}^1)$ we have $\Pi^+ b_-^{-1} \Pi^- b_- \widetilde{h} = 0$. Thus

$$\Pi^+ A_0 \widetilde{R} g_+ = \widetilde{g}_+ + T_1 \widetilde{g}_+, \quad \forall \widetilde{g}_+ \in \widetilde{H}_0^+(\mathbf{R}^1). \tag{1.36}$$

Similarly we have

$$\widetilde{R}\Pi^+ A_0 \widetilde{u}_+ = P_+ A_0^{-1} a_0 P_- a_+^{-1} \Pi^+ A_0 \widetilde{u}_+.$$

We remark that

$$P_- a_+^{-1} \Pi^+ \widetilde{h} = P_- a_+^{-1} \widetilde{h}, \quad \forall \widetilde{h} \in \widetilde{H}_0(\mathbf{R}^1).$$

For,

$$P_- a_+^{-1} \Pi^+ \widetilde{h} = b_-^{-1} \Pi^+ b_- a_+^{-1} (\widetilde{h} - \Pi^- \widetilde{h}) = P_- a_+^{-1} \widetilde{h},$$

because

$$b_- a_+^{-1} \Pi^- \widetilde{h} \in \widetilde{H}^-_{-\frac{1}{2}+\delta_0}(\mathbf{R}^1)$$

and so $\Pi^+ b_- a_+^{-1} \Pi^- \widetilde{h} = 0$. Thus when we use (1.33) we obtain

$$\begin{aligned} \widetilde{R}\Pi^+ A_0 u_+ &= P_+ A_0^{-1} a_+ P_- a_+^{-1} A_0 \widetilde{u}_+ \\ &= P_+ (P_+ A_0^{-1} a_+ - C_1) a_+^{-1} A_0 \widetilde{u}_+ = P_+ \widetilde{u}_+ + T_2 \widetilde{u}_+ = \widetilde{u}_+ + T_2 \widetilde{u}_2, \end{aligned} \tag{1.37}$$

where $T_2 = -P_+ C_1 a_+^{-1} A_0$ is a completely continuous operator in $\widetilde{H}_0^+(\mathbf{R}^1)$. In (1.37) we have used the fact that $P^+ \widetilde{u}_+ = \widetilde{u}_+$ for all $\widetilde{u}_+ \in \widetilde{H}_0^+(\mathbf{R}^1)$. In fact we have

$$b_+ \widetilde{u}_+ \in \widetilde{H}^+_{\frac{1}{2}-\delta_0}(\mathbf{R}^1)$$

and by a result of the author ([20], Lemma 5.4) $b_+ \widetilde{u}_+ = \Pi^+ b_+ \widetilde{u}_+$. Hence $P^+ \widetilde{u}_+ = \widetilde{u}_+$ for every function $\widetilde{u}_+(\xi) \in \widetilde{H}_0^+(\mathbf{R}^1)$. It follows from (1.36) and (1.37) that $\widetilde{R}$ is a regularizer of $\Pi^+ A_0$. Hence $\Pi^+ A_0$ is a Fredholm operator.

5. We now turn from equation (1.10) to (1.6) and explain the conditions for the operator $pA(\omega, D_n)$ defined by (1.6) to be normally solvable. We write

$$A^{(0)}(\xi', \xi_n) = A(\xi', \xi_n)(\xi_n - i|\xi'|)^{-\alpha - i\beta}.$$

The system (1.6) is equivalent to the system (1.7) with

$$A_0(\omega, \xi_n) = A^{(0)}(\omega, \xi_n)\left(\frac{\xi_n - i}{\xi_n + i}\right)^s.$$

Hence we have

$$a_+ = A_0(0, +1) = A^{(0)}(0, +1), \quad a_- = A_0(0, -1) = A^{(0)}(0, -1)\, e^{-2\pi i s}.$$

Let $a_j^{(0)}$, $1 \leq j \leq p$, be the eigenvalues of the matrix $A^{(0)}(0, -1)\,[A^{(0)}(0, +1)]^{-1}$. Then (1.23) is equivalent to the following condition:

(1.38) $$s - \operatorname{Re} \frac{1}{2\pi i} \ln a_j^{(0)} \neq \frac{1}{2} \pmod{k},$$

where k is an arbitrary integer.

Thus it follows from Theorem 1.1 that the following theorem holds.

THEOREM 1.2. *Let* $A(\xi', \xi_n) \in O^{\infty}_{\alpha+i\beta}$ *be an elliptic matrix and* s *a real number satisfying* (1.38). *Then, for any* ω *with* $|\omega| = 1$, $pA(\omega, D_n)$ *is a Fredholm operator from* $H_s^+(\mathbf{R}^1)$ *into* $H_{s-\alpha}(\mathbf{R}^1_+)$.

REMARK 1.1. Let Z_0 be the set of real numbers s for which (1.38) does not hold. We can show that for $s \in Z_0$ the range of the operator $pA(\omega, D_n)$ is not closed in $H_{s-\alpha}(\mathbf{R}^1_+)$, although as before the kernel and cokernel are finite dimensional. Hence $pA(\omega, D_n)$ is not a Fredholm operator for $s \in Z_0$.

We shall now clarify how the kernel and cokernel of the operator $pA(\omega, D_n)$ depend on s.

LEMMA 1.5. *There is a number* $s_+ = s_+(\omega)$ *such that* $\dim \ker pA(\omega, D_n) = 0$ *for* $s > s_+(\omega)$.

PROOF. Let N_s be the kernel of the operator $pA(\omega, D_n)$ that lies in $H_s^+(\mathbf{R}^1)$. The spaces N_s are finite dimensional. Since $H^+_{s_1} \subset H^+_{s_2}$ for $s_1 > s_2$, we have $N_{s_1} \subset N_{s_2}$. Let $\bigcap_s N_s = N_\infty$ be nonempty.* Then there is a function $u_0^+(x_n) \in N_\infty$, $u_0^+ \not\equiv 0$, satisfying the equation $pA(\omega, D_n) u_0^+ = 0$. Because the operators D_n^k and $pA(\omega, D_n)$ commute, $pA(\omega, D_n) D_n^k u_0^+ = 0$ for all $k \geqq 0$. Thus $D_n^k u_0^+ \in N_\infty$ for every k, and the functions $D_n^k u_0^+$, $k = 0, 1, \cdots$, are linearly independent because $u_0^+ \in H_s^+(\mathbf{R}^1)$ and $D_n^k u_0^+ \not\equiv 0$. This contradicts the fact that N_∞ is finite dimensional; hence $N_\infty = 0$.

The following lemma is proved by going over to the conjugate operator.

LEMMA 1.6. *There is an* $s_- = s_-(\omega)$ *such that if* $s < s_-$, *then* $\dim \operatorname{coker} pA(\omega, D_n) = 0$.

§2. An algebra of operators on the half-line that is generated by pseudodifferential operators

1. As we showed in §1, whenever (1.9) holds the regularizer of the operator $\theta^+ A_0(\omega, D_n)$ has the form $\theta^+ A_0^{-1}(\omega, D_n)$. If $a_+ \neq a_-$, then the regularizer of $\theta^+ A(\omega, D_n)$ is not an operator of the same class. In this section we extend the class of operators so that the extended class contains the regularizers of the operators $\theta^+ A_0(\omega, D_n)$.

Let $M(t)$ be a matrix which belongs to C^∞ for $t > 0$ and satisfies the estimates

(2.1) $$\left| t^k \frac{d^k M(t)}{dt^k} \right| \leqslant \begin{cases} C_\varepsilon t^{-\delta-\varepsilon} & \text{for } 0 < t < 1, \quad 0 \leqslant k < \infty, \\ C_\varepsilon t^{-1+\delta+\varepsilon} & \text{for } 1 \leqslant t < \infty, \quad 0 \leqslant k < \infty. \end{cases}$$

---

* *Translator's note.* Clearly $N_\infty \neq 0$ is intended.

In (2.1), $0<\delta<\frac{1}{2}$ and δ is fixed for the given function, and $\epsilon>0$ is arbitrary. Let M be the integral operator defined by

$$(2.2)\qquad Mu_+=\int_0^\infty M\left(\frac{t}{\tau}\right)\frac{u_+(\tau)}{\tau}d\tau,\quad \forall u_+(\tau)\in C_0^\infty(\mathbf{R}^1_+).$$

Let $\mathfrak{M}$ denote the class of integral operators of the form (2.2) with the kernels $M(t/\tau)/\tau$, where $M(t)$ satisfies the estimates (2.1). We apply the Mellin transformation to (2.2). When we change the order of integration and make the substitution $t=t'\tau$ we find that

$$(2.3)\qquad \widehat{Mu_+}=\int_0^\infty t^{s-1}\left(\int_0^\infty M\left(\frac{t}{\tau}\right)\frac{u_+(\tau)}{\tau}d\tau\right)dt=\hat{M}(s)\,\hat{u}_+(s),$$

where $\hat{M}(s)=\int_0^\infty M(t)t^{s-1}dt$ is the Mellin transform of the matrix $M(t)$. The matrix $\hat{M}(s)$ will be called the *symbol of the operator M*. It is not hard to verify that $\hat{M}(s)$ is analytic in the strip $\delta<\operatorname{Re}s<1-\delta$, and it satisfies the estimates

$$(2.3')\qquad |s^m\hat{M}(s)|\leqslant C_{\varepsilon m},\quad \delta+\varepsilon\leqslant\operatorname{Re}s\leqslant 1-\delta-\varepsilon$$

for all $\epsilon>0$ and $m\geqq 0$. Conversely, if $\hat{M}(s)$ is analytic in the strip $\delta<\operatorname{Re}s<1-\delta$ and satisfies (2.3′), then $M(t)\in C^\infty$ for $t>0$, and the estimates (1.1) hold.

Let $\chi(t)\in C_0^\infty(\mathbf{R}^1)$, $\chi(t)\equiv 1$ for $|t|\leqq 1$. We consider the following class of operators on the half-line $\mathbf{R}^1_+$:

$$(2.4)\qquad \begin{gathered}\Theta^+A_0(\omega,D_n)u_+ +\Theta^+\chi(x_n)Mu_+ +\Theta^+Tu_+,\\ \forall u_+\in C_0^\infty(\mathbf{R}^1_+),\end{gathered}$$

where $A_0(\omega,D_n)$ is the p.d.o. in $\mathbf{R}^1$ with the symbol $A(\omega,\xi_n)$, M is an integral operator of the class $\mathfrak{M}$, and T is a completely continuous operator in $H_0(\mathbf{R}^1)=L_2(\mathbf{R}^1)$.

In this section we show that the operators of the form (2.4) form an algebra; that is, sums and products of operators of the form (2.4) are also of this form.

EXAMPLE 2.1. We find the form of the operators

$$\Pi^\pm u_+=\pm\frac{i}{2\pi}\int_0^\infty\frac{u_+(y_n)\,dy_n}{x_n\pm i0-y_n}\quad\text{and}\quad M_0u_+=\frac{i}{2\pi}\int_0^\infty\frac{u_+(y_n)\,dy_n}{x_n+y_n},\ x_n>0,$$

after applying the Mellin transformation.

By analogy with (2.3), for $0<\operatorname{Re}s<1$ we obtain

$$(2.5)\qquad \begin{aligned}\widehat{\Pi^+u_+}&=\lim_{\varepsilon\to 0}\int_0^\infty x_n^{s-1}\left(+\frac{i}{2\pi}\int_0^\infty\frac{u_+(y_n)\,dy_n}{x_n+i\varepsilon-y_n}\right)dx_n\\&=-\frac{e^{2\pi is}}{1-e^{2\pi is}}\hat{u}_+(s),\end{aligned}$$

$$\hat{\Pi}^- u_+ = \lim_{\varepsilon\to 0}\int_0^\infty x_n^{s-1}\left(-\frac{i}{2\pi}\int_0^\infty \frac{u_+(y_n)\,dy_n}{x_n - i\varepsilon - y_n}\right)dx_n$$

$$= \frac{1}{1-e^{2\pi i s}}\,\hat{u}_+(s), \tag{2.6}$$

$$\hat{M}_0 u_+ = \frac{e^{i\pi s}}{1-e^{2\pi i s}}. \tag{2.7}$$

We note that $e^{2\pi is}/(1-e^{2\pi is})$ and $1/(1-e^{2\pi is})$ are analytic in the strip $0 < \mathrm{Re}\, s < 1$ and bounded for $\epsilon \leqq \mathrm{Re}\, s \leqq 1-\epsilon$ for all $\epsilon > 0$, and

$$\frac{e^{2\pi is}}{1-e^{2\pi is}} = o\left(\frac{1}{|s|^k}\right),\quad \frac{1}{1-e^{2\pi is}} = 1 + o\left(\frac{1}{|s|^k}\right) \quad \text{as } \mathrm{Im}\, s \to +\infty, \tag{2.8}$$
$$\varepsilon \leqslant \mathrm{Re}\, s \leqslant 1-\varepsilon,\quad \forall\, k \geqslant 0,$$

$$\frac{e^{2\pi is}}{1-e^{2\pi is}} = -1 + o\left(\frac{1}{|s|^k}\right),\quad \frac{1}{1-e^{2\pi is}} = o\left(\frac{1}{|s|^k}\right) \tag{2.9}$$
$$\text{as } \mathrm{Im}\, s \to -\infty,\quad \varepsilon \leqslant \mathrm{Re}\, s \leqslant 1-\varepsilon,\quad \forall\, k \geqslant 0.$$

By (2.5) – (2.9) we have the following lemma:

LEMMA 2.1. *Let M_1 and M_2 be operators of the class $\mathfrak{M}$. Then the operators M_1M_2, $\Pi^{\pm}M_1$ and $M_1\Pi^{\pm}$ are also in $\mathfrak{M}$, and they have the symbols $\hat{M}_1(s)\hat{M}_2(s)$, $\hat{M}_1(s)/(1-e^{2\pi is})$ and $-e^{2\pi is}\hat{M}_1(s)/(1-e^{2\pi is})$.*

2. Let $A_0(\omega, D_n)$ be a p.d.o. in $\mathbf{R}^1$. We find an expression for $A_0(\omega, D_n)$ as an integral operator of convolution type.

Let $a_0 = F_{\xi_n}^{-1}A_0(\omega, \xi_n)$. Then the following lemma holds:

LEMMA 2.2. *Let $A_0(\omega, \xi_n) \in O_0$. Then $a_0 \in C^\infty$ for $x_n \neq 0$, and it has the following properties:*

$$\text{a) } |x_n^k a_0(x_n)| \leqslant C_k,\quad \forall\, k \geqslant 0,\quad |x_n| \geqslant 1; \tag{2.10}$$

$$\text{b) } A_0(\omega, D_n)u_+ = a_0 * u_+ = a_+\Pi^- u_+ + a_-\Pi^+ u_+ + b * u_+$$
$$= \frac{a_+ + a_-}{2}u_+ + (a_- - a_+)\Pi^\circ u_+ + b * u_+, \tag{2.11}$$

where

$$b(x_n) \in L_2(\mathbf{R}^1),\quad a_+ = A_0(0, +1),\quad a_- = A_0(0, -1),$$

$$\Pi^\circ u_+ = \frac{i}{2\pi} v.\,p. \int_0^\infty \frac{u_+(y_n)\,dy_n}{x_n - y_n}. \tag{2.12}$$

and the integral is taken in the sense of principal value.

PROOF. Because $A_0(\omega, \xi_n) \in O_0$, we have

$$\left|\frac{\partial^k A_0(\omega, \xi_n)}{\partial \xi_n^k}\right| \leqslant C_k(1+|\xi_n|)^{-k},\quad \forall\, k \geqslant 0. \tag{2.13}$$

It follows from (2.13) that $a_0(x_n) \in C^\infty$ for $x_n \neq 0$ and that (2.10) holds.

We now put $B(\omega,\xi_n) = A_0(\omega,\xi_n) - a_+\theta(\xi_n) - a_-(1-\theta(\xi_n))$. Then $|B(\omega,\xi_n)| \leqq C(1+|\xi_n|)^{-1}$, and so $b(x_n) = F_{\xi_n}^{-1}B(\omega,\xi_n) \in L_2(\mathbf{R}^1)$. We have

$$F_{\xi_n}^{-1}\Theta(\xi_n)e^{-\varepsilon\xi_n} = \frac{1}{2\pi}\int_0^\infty e^{-\varepsilon\xi_n - ix_n\xi_n}\,d\xi_n = \frac{1}{2\pi}\frac{1}{ix_n+\varepsilon} = -\frac{i}{2\pi}\frac{1}{x_n - i\varepsilon}.$$

Hence

$$F_{\xi_n}^{-1}\Theta(\xi_n) = -\frac{i}{2\pi}\frac{1}{x_n - i0}. \tag{2.14}$$

Similarly

$$F_{\xi_n}^{-1}(1-\Theta(\xi_n)) = \frac{i}{2\pi}\frac{1}{x_n + i0}. \tag{2.15}$$

Thus the decomposition (2.11) follows from (2.14) and (2.15).

3. Let $A_1(\omega,\xi_n)$ and $A_2(\omega,\xi_n)$ belong to the class O_0, $a_1^\pm = A_1(0,\pm 1)$, and $a_2^\pm = A_2(0,\pm 1)$. The product of two p.d.o.'s in $\mathbf{R}_+^1$ is not necessarily a p.d.o. In the following lemma we show that the product of the operators $\theta^+A_1(\omega,D_n)$ and $\theta^+A_2(\omega,D_n)$ can be put in the form (2.4).

LEMMA 2.3. *We have*

$$\begin{gathered}\Theta^+A_1(\omega,D_n)\Theta^+A_2(\omega,D_n)u_+ \\ = \Theta^+A_3(\omega,D_n)u_+ + \Theta^+\chi(x_n)M_{12}u_+ + \Theta^+T_1u_+, \\ \forall\, u_+ \in C_0^\infty(\mathbf{R}_+^1),\end{gathered} \tag{2.16}$$

where $A_3(\omega,D_n)$ is the p.d.o. in $\mathbf{R}^1$ with $A_3(\omega,\xi_n) = A_1(\omega,\xi_n)A_2(\omega,\xi_n)$, M_{12} is an integral operator of the class $\mathfrak{M}$ with the symbol

$$\hat{M}_{12}(s) = (a_1^+ - a_1^-)(a_2^+ - a_2^-)\frac{e^{2\pi is}}{(1-e^{2\pi is})^2}, \tag{2.17}$$

and T is a completely continuous operator in $H_0(\mathbf{R}^1)$.

PROOF. We have

$$\begin{aligned}\Theta^+A_1(\omega,D_n)\Theta^+A_2(\omega,D_n)u_+ &= \Theta^+A_1(\omega,D_n)A_2(\omega,D_n)u_+ \\ &\quad - \Theta^-A_1(\omega,D_n)\Theta^-A_2(\omega,D_n)u_+,\end{aligned} \tag{2.18}$$

where θ^- denotes the operator of multiplication by $(1-\theta(x_n)) = \theta(-x_n)$. Because $\theta(x_n)\theta(-x_n) = 0$, by (2.11) we have

$$\begin{gathered}\Theta^+A_1(\omega,D_n)\Theta^-A_2(\omega,D_n)u_+ \\ = \Theta^+[(a_1^- - a_1^+)\Pi^\circ + B_1]\Theta^-[(a_2^- - a_2^+)\Pi^\circ + B_2]u_+,\end{gathered} \tag{2.19}$$

where $B_iu_+ = b_i * u_+$ and $b_i(x_n) \in L_2(\mathbf{R}^1)$, $i = 1,2$.

The integral operator $\chi(x_n)B_1v = \int_{-\infty}^{\infty}\chi(x_n)b_1(x_n-y_n)v(y_n)dy_n$ is clearly an integral operator of Hilbert-Schmidt type. By (2.10) we have

$$|(1-\chi(x_n))a_1(x_n-t)| \leqq C_N(x+|t|)^{-N}$$

for all N for $x_n \geqq 1$ and $t \leqq 0$. Hence $(1-\chi(x_n))\Theta^+ A_1(\omega, D_n)\Theta^-$ is also an integral operator of Hilbert-Schmidt type. Thus

$$\begin{aligned}\Theta^+ A_1(\omega, D_n)\Theta^- A_2(\omega, D_n)u_+ &= \chi(x_n)\Theta^+ A_1(\omega, D_n)\Theta^- A_2(\omega, D_n)u_+ \\ &+ (1-\chi(x_n))\Theta^+ A_1\Theta^- A_2 u_+ = \\ &= \chi(x_n)\Theta^+(a_1^- - a_1^+)\Pi^\circ\Theta^-[(a_2^- - a_2^+)\Pi^\circ + B_2]u_+ + T_1 u_+,\end{aligned} \tag{2.20}$$

where T_1 is a completely continuous operator in $H_0(\mathbf{R}^1)$. We have

$$\Theta^+\Pi^\circ\Theta^-\Pi^\circ u_+ = \Theta(x_n)\left(\frac{i}{2\pi}\right)^2 \int_{-\infty}^{0} \frac{1}{x_n - y_n}\left(\Theta(-y_n)\int_0^\infty \frac{u_+(z_n)\,dz_n}{y_n - z_n}\right) dy_n. \tag{2.21}$$

When we make the change of variables $y_n = -t$ in (2.21) we obtain

$$\Theta^+\Pi^\circ\Theta^-\Pi^\circ u_+ = -M_0^2 u_+, \quad \text{where} \quad M_0 u_+ = \frac{i}{2\pi}\int_0^\infty \frac{u_+(y_n)\,dy_n}{x_n + y_n}$$

(see Example 2.1). Hence, bearing in mind that the commutator of $\chi(x_n)$ and Π° and also the product $\chi(x_n)B_2$ are Hilbert-Schmidt operators, we obtain

$$\begin{aligned}&\Theta^+ A_1(\omega, D_n)\Theta^- A_2(\omega, D_n)u_+ \\ &= -\chi(x_n)(a_1^- - a_1^+)(a_2^- - a_2^+)M_0^2 u_+ + T_2 u_+,\end{aligned} \tag{2.22}$$

where T_2 is a completely continuous operator. The validity of the representation (2.16) follows from (2.18) and (2.22).

The following lemma is proved similarly:

LEMMA 2.4. *Let $A_0(\omega, \xi_n) \in O_0$, and let M be an integral operator of the class $\mathfrak{M}$. Then*

$$\Theta^+ A_0(\omega, D_n)\Theta^+\chi(x_n)Mu_+ = \Theta^+\chi(x_n)M_1u_+ + \Theta^+ T_1u_+, \tag{2.23}$$

$$\Theta^+\chi(x_n)M\Theta^+ A_0(\omega, D_n)u_+ = \Theta^+\chi(x_n)M_2u_+ + \Theta^+ T_2u_+, \tag{2.24}$$

where M_1 and M_2 are integral operators of the class $\mathfrak{M}$ with the symbols

$$\widehat{M}_1(s) = \left(a_+\frac{1}{1-e^{2\pi is}} - a_-\frac{e^{2\pi is}}{1-e^{2\pi is}}\right)\widehat{M}(s), \tag{2.25}$$

$$\widehat{M}_2(s) = \widehat{M}(s)\left(a_+\frac{1}{1-e^{2\pi is}} - a_-\frac{e^{2\pi is}}{1-e^{2\pi is}}\right), \tag{2.26}$$

and T_1 and T_2 are completely continuous operators.

LEMMA 2.5. *Let M_1 and M_2 be operators of the class $\mathfrak{M}$. Then*

$$\chi M_1\chi M_2 u_+ = \chi M_3 u_+ + \Theta^+ Tu_+, \tag{2.27}$$

where M_3 is the operator of the class $\mathfrak{M}$ with the symbol $\hat{M}_3(s) = \hat{M}_1(s)\hat{M}_2(s)$ and T is a completely continuous operator in $L_2(\mathbf{R}^1)$.

4. By the symbol of the operator (2.4) we mean the pair of matrices

$$A_0(\omega,\ \xi_n),\ -\infty\leqslant\xi_n\leqslant+\infty \text{ and } a_+\frac{1}{1-e^{2\pi is}}-a_-\frac{e^{2\pi is}}{1-e^{2\pi is}}+\widehat{M}(s),$$

$$\operatorname{Re}s=\frac{1}{2}, \tag{2.28}$$

where $\widehat{M}(s)$ is the symbol of the operator M, and $a_\pm=A_0(0,\pm 1)=\lim_{\xi_n\to\pm\infty}A_0(\omega,\xi_n)$. By Lemmas 2.3, 2.4 and 2.5 the following theorem holds.

THEOREM 2.1. *Let* $\mathscr{A}_1=\theta^+A_1(\omega,D_n)+\theta^+\chi(x_n)M_1+\theta^+T_1$ *and* $\mathscr{A}_2=\theta^+A_2(\omega,D_n)+\theta^+\chi(x_n)M_1+\theta^+T_2$ *be two operators of the form* (2.4) *with the symbols*

$$A_1(\omega,\ \xi_n),\ a_1^+\frac{1}{1-e^{2\pi is}}-a_1^-\frac{e^{2\pi is}}{1-e^{2\pi is}}+\widehat{M}_1(s)$$

and

$$A_2(\omega,\ \xi_n),\ a_2^+\frac{1}{1-e^{2\pi is}}-a_2^-\frac{e^{2\pi is}}{1-e^{2\pi is}}+\widehat{M}_2(s),$$

where $a_1^\pm=A_1(0,\pm 1)$ *and* $a_2^\pm=A_2(0,\pm 1)$. *Then their product* $\mathscr{A}_3$ *can be put in the form* (2.4)

$$\mathscr{A}_3=\mathscr{A}_1\mathscr{A}_2=\Theta^+A_3(\omega,\ D_n)+\Theta^+\chi(x_n)M_3+\Theta^+T_3, \tag{2.29}$$

and $\mathscr{A}_3$ *has the symbol*

$$A_3(\omega,\ \xi_n),\ a_3^+\frac{1}{1-e^{2\pi is}}-a_3^-\frac{e^{2\pi is}}{1-e^{2\pi is}}+\widehat{M}_3(s),$$

which is equal to the product of the symbols of $\mathscr{A}_1$ *and* $\mathscr{A}_2$; *that is,*

$$A_3(\omega,\ \xi_n)=A_1(\omega,\ \xi_n)A_2(\omega,\ \xi_n), \tag{2.30}$$

$$a_3^+\frac{1}{1-e^{2\pi is}}-a_3^-\frac{e^{2\pi is}}{1-e^{2\pi is}}+\widehat{M}_3(s)$$

$$=\left(a_1^+\frac{1}{1-e^{2\pi is}}-a_1^-\frac{e^{2\pi is}}{1-e^{2\pi is}}+\widehat{M}_1(s)\right)\left(a_2^+\frac{1}{1-e^{2\pi is}}-a_2^-\frac{e^{2\pi is}}{1-e^{2\pi is}}+\widehat{M}_2(s)\right). \tag{2.31}$$

For, by Lemmas 2.3, 2.4 and 2.5 the operator $\mathscr{A}_3=\mathscr{A}_1\mathscr{A}_2$ can be put in the form (2.31), with $A_3(\omega,\xi_n)=A_1(\omega,\xi_n)A_2(\omega,\xi_n)$,

$$\begin{aligned}\widehat{M}_3(s)&=(a_1^+-a_1^-)(a_2^+-a_2^-)\frac{e^{2\pi is}}{(1-e^{2\pi is})^2}\\&+\widehat{M}_1(s)\left(a_2^+\frac{1}{1-e^{2\pi is}}-a_2^-\frac{e^{2\pi is}}{1-e^{2\pi is}}\right)\\&+\left(a_1^+\frac{1}{1-e^{2\pi is}}-a_1^-\frac{e^{2\pi is}}{1-e^{2\pi is}}\right)\widehat{M}_2(s)+\widehat{M}_1(s)\widehat{M}_2(s).\end{aligned} \tag{2.32}$$

Because $a_3^\pm=a_1^\pm a_2^\pm$ and

$$\left(\frac{a_1^+}{1-e^{2\pi is}} - \frac{a_1^- e^{2\pi is}}{1-e^{2\pi is}}\right)\left(\frac{a_2^+}{1-e^{2\pi is}} - \frac{a_2^- e^{2\pi is}}{1-e^{2\pi is}}\right)$$
$$= \frac{a_1^+ a_2^+}{1-e^{2\pi is}} - \frac{a_1^- a_2^- e^{2\pi is}}{1-e^{2\pi is}} + (a_1^+ - a_1^-)(a_2^+ - a_2^-)\frac{e^{2\pi is}}{(1-e^{2\pi is})^2},$$

(2.32) implies (2.31).

Theorem 2.1 allows one to find a condition for the normal solvability of an operator $\mathscr{A}$ of form (2.4).

COROLLARY 2.1. *For $\mathscr{A}$ to be a Fredholm operator it is sufficient that the determinant of its symbol* (2.30) *is nonzero; that is, that*

(2.33) $$\det A_0(\omega, \xi_n) \neq 0, \quad -\infty \leqslant \xi_n \leqslant +\infty,$$

(2.34) $$\det\left(\frac{a_+}{1-e^{2\pi is}} - \frac{a_- e^{2\pi is}}{1-e^{2\pi is}} + \hat{M}(s)\right) \neq 0, \quad \operatorname{Re} s = \frac{1}{2}.$$

For let R be the operator of form (2.3) whose symbol is the reciprocal of that of $\mathscr{A}$; that is,

(2.35) $$Ru_+ = \Theta^+ A_0^{-1}(\omega, D_n) u_+ + \Theta^+ \chi(x_n) M_1 u_+,$$

where

(2.36) $$\hat{M}_1(s) = \left(\frac{a_+}{1-e^{2\pi is}} - \frac{a_- e^{2\pi is}}{1-e^{2\pi is}} + \hat{M}(s)\right)^{-1} - \left(\frac{a_+^{-1}}{1-e^{2\pi is}} - \frac{a_-^{-1} e^{2\pi is}}{1-e^{2\pi is}}\right).$$

We note that by (2.8) and (2.9) the operator M_1 belongs to the class $\mathfrak{M}$.

By Theorem 2.1 the operator R is a left and right regularizer of the operator $\mathscr{A}$, and so $\mathscr{A}$ is a Fredholm operator.

In particular, let $\mathscr{A} = \Theta^+ A_0(\omega, D_n)$; that is, $M = 0$, and $T = 0$. Then $\Theta^+ A_0(\omega, D_n)$ is a Fredholm operator if

(2.37) $$\det A_0(\omega, \xi_n) \neq 0, \quad -\infty \leqslant \xi_n \leqslant +\infty,$$

(2.38) $$\det\left(\frac{a_+}{1-e^{2\pi is}} - \frac{a_- e^{2\pi is}}{1-e^{2\pi is}}\right) \neq 0, \quad \operatorname{Re} s = \frac{1}{2}.$$

Condition (2.38) can be given a more convenient form. Since by (2.37) $\det a_+ \neq 0$, and since for $s = \frac{1}{2} + i\tau$ we have $e^{2\pi is} = -e^{-2\pi\tau}$ and $1 - e^{2\pi is} = 1 + e^{2\pi\tau}$, $-\infty < \tau < +\infty$, condition (2.38) is equivalent to the following:

(2.39) $$\det(e^{2\pi\tau} I + a_+^{-1} a_-) \neq 0 \quad \text{for} \quad -\infty < \tau < +\infty.$$

When τ ranges over the real axis $e^{2\pi\tau}$ runs through the set of all positive numbers. Thus (2.39) means that $a_+^{-1} a_-$ has no negative eigenvalues. Hence the conditions (2.38) and (1.19) are equivalent.

REMARK 2.1. One can show that the condition (2.35) is also necessary for an operator of type (2.4) to be a Fredholm operator.

5. We find the index of a Fredholm operator $\mathscr{A}$ of the form (2.4).

THEOREM 2.2. *When* (2.35) *holds the index of the operator* $\mathscr{A}$ *is given by*

$$\operatorname{ind}\mathscr{A} = \frac{1}{2\pi}\,\Delta\arg\det A_0(\omega,\ \xi_n)\Big|_{\xi_n=+\infty}^{\xi_n=-\infty}$$

$$+\frac{1}{2\pi}\,\Delta\arg\det\left(\frac{a_+}{1-e^{2\pi is}}-\frac{a_-\,e^{2\pi is}}{1-e^{2\pi is}}+\widehat{M}(s)\right)\Bigg|_{s=\frac{1}{2}-i\infty}^{s=\frac{1}{2}+i\infty}. \tag{2.40}$$

We note that the right-hand side of (2.40) is an integer, because, for $\xi_n = -\infty$, the matrix $A_0(\omega,\xi_n)$ is equal to the value of

$$N(s)=\frac{a_+}{1-e^{2\pi is}}-\frac{a_-\,e^{2\pi is}}{1-e^{2\pi is}}+M(s)$$

when $s=\frac{1}{2}-i\infty$, and, for $\xi_n=+\infty$, $A_0(\omega,\xi_n)$ is equal to the value of $N(s)$ when $s=\frac{1}{2}+i\infty$ (see (2.8) and (2.9)).

PROOF OF THEOREM 2.2. To begin with we deform the operator $\mathscr{A}$ homotopically with the conditions (2.33) and (2.34) preserved into an operator of the form (2.40) for which

$$a_+ = a_- = I. \tag{2.41}$$

Because the left and right sides of (2.40) are unchanged under this deformation, it is sufficient to prove (2.40) when (2.41) holds. We note that in this case each of the two terms on the right-hand side of (2.40) are separately equal to integers.

If (2.41) holds, then both the matrix $A_0(\omega,\xi_n)$ and the matrix $N(s)=I+\widehat{M}(s)$ can be deformed homotopically into diagonal matrices that have all their elements save one equal to unity. The homotopic deformation can be carried out in such a way that the operators remain in the class (2.4), and the conditions (2.33) and (2.34) are preserved. Hence the system of equations $Au_+=f$ can be deformed homotopically into a diagonal system of the form

$$\begin{aligned}\Theta^+ a_1(\omega,\ D_n)\,u_1^+ &= f_1^+\\ u_2^+ + \Theta^+\chi m_2 u_2^+ &= f_2^+\\ u_3^+ = f_3^+,\ \cdots,\ u_p^+ &= f_p^+,\end{aligned} \tag{2.42}$$

where $a_1(\omega,\xi_n)$ is a scalar symbol of the class O_0, $a_1(0,\pm 1)=1$, and m_2 is a scalar operator of the class $\mathfrak{M}$. Obviously the index of the system (2.42) is equal to the sum of the indices of the first and second equations of the system, so that it is sufficient to prove (2.40) separately for each equation. The validity of (2.40) for the first equation of (2.42) follows from Lemma 1.2.

We show that it is also true for the second equation of (2.42). Let $\chi_1(x_n)$ be the function that is 1 for $0 \leq x_n \leq 1$ and zero for $x_n \notin [0,1]$. By (2.1) the operator $\theta^+\chi m_2 u_2^+$ differs from $\chi_1 m_2 u_2^+$ by an operator which is completely continuous on $H_0(\mathbf{R}^1)$. The equation $u_2^+ + \chi_1 m_2 u_2^+ = f_2^+$ is clearly equivalent to the following equation on $[0,1]$:

$$v_2^+ + \chi_1 m_2 v_2^+ = g_2^+, \tag{2.43}$$

where $v_2^+ = u_2^+$ for $0 \leqq x_n \leqq 1$, and $g_2^+ = f_2^+ - \chi_1 m_2 (1-\chi_1) f_2^+$, $0 \leqq x_n \leqq 1$.

We find the index of the equation (2.43). If in (2.43) we make the change of variables $x_n = e^{-t}$, $0 \leqq t < \infty$, and write $w_2(t) = v_2(e^{-t})e^{-t/2}$, then the equation (2.43) becomes an integral equation of convolution type on the half-line:

$$w_2(t) + \int_0^\infty e^{-\frac{t-\tau}{2}} m_2(e^{-(t-\tau)}) w_2(\tau)\, d\tau = h_2(t), \tag{2.44}$$

and this has a form similar to a pseudodifferential equation of the form (1.7). The symbol of the equation (2.44) is

$$1 + \widetilde{m}_2(\xi_n) = 1 + \int_{-\infty}^{\infty} e^{-\frac{t}{2}} m_2(e^{-t}) e^{it\xi_n}\, dt. \tag{2.45}$$

When we make in (2.45) the change of variable $e^{-t} = x_n$, we obtain

$$\widetilde{m}_2(\xi_n) = \int_0^\infty m_2(x_n) x_n^{-\frac{1}{2} - i\xi_n} dx_n = \widehat{m}_2\left(\frac{1}{2} - i\xi_n\right),$$

where $\hat{m}_2(s)$ is the Mellin transform of $m_2(x_n)$. Thus by Lemma 1.2 the index of the equation (2.43) is

$$\frac{1}{2\pi} \Delta \arg (1 + \widetilde{m}_2(\xi_n)) \Big|_{\xi_n=+\infty}^{\xi_n=-\infty} = \frac{1}{2\pi} \Delta \arg (1 + \widehat{m}_2(s)) \Big|_{s=\frac{1}{2}-i\infty}^{s=\frac{1}{2}+i\infty},$$

as we were required to prove.

COROLLARY 2.2. We deduce the formula for the index of the operator $\mathscr{A}_0 = \theta^+ A_0(\omega, D_n)$. By (2.40) we have

$$\begin{aligned} \operatorname{ind} \mathscr{A}_0 = {} & \frac{1}{2\pi} \Delta \arg\det A_0(\omega, \xi_n) \Big|_{\xi_n=+\infty}^{\xi_n=-\infty} \\ & + \frac{1}{2\pi} \Delta \arg\det \left(\frac{a_+}{1-e^{2\pi i s}} - \frac{a_- e^{2\pi i s}}{1-e^{2\pi i s}} \right) \Big|_{s=\frac{1}{2}-i\infty}^{s=\frac{1}{2}+i\infty} \end{aligned} \tag{2.46}$$

If $s = \frac{1}{2} + i\tau$, then $e^{2\pi i s} = e^{-2\pi\tau}$. We write $t = 1/(1-e^{2\pi i s}) = 1/(1+e^{-2\pi\tau})$. As τ ranges over the real axis, t ranges over $[0,1]$. Hence (2.46) can be written as

$$\begin{aligned} \operatorname{ind} \mathscr{A}_0 = {} & \frac{1}{2\pi} \Delta \arg\det A_0(\omega, \xi_n) \Big|_{\xi_n=+\infty}^{\xi_n=-\infty} \\ & + \frac{1}{2\pi} \Delta \arg\det [t a_+ + (1-t) a_-] \Big|_{t=0}^{t=1}. \end{aligned} \tag{2.47}$$

The formula (2.47) can be given the following geometrical interpretation (see [**9**]):

As ξ_n varies from $+\infty$ to $-\infty$, the matrix $A_0(\omega,\xi_n)$ describes a curve, which in general is not closed, in the space of nonsingular matrices of order p. We connect the matrices $a_+ = A_0(0,+1)$ and $a_- = A_0(0,-1)$ by the segment $ta_+ + (1-t)a_-$ in the space of matrices (when $A_0(\omega,\xi_n)$ is a scalar symbol this is the segment in the complex plane that joins the points a_+ and a_-). We remark that the condition that the matrices $ta_+ + (1-t)a_-$ are nonsingular for all t in $[0,1]$ is one of the conditions for the normal solvability of $\mathscr{A}_0$.

As a result we obtain a closed curve in the space of nonsingular matrices. The increase in the argument of the determinant of this curve, divided by 2π, is the index of the operator $\mathscr{A}_0$.

REMARK 2.2. Theorem 1.1 and Corollary 2.1 give two different methods for constructing a regularizer of the operator $\mathscr{A}_0 = \theta^+ A_0(\omega, D_n)$. We give yet another method of constructing a regularizer.

Let $b_\pm(\xi_n)$ be the same matrices as in §1.4. We suppose that (1.22) holds. Then by Lemma 1.3 the operator $b_+^{-1}(\xi_n)\Pi^+ b_-(\xi_n)a_+^{-1}$ is bounded in $H_0^+(\mathbf{R}^1)$. We note that the operator $b_+^{-1}(\xi_n)\Pi^+ b_-(\xi_n)a_+^{-1}$ is the inverse of $\Pi^+ a_+ b_-^{-1}(\xi_n)b_+(\xi_n)$.

Let R_0 denote the operator defined by

$$
\begin{aligned}
R_0 g_+ &= \chi(x_n) F_{\xi_n}^{-1} b_+^{-1}(\xi_n)\Pi^+ b_-(\xi_n) a_+ \tilde{g}_+(\xi_n) \\
&\quad + (1-\chi(x_n))\Theta^+ F_{\xi_n}^{-1} A_0^{-1}(\omega,\xi_n)\tilde{g}_+(\xi_n).
\end{aligned} \tag{2.48}
$$

We note that $(1-\chi(x_n))\theta(x_n) \in C^\infty(\mathbf{R}^1)$.

We show that R_0 is a left and right regularizer of $\theta^+ A_0(\omega, D_n)$. We have $A_0(\omega,\xi_n) = a_+ b_-^{-1}(\xi_n) b_+(\xi_n) + B(\omega,\xi_n)$, where $|B(\omega,\xi_n)| \leq C(1+|\xi_n|)^{-1}$.

We have already noted that the operators $\chi(x_n)B(\omega,D_n)$ and $B(\omega,D_n)\chi(x_n)$, and the commutators of $\chi(x_n)$ with $A_0(\omega,D_n)$ and $a_+ b_-^{-1}(D_n)b_+(D_n)$, are Hilbert-Schmidt operators. Hence

$$
\begin{aligned}
\Theta^+ A_0(\omega, D_n) R_0 g_+ &= \Theta^+ (a_+ b_-^{-1}(D_n) b_+(D_n) \\
&\quad + B(\omega, D_n)) R_0 g_+ = \chi(x_n) g_+ + \Theta^+ T_1 g_+ \\
&\quad + (1-\chi(x_n))\Theta^+ A_0(\omega, D_n) F_{\xi_n}^{-1} A_0^{-1}(\omega,\xi_n)\tilde{g}_+(\xi_n) + \Theta^+ T_2 g_+ \\
&= g_2 + \Theta^+ T_3 g,
\end{aligned}
$$

where T_i is a completely continuous operator in $H_0(\mathbf{R}^1)$, $i = 1,2,3$. Similarly we have

$$R_0\theta^+ A_0(\omega,D_n)u_+ = u_+ + \theta^+ T_4 u_+,$$

where T_4 is a completely continuous operator.

§3. Boundary value problems for systems of elliptic pseudodifferential equations in a half-space

Let $\hat{A}(D)$ be a p.d.o. with $\hat{A}(\xi) = \|\hat{a}_{ij}(\xi)\|_{i,j=1}^p$, where $\hat{A}(\xi) \in O_{\alpha+i\beta}^\infty$ and the ellipticity condition holds:

$$\det \hat{A}(\xi) \neq 0 \quad \text{for} \quad \xi \neq 0. \tag{3.1}$$

Let s be a real number that satisfies (1.38). Then by Theorem 1.2 the operator $pA(\omega, D_n)$ is a Fredholm operator from $H_s^+(\mathbf{R}^1)$ into $H_{s-\alpha}(\mathbf{R}_+^1)$ for every $\omega \in S^{n-2}$, where S^{n-2} is the unit sphere in $\mathbf{R}^{n-1}$. We denote the dimension of the kernel of the operator $pA(\omega, D_n)$ by $n_+(\omega)$, and the dimension of its cokernel by $n_-(\omega)$. As is well known, the functions $n_+(\omega)$ and $n_-(\omega)$ are bounded on S^{n-2}: $\max_\omega n_+(\omega) < \infty$ and $\max_\omega n_-(\omega) < \infty$.

By analogy with the scalar case (see [**21**]), to make the boundary value problem on $\mathbf{R}_+^n$ correctly posed we add to the equation $p\hat{A}(D)u_+ = f$ at least $\max_\omega n_-(\omega)$ potentials so as to eliminate the cokernel of the operator $pA(\omega, D_n)$, and we specify at least $\max_\omega n_+(\omega)$ boundary conditions at $x_n = 0$ so as to eliminate the kernel of $pA(\omega, D_n)$.

Let m_+ and m_- be any integers such that $m_+ \geqq \max_\omega n_+(\omega)$, $m_- \geqq \max_\omega n_-(\omega)$ and $m_+ - m_- = n_+(\omega) - n_-(\omega) = m$, where m is the index of $pA(\omega, D_n)$. We note that m does not depend on ω for $n \geqq 3$, since the sphere S^{n-2} is connected and $pA(\omega, D_n)$ depends continuously on ω in the operator norm. For $n = 2$ we make the additional requirement that $m = n_+(\omega) - n_-(\omega)$ does not depend on ω.

Let $B_j(\xi', \xi_n) \in O'_{\beta_j}$ and $C_k(\xi', \xi_n) \in O'_{\gamma_k}$, $1 \leqq j \leqq m_+$, $1 \leqq k \leqq m_-$; that is, $B_j(\xi', \xi_n)$ and $C_k(\xi', \xi_n)$ are homogeneous symbols continuous for $\xi' \neq 0$, $B_j(\xi)$ and $C_k(\xi)$ are p-dimensional vectors, and

$$\operatorname{Re}\beta_j < s - \tfrac{1}{2}, \qquad \operatorname{Re}\gamma_k < \alpha - s - \tfrac{1}{2}. \tag{3.2}$$

Let $E_{jk}(\xi')$ be homogeneous functions of order $\beta_{jk} = \beta_j + \gamma_k - \alpha - i\beta + 1$ with respect to ξ' that are continuous for $\xi' \neq 0$.

We consider in $\mathbf{R}_+^n$ the boundary value problem

$$p\hat{A}u_+ + \sum_{k=1}^{m_-} p\hat{G}_k (v_k(x') \times \delta(x_n)) = f(x), \tag{3.3}$$

$$p'\hat{B}_j u_+ + \sum_{k=1}^{m_-} \hat{E}_{jk}(D') v_k(x') = g_j(x'), \qquad 1 \leqslant j \leqslant m_+, \tag{3.4}$$

where $\hat{A}$, $\hat{G}_k$, and B_j are p.d.o.'s in $\mathbf{R}^n$ with the symbols

$$\hat{A}(\xi', \xi_n), \quad C_k((1 + |\xi'|)\omega, \xi_n), \quad B_j((1 + |\xi'|)\omega, \xi_n),$$

$\hat{E}_{jk}(\hat{D}')$ is a p.d.o. in $\mathbf{R}^{n-1}$ with the symbol $E_{jk}((1 + |\xi'|)\omega)$, and p' is the restriction operator to $\mathbf{R}^{n-1}$. We write

$$\mathscr{H}_s^{(1)}(\mathbf{R}^n) = H_s^+(\mathbf{R}^n) \times \prod_{k=1}^{m_-} H_{s-\alpha+\operatorname{Re}\gamma_k+\frac{1}{2}}(\mathbf{R}^{n-1}),$$

$$\mathscr{H}_s^{(2)}(\mathbf{R}^n) = H_{s-\alpha}(\mathbf{R}_+^n) \times \prod_{j=1}^{m_+} H_{s-\operatorname{Re}\beta_j-\frac{1}{2}}(\mathbf{R}^{n-1}).$$

Let $\mathfrak{A}$ be the operator corresponding to the equations (3.3) and (3.4).

By the relations (3.2) $\mathfrak{A}$ is a bounded operator from $\mathscr{H}_s^{(1)}(\mathbf{R}^n)$ into $\mathscr{H}_s^{(2)}(\mathbf{R}^n)$.

We find conditions for the boundary value problem (3.3), (3.4) to be uniquely solvable.

We connect with (3.3) and (3.4) the following family of boundary value problems on the half-line that depends on a parameter $\omega \in S^{n-2}$:

$$pA(\omega, D_n) v_+(x_n) + \sum_{k=1}^{m_-} \rho_k p C_k(\omega, D_n)\delta(x_n) = g(x_n), \tag{3.5}$$

$$pB_j(\omega, D_n) v_+(x_n) + \sum_{k=1}^{m_+} E_{jk}(\omega)\rho_k = h_j, \quad 1 \leqslant j \leqslant m_+, \tag{3.6}$$

where ρ_k and h_j are complex numbers, $v_+(x_n) \in H_s^+(\mathbf{R}^1)$, and $g(x_n) \in H_{s-\alpha}(\mathbf{R}_+^1)$. Let $\mathscr{H}_s^{(1)}(\mathbf{R}^1) = H_s^+(\mathbf{R}^1) \times \mathbf{C}_-^m$ and $\mathscr{H}_s^{(2)}(\mathbf{R}^1) = H_{s-\alpha}(\mathbf{R}_+^1) \times \mathbf{C}_+^m$, where $\mathbf{C}^k$ is k-dimensional vector space. Let $\mathfrak{A}(\omega)$ be the operator corresponding to the system (3.5), (3.6). The operator $\mathfrak{A}(\omega)$ is bounded from $\mathscr{H}_s^{(1)}(\mathbf{R}^1)$ into $\mathscr{H}_s^{(2)}(\mathbf{R}^1)$, and depends continuously on $\omega \in S^{n-2}$ in the operator norm. Let us explain in detail the connection between the operators $\mathfrak{A}$ and $\mathfrak{A}(\omega)$.

We take the Fourier transformation with respect to x' in (3.3), (3.4). We obtain

$$p\hat{A}(\xi', D_n)\tilde{w}_+(\xi', x_n) + \sum_{k=1}^{m_-} p\,\hat{C}_k(\xi', D_n)(\tilde{v}_k(\xi') \times \delta(x_n)) = \tilde{h}(\xi', x_n), \tag{3.7}$$

$$p'\hat{B}_j(\xi', D_n)\tilde{w}_+(\xi', x_n) + \sum_{k=1}^{m_-} \hat{E}_{jk}(\xi')\tilde{v}_k(\xi') = \tilde{g}_j(\xi'), \quad 1 \leqslant j \leqslant m_+, \tag{3.8}$$

where $\tilde{w}(\xi', x_n) = F_{x'} u_+(x', x_n)$, and $\tilde{h}(\xi', x_n) = F_{x'} f(x', x_n)$. We now make in (3.7) and (3.8) the change of variables $x_n = y_n/(1+|\xi'|)$. When we put $\xi_n = (1+|\xi'|)\eta_n$ we obtain

$$\tilde{w}_+\left(\xi', \frac{y_n}{1+|\xi'|}\right) = F_{\eta_n}^{-1}[(1+|\xi'|)\tilde{u}_+(\xi', (1+|\xi'|)\eta_n)].$$

Similarly we find that

$$\begin{aligned} \hat{A}(\xi', D_n)\tilde{w}_+(\xi', x_n) &= F_{\xi_n}^{-1}\hat{A}(\xi', \xi_n)\tilde{u}_+(\xi', \xi_n) \\ &= (1+|\xi'|)^{\alpha+i\beta} F_{\eta_n}^{-1} A(\omega, \eta_n)(1+|\xi'|)\tilde{u}_+(\xi', (1+|\xi'|)\eta_n) \\ &= (1+|\xi'|)^{\alpha+i\beta}\tilde{A}(\omega, D_n)\tilde{w}_+\left(\xi', \frac{y_n}{1+|\xi'|}\right). \end{aligned}$$

Hence from (3.7) and (3.8) we obtain

$$pA(\omega, D_n)\tilde{w}_+\left(\xi', \frac{y_n}{1+|\xi'|}\right) + \sum_{k=1}^{m_-} pC_k(\omega, D_n)\delta(x_n)\tilde{\rho}_k(\xi') = \tilde{g}\left(\xi', \frac{y_n}{1+|\xi'|}\right), \tag{3.9}$$

$$p'B_j(\omega, D_n)\widetilde{w}_+\left(\xi', \frac{y_n}{1+|\xi'|}\right) + \sum_{k=1}^{m_-} E_{jk}(\omega)\widetilde{\rho}_k(\xi') = \widetilde{h}_j(\xi'), \tag{3.10}$$
$$1 \leqslant j \leqslant m_+,$$

where $\tilde{\rho}(\xi') = \tilde{v}_k(\xi')(1+|\xi'|)^{\gamma_k - \alpha - i\beta - 1}$ and

$$\widetilde{g}\left(\xi', \frac{y_n}{1+|\xi'|}\right) = (1+|\xi'|)^{-\alpha - i\beta}\widetilde{h}\left(\xi', \frac{y_n}{1+|\xi'|}\right),$$
$$\widetilde{h}_j(\xi') = (1+|\xi'|)^{-\beta_j}\widetilde{g}_j(\xi').$$

We note that $\tilde{\rho}_k(\xi') \in \widetilde{H}_{s-1/2}(\mathbf{R}^{n-1})$ and $\tilde{h}_j(\xi') \in H_{s-1/2}(\mathbf{R}^{n-1})$, $1 \leq k \leq m_-$, $1 \leq j \leq m_+$. Because

$$\int_{-\infty}^{\infty} (1+|\xi'|+|\xi_n|)^{2s}|\widetilde{u}_+(\xi', \xi_n)|^2 d\xi' d\xi_n \tag{3.11}$$
$$= \int_{-\infty}^{\infty}\left(\int_{-\infty}^{\infty}(1+|\eta_n|)^{2s}|(1+|\xi'|)\widetilde{u}_+(\xi', (1+|\xi'|)\eta_n)|^2 d\eta_n\right)(1+|\xi'|^{2s-1}d\xi',$$

the function $\tilde{w}_+(\xi', y_n/(1+|\xi'|)) = F_{\eta_n}^{-1}((1+|\xi'|)\tilde{u}_+(\xi', (1+|\xi'|)\eta_n))$ can be considered as a function of $\xi' \in \mathbf{R}^{n-1}$ with values in $H_s^+(\mathbf{R}^1)$ for almost all ξ', and $\tilde{w}_+(\xi', y_n/(1+|\xi'|))$ belongs to $H_{s-1/2}(\mathbf{R}^{n-1}, H_s^+(\mathbf{R}^1))$ (the Sobolev-Slobodeckiĭ space of order $s-\frac{1}{2}$ that consists of functions defined in $\mathbf{R}^{n-1}$ with values in $H_s^+(\mathbf{R}^1)$ for almost all $\xi' \in \mathbf{R}^{n-1}$, and for which the norm (3.11) is finite).

Let V and Φ stand for the following vectors:

$$\widetilde{V}(\xi') = \left(\widetilde{w}_+\left(\xi', \frac{y_n}{1+|\xi'|}\right), \widetilde{\rho}_1(\xi'), \ldots, \widetilde{\rho}_{m_-}(\xi')\right),$$
$$\widetilde{\Phi}(\xi') = \left(\widetilde{g}\left(\xi', \frac{y_n}{1+|\xi'|}\right), \widetilde{h}_1(\xi'), \ldots, \widetilde{h}_{m_+}(\xi')\right).$$

The vector $\widetilde{V}(\xi')$ belongs to $\mathscr{A}_s^{(3)}(\mathbf{R}^n) = H_{s-1/2}(\mathbf{R}^{n-1}, H_s^1(\mathbf{R}^1))$ (the Sobolev-Slobodeckiĭ space of order $s-\frac{1}{2}$ that consists of functions defined in $\mathbf{R}^{n-1}$ with values in $\mathscr{A}_s^{(1)}(\mathbf{R}^1)$ for almost all $\xi' \in \mathbf{R}^{n-1}$). Similarly the vector $\tilde{\Phi}(\xi')$ belongs to the space $\mathscr{A}_s^{(4)}(\mathbf{R}^n) = H_{s-1/2}(\mathbf{R}^{n-1}, \mathscr{A}_s^{(2)}(\mathbf{R}^1))$ of functions defined for $\xi' \in \mathbf{R}^{n-1}$ with values in $\mathscr{A}_s^{(2)}(\mathbf{R}^1)$.

Let $\mathfrak{A}^{(1)}$ denote the operator corresponding to the system (3.9), (3.10), so that this system can be written as

$$\mathfrak{A}^{(1)}\widetilde{V} = \tilde{\Phi}. \tag{3.12}$$

The system (3.9), (3.10) is equivalent to the system (3.3), (3.4), and so it has a unique solution if and only if (3.3), (3.4) does.

We suppose that the following condition holds:

(3.13) *The system of equations* (3.5), (3.6) *is uniquely solvable for any* $\omega \in S^{n-2}$.

Then we have the following theorem.

Theorem 3.1. *Let the conditions* (3.1), (1.38) *and* (3.13) *hold. Then the operator* $\mathfrak{A}$ *maps the space* $\mathscr{H}_s^{(1)}(\mathbf{R}^n)$ *onto* $\mathscr{H}_s^{(2)}(\mathbf{R}^n)$ *in a continuous and one-to-one manner.*

Proof. Let $R(\omega)$ be the operator inverse to $\mathfrak{A}(\omega)$. Because $\mathfrak{A}(\omega)$ depends continuously on ω in the operator norm and it has an inverse for any $\omega \in S^{n-2}$, $R(\omega)$ also depends continuously on ω in the operator norm and is bounded from $\mathscr{H}_s^{(2)}(\mathbf{R}^1)$ into $\mathscr{H}_s^{(1)}(\mathbf{R}^1)$. In particular, the norm of $R(\omega)$ is bounded on S^{n-2}.

For a fixed $\xi' \in \mathbf{R}^{n-1}$ the operator $\mathfrak{A}^{(1)}$ takes the vector $\widetilde{V}(\xi') \in \mathscr{H}_s^{(1)}(\mathbf{R}^1)$ to the vector

$$\mathfrak{A}(\omega)\widetilde{V}(\xi') \in \mathscr{H}_s^{(2)}(\mathbf{R}^1), \quad \text{where } \omega = \xi'/|\xi'|.$$

Let $\tilde{\Phi}(\xi')$ be any vector in $\mathscr{H}_s^{(4)}(\mathbf{R}^n)$. Let $R^{(1)}$ stand for the operator from $\mathscr{H}_s^{(4)}(\mathbf{R}^n)$ into $\mathscr{H}_s^{(3)}(\mathbf{R}^n)$ that is defined as follows: for a fixed $\xi' \in \mathbf{R}^{n-1}$, $R^{(1)}$ takes $\tilde{\Phi}(\xi')$ to the vector $R(\omega)\tilde{\Phi}(\xi')$ in $\mathscr{H}_s^{(3)}(\mathbf{R}^1)$. Because $R(\omega)$ is continuous with respect to ω, we have $R(\omega)\tilde{\Phi}(\xi') \in \mathscr{H}_s^{(3)}(\mathbf{R}^n)$ and

$$\| R(\omega)\tilde{\Phi}(\xi') \|_1 \leq C \| \tilde{\Phi}(\xi') \|_2, \tag{3.14}$$

where $\| \widetilde{V}(\xi') \|_1$ and $\| \tilde{\Phi}(\xi') \|_2$ denote the norms in $\mathscr{H}_s^{(3)}(\mathbf{R}^n)$ and $\mathscr{H}_s^{(4)}(\mathbf{R}^n)$ respectively.

Hence the operator $R^{(1)}$ is bounded from $\mathscr{H}_s^{(4)}(\mathbf{R}^n)$ into $\mathscr{H}_s^{(3)}(\mathbf{R}^n)$, and obviously

$$\mathfrak{A}^{(1)}R^{(1)} = I_4, \qquad R^{(1)}\mathfrak{A}^{(1)} = I_3,$$

where I_3 (respectively, I_4) is the identity operator in $\mathscr{H}_s^{(3)}(\mathbf{R}^n)$ (in $\mathscr{H}_s^{(4)}(\mathbf{R}^n)$).

Thus if the conditions of Theorem 3.1 hold, then the system (3.9), (3.10), and consequently the system (3.3), (3.4), is uniquely solvable.

It follows from Theorem 3.1 that there is an operator R which is bounded from $\mathscr{H}_s^{(2)}(\mathbf{R}^n)$ into $\mathscr{H}_s^{(1)}(\mathbf{R}^n)$ and inverse to $\mathfrak{A}$; that is,

$$R\mathfrak{A} = I_1, \qquad \mathfrak{A}R = I_2, \tag{3.15}$$

where I_1 (respectively, I_2) is the identity operator in $\mathscr{H}_s^{(1)}(\mathbf{R}^n)$ (in $\mathscr{H}_s^{(2)}(\mathbf{R}^n)$).

Remark 3.1. The operators $\mathfrak{A}^{(1)}$ and $R^{(1)}$ can be considered as pseudo-differential operators in $\mathbf{R}^{n-1}$ with the operator-valued symbols $\mathfrak{A}(\omega)$ and $R(\omega)$ (see [**10**]).

2. Here we give a generalization of Theorem 3.1 to the case of pseudo-differential operators with symbols depending on x. The class of symbols $A_0(x, \xi)$ which are homogeneous of order $\alpha + i\beta$, are infinitely differentiable for $\xi \neq 0$ and are stabilizing with respect to x for $|x| \geqq N$ will be denoted as before by $O_{\alpha+i\beta}^{\infty}$. Let

$$\chi(\xi) \in C_0^{\infty}(\mathbf{R}^n), \quad \chi(\xi) = 1 \quad \text{for } |\xi| \leqq 1.$$

The (matrix) function $A(x,\xi)$ belongs to the class $\mathring{O}^{\infty}_{\alpha+i\beta}$ if the following conditions hold:

a) $A(x,\xi) \in C^{\infty}(\mathbf{R}^n \times \mathbf{R}^n)$, and

$$|D_x^p D_\xi^k A(x, \xi)| \leqslant C_{pk}(1+|\xi|)^{\alpha-|k|}, \quad |k| \geqslant 0, \ |p| \geqslant 0; \tag{3.16}$$

b) $A(x,\xi) = A(\infty,\xi)$ for $|x| \geqq N$.

c) There is a homogeneous symbol $A_0(x,\xi)$ of the class $O^{\infty}_{\alpha+i\beta}$ such that $A_1(x,\xi) - A_0(x,\xi)(1-\chi(\xi))$ satisfies estimates of the form (3.16) with α changed to $\alpha - 1$.

Let $A(x,D)$ be the p.d.o. with the symbol $A(x,\xi) \in \mathring{O}^{\infty}_{\alpha+i\beta}$:

$$A(x, D)\,u = \frac{1}{(2\pi)^n} \int_{-\infty}^{\infty} A(x, \xi)\,\tilde{u}(\xi)\,e^{-i(x,\xi)}\,d\xi, \quad \forall u \in S(\mathbf{R}^n).$$

The symbol $A_0(x,\xi)$ will be called the *principal homogeneous symbol* of the p.d.o. $A(x,D)$. Let G be a bounded domain in $\mathbf{R}^n$ with the smooth boundary Γ, let $\mathring{H}_s(G)$ be the subspace of $H_s(\mathbf{R}^n)$ consisting of functions with supports in $\overline{G}$, $H_s(G)$ and $H_s(\Gamma)$ the Sobolev-Slobodeckiĭ spaces on G and Γ, p the restriction operator to G, and p' the restriction operator to Γ.

We consider the boundary value problem

$$p\left(A(x, D)\,u_0 + \sum_{k=1}^{m-} C_k(x, D)\,(v_k(x') \times \delta(\Gamma))\right) = f(x), \tag{3.17}$$

$$p' B_j(x, D)\,u_0 + \sum_{k=1}^{m-} E_{jk}(x', D')\,v_k(x') = g_j(x'), \quad 1 \leqslant j \leqslant m_+, \tag{3.18}$$

where $A(x,D) \in \mathring{O}^{\infty}_{\alpha+i\beta}$, $C_k(x,\xi) \in \mathring{O}^{\infty}_{\gamma_k}$, $B_j(x,\xi) \in \mathring{O}^{\infty}_{\beta_j}$, the conditions (3.2) hold, and $E_{jk}(x',D')$ is a pseudodifferential operator of class $\mathring{O}^{\infty}_{\beta_j+\gamma_k-\alpha-i\beta+1}$ on Γ. In (3.17) and (3.18)

$$u_0 \in \mathring{H}_s(G), \quad v_k \in H_{s-\operatorname{Re}\gamma_k+\alpha+\frac{1}{2}}(\Gamma), \quad f \in H_{s-\alpha}(G),$$

$$g_j \in H_{s-\operatorname{Re}\beta_j-\frac{1}{2}}(\Gamma).$$

Let $\{U_j\}$, $1 \leqq j \leqq N$, be a covering of $\overline{G}$. For those U_j that meet Γ we introduce a system of coordinates $(x_1^{(j)}, \cdots, x_n^{(j)})$ in which the equation of Γ has the form $x_n^{(j)} = 0$. We call such a system of coordinates in U_j the jth *local system of coordinates*.

The following theorem holds.

THEOREM 3.2. *Let* $\det A_0(x,\xi) \neq 0$ *for* $\xi \neq 0$, *and let the conditions* (3.2) *hold. Assume that there exists a real number s such that for any fixed* $x \in \Gamma$ *the conditions* (1.38) *and* (3.13) *hold in the corresponding local system of coordinates. Then the equations* (3.17) *and* (3.18) *define a Fredholm operator*

$\mathfrak{A}$, *which acts from the space*

$$\mathcal{H}_s^{(1)}(G) = \overset{\circ}{H}_s(G) \times \prod_{k=1}^{m_-} H_{s-\operatorname{Re}\gamma_k+\alpha+\frac{1}{2}}(\Gamma)$$

into the space

$$\mathcal{H}_s^{(2)}(G) = H_{s-\alpha}(G) \times \prod_{j=1}^{m_+} H_{s-\operatorname{Re}\beta_j-\frac{1}{2}}(\Gamma).$$

Theorem 3.2 is proved by means of Theorem 3.1 and by using the standard technique of "majorizing" the coefficients (see, for example, [4]).

§4. The factorization of elliptic matrices

1. In the scalar case the explicit form of a solution of a boundary value problem for an elliptic pseudodifferential operator in $\mathbf{R}_+^n$ can be found by factoring the scalar symbol. Factorization is also possible for elliptic matrices for a fixed ξ', but in general the factors in the factorization of elliptic matrices are discontinuous functions of ξ'. Hence factorization does not have the same fundamental role for systems of pseudodifferential operators as it does for the scalar case. In addition, the factorization of matrices is not as effective as that of a scalar symbol.

Here we distinguish a class of elliptic matrices for which the factors in the factorization depend smoothly on the parameters. We first mention the following lemma on Cauchy type integrals.

LEMMA 4.1. *Let* $b(\omega, \eta_n) \in C^\infty$ *with respect to* (ω, η_n), *and let*

$$|D_\omega^p D_{\eta_n}^k b(\omega, \eta_n)| \leqslant C_{pk}(1 + |\eta_n|)^{-\gamma-k}, \tag{4.1}$$
$$0 < \gamma < 1, \quad k \geqslant 0, \quad |p| \geqslant 0.$$

Then the function

$$B_+(\omega, \xi_n + i\tau) = \frac{i}{2\pi} \int_{-\infty}^{\infty} \frac{b(\omega, \eta_n)}{\xi_n + i\tau - \eta_n} d\eta_n$$

is analytic in $\xi_n + i\tau$ *for* $\tau > 0$, *infinitely differentiable with respect to* $(\omega, \xi_n + i\tau)$ *for* $\tau \geqq 0$, *and satisfies the estimates*

$$|D_\omega^p D_{\zeta_n}^k B_+(\omega, \zeta_n)| \leqslant C_{pk}(1 + |\xi_n| + |\tau|)^{-\gamma-k}, \quad \zeta_n = \xi_n + i\tau. \tag{4.2}$$

COROLLARY 4.1. *Let* $P_+ = b_+^{-1}(\xi_n)\Pi^+ b_+(\xi_n)$ *be the operator defined by* (1.24), *and let the function* $g(\omega, \eta_n)$ *satisfy the estimates* (4.1) *for* $\gamma > \frac{1}{2} + \delta_0$, *where*

$$0 < \delta_0 < \frac{1}{2} - \max_{1 \leqslant j \leqslant p} |\operatorname{Re} b_j|$$

(b_j *are the eigenvalues of the matrix* b (*see* (1.21)). *Then* $P_+ g$ *satisfies the estimates*

$$|D_\omega^p D_{\zeta_n}^k P_+ g| \leqslant C_{pk}(1 + |\xi_n| + |\tau|)^{-\frac{1}{2}-\delta_0-k}, \tag{4.3}$$
$$\forall p, \ \forall k, \ \zeta_n = \xi_n + i\tau, \ \tau \geqslant 0.$$

Lemma 4.2. *Let $B(\omega,\eta_n)\in O_0$ and $B(0,\pm 1)=I$. Then the operator $C=P^+B-BP^+$ is an integral operator with the kernel $C(\omega,\xi_n,\eta_n)$, which is infinitely differentiable with respect to (ω,ξ_n,η_n) and for any p and k satisfies the estimates*

$$|D^p_\omega D^k_{\xi_n} C(\omega,\xi_n,\eta_n)| \leqslant C_{pk}(1+|\xi_n|)^{-\frac{1}{2}-\delta_0-k}(1+|\eta_n|)^{-\frac{1}{2}-\delta_0}. \tag{4.4}$$

We can now prove the main lemma of this subsection.

Lemma 4.3. *Let $A_0(\omega,\eta_n)\in O_0$ be an elliptic matrix that satisfies the condition* (1.22). *Assume that for all $\omega\in S^{n-2}$ the operator $\Theta^+A_0(\omega,D_n)$ has a unique inverse in $H_0^+(\mathbf{R}^1)$. Then the matrix $A_0(\omega,\eta_n)$ has the following factorization:*

$$A_0(\omega,\eta_n)=A_-(\omega,\eta_n)\,A_+(\omega,\eta_n), \tag{4.5}$$

where

$$A_+(\omega,\eta_n)=b_+^{-1}(\eta_n)+c_+(\omega,\eta_n),\quad A_-(\omega,\eta_n)=a_+b_-^{-1}(\eta_n)+c_-(\omega,\eta_n), \tag{4.6}$$

$b_\pm(\omega_n)$ are the matrices in (1.21), *and $c_+(\omega,\eta_n+i\tau)$ is analytic in $\eta_n+i\tau$ for $\tau>0$, infinitely differentiable with respect to $(\omega,\eta_n+i\tau)$ for $\tau\geqq 0$, and satisfies the estimate*

$$|D^p_\omega D^k_{\zeta_n} c_+(\omega,\eta_n+i\tau)| \leqslant \frac{C_{pk}}{(1+|\eta_n|+|\tau|)^{\frac{1}{2}+\delta_0+k}}, \tag{4.7}$$
$$\tau\geqslant 0,\ \zeta_n=\xi_n+i\tau,\ k\geqslant 0,\ |p|\geqslant 0.$$

Similarly the matrix $c_-(\omega,\eta_n+i\tau)$ is analytic in $\xi_n+i\tau$ for $\tau<0$, of class C^∞ in $(\omega,\eta_n+i\tau)$ for $\tau\leqq 0$, and satisfies the estimate (4.7) *for $\tau<0$. In addition,*

$$\det A_+(\omega,\xi_n+i\tau)\neq 0 \ \text{ for } \tau\geqslant 0,\quad \det A_-(\omega,\eta_n+i\tau)\neq 0 \ \text{ for } \tau\leqslant 0. \tag{4.8}$$

Proof. The matrix $A_0(\omega,\eta_n)$ can be represented as

$$A_0(\omega,\eta_n)=a_+b_-^{-1}(\eta_n)\,b_+(\eta_n)+a_0(\omega,\eta_n), \tag{4.9}$$

where $a_0(\omega,\eta_n)=O(1/(1+|\eta_n|))$.

We find matrices $c_+(\omega,\eta_n)$ and $c_-(\omega,\eta_n)$ such that

$$A_0(\omega,\eta_n)\,(b_+^{-1}(\eta_n)+c_+(\omega,\eta_n))=a_+b_-^{-1}(\eta_n)+c_-(\omega,\eta_n). \tag{4.10}$$

By (4.9) we have

$$A_0(\omega,\eta_n)\,c_+(\omega,\eta_n)=c_-(\omega,\eta_n)-a_0(\omega,\eta_n)\,b_+^{-1}(\eta_n). \tag{4.11}$$

Because

$$a_0(\omega,\eta_n)\,b_+^{-1}(\eta_n)=O\left(\frac{1}{(1+|\eta_n|)^{\frac{1}{2}+\delta_0}}\right)$$

and so belongs to $H_0(\mathbf{R}^1)$, we have

$$\Pi^{\pm} a_0(\omega, \eta_n) b_{\mp}^{-1}(\eta_n) \in H_0^{\pm}(\mathbf{R}^1).$$

When we apply the operator Π^+ to (4.11) we obtain

$$\Pi^+ A_0(\omega, \eta_n) c_+(\omega, \eta_n) = -\Pi^+ a_0(\omega, \eta_n) b_{+}^{-1}(\eta_n). \tag{4.12}$$

By hypothesis, the equation (4.12) and the equivalent equation (4.11) are uniquely solvable. Hence there are unique matrices $c_+(\omega, \eta_n) \in H_0^+(\mathbf{R}^1)$ and $c_-(\omega, \eta_n) \in H_0^-(\mathbf{R}^1)$ satisfying (4.11). Because the matrix $\Pi^+ a_0(\omega, \eta_n) b_{\pm}^{-1}(\eta_n)$ is obviously infinitely differentiable with respect to ω in the norm of $H_0^+(\mathbf{R}^1)$, and the operator $R(\omega)$ inverse to $\Pi^+ A_0(\omega, \eta_n)$ also belongs to C^∞ with respect to ω, it follows that $c_\pm(\omega, \eta_n)$ are infinitely differentiable with respect to ω with values in $H_0^{\pm}(\mathbf{R}^1)$.

We show that the matrices $c_\pm(\omega, \eta_n)$ have the properties stated in Lemma 4.3. We apply to the left of (4.12) the regularizer

$$\widetilde{R} = P_+ A^{-1}(\omega, \eta_n) a_+ P_- a_+^{-1},$$

constructed in §1. By analogy with (1.37) we obtain

$$c_+(\omega, \eta_n) = P_+ C b_-^{-1} b_+ a_+^{-1} A_0 c_+ + P_+ A_0^{-1} a_+ P_- a_+^{-1} a_0 b_+^{-1}, \tag{4.13}$$

where

$$C = P_+ A_0^{-1} a_+ b_- b_+^{-1} - A_0 a_+ b_- b_+^{-1} P_+$$

is by Lemma 4.2 an integral operator with an infinitely differentiable kernel that satisfies (4.4). The matrix $P_+ A_0^{-1} a_+ P_- a_+^{-1} a_0 b_+^{-1}$ belongs to C^∞ with respect to $(\omega, \xi_n + i\tau)$, $\tau \geqq 0$, by Corollary 4.1, and it satisfies estimates of the form (4.7). Because $a_1 = b_-^{-1} b_+ A_0 c_+$ belongs to C^∞ with respect to ω and has values in $H_0(\mathbf{R}^1)$ with respect to η_n, the matrix Ca_1 is infinitely differentiable with respect to (ω, ξ_n) and it satisfies estimates of the form (4.1) for $\gamma > \frac{1}{2} + \delta_0$. Hence, by Corollary 4.1, $P_+ Ca_1 \in C^\infty$ with respect to $(\omega, \xi_n + i\tau)$ for $\tau \geqq 0$, and it satisfies the estimates (4.7).

It follows from (4.11) that $c_-(\omega, \eta_n)$ also has the properties stated in Lemma 4.2.

We now show that the inequalities (4.8) hold. Let us suppose the contrary. For some ω_0 and $\xi_{n0} + i\tau_0$, $\tau_0 \geq 0$, let

$$\det\,(b_+^{-1}(\xi_{n0} + i\tau_0) + c_+(\omega_0, \xi_{n0} + i\tau_0)) = 0.$$

Then one can find numbers $\alpha_1, \cdots, \alpha_p$, not all zero, such that the linear combination $\sum_{k=1}^p \alpha_k s_k^+(\omega_0, \xi_n + i\tau)$ of the columns of the matrix

$$b_+^{-1}(\omega_0, \xi_n + i\tau) + c_+(\omega_0, \xi_n + i\tau)$$

is zero for $\xi_n + i\tau = \xi_{n0} + i\tau_0$.

We write

$$s_+(\omega_0, \xi_n + i\tau) = \sum_{k=1}^{p} \frac{\alpha_k s_k^+(\omega_0, \xi_n + i\tau)}{(\xi_n + i\tau) - (\xi_{n0} + i\tau_0)}.$$

Then $s_+(\omega_0, \xi_n) \in \widetilde{H}_0^+(\mathbf{R}^1)$, and by (4.10) it satisfies the equation

$$A_0(\omega_0, \xi_n) s_+(\omega_0, \xi_n) = s_-(\omega_0, \xi_n), \tag{4.14}$$

where

$$s_-(\omega_0, \xi_n) = \sum_{k=1}^{p} \frac{\alpha_k \bar{s}_k^-(\omega_0, \xi_n)}{\xi_n - (\xi_{n0} + i\tau_0)},$$

and $\bar{s}_1, \cdots, \bar{s}_p$ are the columns of the matrix $a_+ b_-^{-1}(\xi_n) + c_-(\omega_0, \xi_n)$. Because $\tau \geq 0$, we have $s_-(\omega_0, \xi_n) \in \widetilde{H}_0^-(\mathbf{R}^1)$. But (4.14) contradicts the hypothesis that the operator $\Pi^+ A_0(\omega, \xi_n)$ has no kernel. Hence

$$\det (b_+^{-1}(\xi_n + i\tau) + c_+(\omega, \xi_n + i\tau)) \neq 0 \quad \text{for } \tau \geqslant 0, \ \omega \in S^{n-2}.$$

If we suppose that

$$\det (a_+ b_-^{-1}(\xi_n^{(0)} + i\tau^{(0)}) + c_-(\omega^{(0)}, \xi_n^{(0)} + i\tau^{(0)})) = 0 \quad \text{for } \tau^{(0)} \leqslant 0,$$

then we can similarly construct a nontrivial solution of the equation of the form (4.14), which is impossible. We note that Lemma 4.3 obviously remains valid for the matrix $A_0(x, \omega, \xi_n)$, which depends on x, if the conditions of Lemma 4.3 hold for all $x \in \mathbf{R}^n$. Then the matrices (4.6) belong to C^∞ with respect to x. The following theorem follows from Lemma 4.3:

THEOREM 4.1. *Let $A(x, \xi', \xi_n) \in O^\infty_{\alpha+i\beta}$, and let the matrix*

$$A_0(x, \xi', \xi_n) = \Lambda_-^{-\alpha-i\beta+s}(\xi)\, \Lambda_+^{-s}(\xi)\, A(x, \xi)$$

satisfy the conditions of Lemma 4.3 for all $x \in \mathbf{R}^n$. Then the matrix $A(x, \xi)$ has the following factorization:

$$A(x, \xi) = A_-(x, \xi)\, A_+(x, \xi), \tag{4.15}$$

$$A_+(x, \xi', \xi_n) = \Lambda_+^s(\xi)\, b_+(x, \xi) \left(I + c_+^{(1)}\left(x, \frac{\xi'}{|\xi'|}, \frac{\xi_n}{|\xi'|}\right)\right), \tag{4.16}$$

$$A_-(x, \xi', \xi_n) = \left(I + c_-^{(1)}\left(x, \frac{\xi'}{|\xi'|}, \frac{\xi_n}{|\xi'|}\right)\right) a_+(x)\, b_-^{-1}(x, \xi)\, \Lambda_-^{\alpha+i\beta-s}, \tag{4.17}$$

where

$$a_+(x) = A(x, 0, +1), \quad b_\pm(x, \xi) = \exp\{\ln \Lambda_\pm(\xi) \cdot b(x)\},$$

$$b(x) = \frac{1}{2\pi i} \ln (A^{-1}(x, 0, +1)\, A(x, 0, -1)\, e^{\pi i(\alpha+i\beta-2s)}),$$

the matrix $c_+^{(1)}(x, \xi'/|\xi'|, \xi_n/|\xi'|)$ (respectively, $c_-^{(1)}(x, \xi'/|\xi'|, \xi_n/|\xi'|)$) is analytic in $\xi_n + i\tau$ for $\tau > 0$ (respectively, $\tau < 0$), is of class C^∞ for $x \in \mathbf{R}^n$, $\xi' \neq 0$, $\tau \geq 0$ (respectively, $\tau \leq 0$), and satisfies the estimates

$$\left| D_x^p D_{\xi'}^{k'} D_{\xi_n}^{k_n} c_\pm^{(1)}\left(x, \frac{\xi'}{|\xi'|}, \frac{\xi_n + i\tau}{|\xi'|}\right)\right| \leqslant C_{pk'k_n} \frac{|\xi'|^{2\delta_0 - |k'|}}{(|\xi'| + |\xi_n| + |\tau|)^{2\delta_0 + k_n}}, \quad \forall p, \ \forall k, \tag{4.18}$$

and

$$\det(I + c_+^{(1)}) \neq 0 \quad for \quad |\xi'| + |\xi_n| + \tau > 0, \ \tau > 0,$$

(4.19)

$$\det(I + c_-^{(1)}) \neq 0 \quad for \quad |\xi'| + |\xi_n| + |\tau| \geqslant 0, \ \tau \leqslant 0.$$

EXAMPLE 4.1. The matrix $A(x,\xi) \in O_\alpha^\infty$ is called *strongly elliptic* if the matrix $A_R(x,\xi) = \frac{1}{2}[A(x,\xi) + A^*(x,\xi)]$ is positive definite for $x \in \mathbf{R}^n$, $|\xi| > 0$. The matrix $A_0(x,\xi) = \Lambda_-^{-\alpha+s}\Lambda_+^{-s}A(x,\xi)$ satisfies the hypotheses of Lemma 4.3 for $s = \alpha/2$. Hence a strongly elliptic matrix has a factorization of the form (4.15), (4.16), (4.17), where $s = \alpha/2$.

Let the matrix $A(x,\xi) \in O_0$ be close to the unit matrix:

$$|A(x,\xi) - I| < \epsilon, \tag{4.20}$$

where ϵ is sufficiently small. Then $A(x,\xi)$ is clearly a strongly elliptic matrix and so it has a factorization of the form (4.15) for $s = 0$, and $\alpha + i\beta = 0$. We note that in this case the factors in the factorization of $A(x,\xi)$ can be found by the method of successive approximations (see §5).

3. We now consider the general case.

THEOREM 4.2. *Let $A(x,\xi',\xi_n) \in O_{\alpha+i\beta}^\infty$ be an elliptic matrix, and s a real number satisfying* (1.38). *Then there are an integer N and matrices $Q_N^+(x,\xi)$ and $Q_N^-(x,\xi)$ such that*

$$A(x,\xi) = \Lambda_-^{\alpha+i\beta-s+N}(\xi)\, Q_N^-(x,\xi)\, Q_N^+(x,\xi)\, P_{2N}^{-1}(x,\xi)\, \Lambda_+^{s+N}(\xi), \tag{4.21}$$

$$Q_N^-(x,\xi) = (I + q_N^-(x,\xi))\, a_+(x)\, b_-^{-1}(x,\xi), \tag{4.22}$$

$$Q_N^+(x,\xi) = b_+(x,\xi)\,(I + q_N^+(x,\xi)), \tag{4.23}$$

where $a_+(x)$ and $b_\pm^{-1}(x,\xi)$ are the matrices in (4.16), (4.17), *and the matrices $q_N^\pm(x,\xi)$ and $I + q_N^\pm(x,\xi)$ have the same properties as the matrices $c_\pm^{(1)}$ and $I + c_\pm^{(1)}$ in* (4.16), (4.17), *and $P_{2N}(x,\xi)$ is a homogeneous matrix of order $2N$ that is polynomial in ξ_n,*

$$\det P_{2N}(x,\xi) \neq 0 \quad for \quad \xi \neq 0, \quad P_{2N}(x, 0, \pm 1) = I.$$

PROOF. Let $A_1(x,\xi)$ denote the following matrix:

$$A_1(x,\xi) = b_+^{-1}(x,\xi)\, b_-(x,\xi)\, a_+^{-1}(x)\, A(x,\xi)\, \Lambda_-^{-\alpha-i\beta+s}(\xi)\, \Lambda_+^{-s}(\xi). \tag{4.24}$$

We put $z = \Lambda_+(\xi)/\Lambda_-(\xi)$. Then $\xi_n = i|\xi'|(z+1)/(z-1)$. Because

$$A_1(x, 0, +1) = A_1(x, 0, -1) = I,$$

the matrix

$$A_1\left(x, \xi', i|\xi'|\frac{z+1}{z-1}\right) = A_1\left(x, \omega, i\frac{z+1}{z-1}\right)$$

can be represented by a uniformly convergent Fourier series

$$A_1\left(x, \omega, i\frac{z+1}{z-1}\right) = \sum_{k=-\infty}^{\infty} a_k(x,\omega)\frac{\Lambda_+^k}{\Lambda_-^k}. \tag{4.25}$$

If we take a sufficiently large segment of the Fourier series, we obtain

$$A_1(x,\xi) = B_N(x,\xi)\, R_N(x,\xi), \tag{4.26}$$

where

$$R_N(x,\xi) = \left(\sum_{k=-N}^{N} a_k(x,\omega)\frac{\Lambda_+^k}{\Lambda_-^k}\right)\left(\sum_{k=-N}^{N} a_k(x,\omega)\right)^{-1}$$

is a matrix rational in ξ_n,

$$\det R_N(x,\xi) \neq 0 \quad \text{for} \quad \xi \neq 0, \qquad R_N(x,0,\pm 1) = I,$$

and $B_N(x,\xi)$ is a matrix close to the unit matrix; that is,

$$|B_N(x,\xi) - I| < \epsilon, \tag{4.27}$$

where ϵ is sufficiently small. By Theorem 4.1 the matrix

$$a_+(x)b_-^{-1}(x,\xi)b_+(x,\xi)B_N(x,\xi)$$

has the factorization

$$\begin{aligned} &a_+(x)\,b_-^{-1}(x,\xi)\,b_+(x,\xi)\,B_N(x,\xi) \\ &= (I + q_N^-(x,\xi))\,a_+(x)b_-^{-1}(x,\xi)\,b_+(x,\xi)\,(I + q_N^+(x,\xi)). \end{aligned} \tag{4.28}$$

We represent the matrix $R_N^{-1}(x,\xi)$ as

$$R_N^{-1}(x,\xi) = \Lambda_+^{-N}(\xi)\,\Lambda_-^{-N}(\xi)\,P_{2N}(x,\xi), \tag{4.29}$$

where $P_{2N}(x,\xi)$ is a matrix polynomial in ξ_n. Thus the factorization (4.21) follows from (4.24), (4.28) and (4.29).

§5. The factorization of elliptic pseudodifferential operators

1. *The class S_α^t of symbols.* Because the factors in the factorization of an elliptic symbol of the class $O_{\alpha+i\beta}^\infty$ are not infinitely differentiable with respect to ξ, in constructing the parametrix we need to widen the class $\hat{O}_{\alpha+i\beta}^\infty$ and consider symbols with finite smoothness with respect to ξ. Let $\widetilde{A}'(\eta,\xi)$ be a function locally summable in the Lebesgue sense and satisfying the estimate

$$(1+|\eta|)^N\,|\widetilde{A}'(\eta,\xi)| \leqslant C_N(1+|\xi|)^\alpha, \quad \forall N. \tag{5.1}$$

Let $S_\alpha^0 = S_\alpha^0(\mathbf{R}^n)$ be the class of functions $A(x,\xi)$ of the form

$$A(x,\xi) = A(\infty,\xi) + A'(x,\xi), \tag{5.2}$$

where $A(\infty,\xi)$ is a locally summable function such that $|A(\infty,\xi)| \leqq C(1+|\xi|)^\alpha$, and $A'(x,\xi) = F_\eta^{-1}\widetilde{A}'(\eta,\xi)$ while $\widetilde{A}'(\eta,\xi)$ satisfies (5.1).

Let $t = m+\gamma$, where $m \geqq 0$ is an integer and $0 < \gamma \leqq 1$. By definition, the function $A(x,\xi)$ belongs to the class S_α^t if $D_\xi^k A(x,\xi)$ is an absolutely continuous function which belongs to $S_{\alpha-|k|}^0$ for $0 \leqq |k| \leqq m$, and $D_\xi^k A(x,\xi) \in S_{\alpha-m-\gamma}^0$ for $|k| = m+1$.

We note the following lemma concerning pseudodifferential operators with symbols that belong to the class S_α^t.

LEMMA 5.1. *Let* $A(x,\xi) \in S^0_\alpha$ *and* $B(x,\xi) \in S^0_\beta$. *Then the product of the p.d.o.'s* $A(x,D)$ *and* $B(x,D)$ *is a p.d.o.* $C(x,D)$ *with the symbol* $C(x,\xi) \in S^0_{\alpha+i\beta}$ *defined by*

$$(5.3) \quad C(x, \xi) = A(x, \xi) B(\infty, \xi) + \frac{1}{(2\pi)^n} \int_{-\infty}^{\infty} A(x, \xi + \zeta) B'(\zeta, \xi) e^{-i(x,\zeta)} d\zeta.$$

If $A(x,\xi) \in S^t_\alpha$ *for* $t = m + \gamma$, *where* m *is a nonnegative integer and* $0 < \gamma \leq 1$, *then the symbol* $C(x,\xi)$ *has the expansion*

$$(5.4) \qquad C(x, \xi) = \sum_{|k|=0}^{m} \frac{1}{k!} \frac{\partial^k A(x, \xi)}{\partial \xi^k} D^k_x B(x, \xi) + C^{(m+1)}(x, \xi),$$

where

$$C^{(m+1)}(x, \xi) \in S^0_{\alpha+\beta-m-\gamma}, \quad C^{(m+1)}(\infty, \xi) = 0.$$

2. *Elliptic pseudodifferential operators*. The p.d.o. $A(x,D)$ with the symbol $A(x,\xi) \in S^t_\alpha$, $t > 0$, is said to be *elliptic* if there is a symbol

$$R_0(x, \xi) \in S^t_{-\alpha-\delta}, \quad 0 \leqslant \delta < t_0, \quad t_0 = \min(t, 1),$$

such that

$$A(x, \xi) R_0(x, \xi) = I + T'_1(x, \xi) \text{ and } R_0(x, \xi) A(x, \xi) = I + T'_2(x, \xi),$$

where $T'_i(x,\xi') \in S^t_{-\delta_1}$, $\delta_1 > 0$.

By Lemma 5.1 we have

$$(5.5) \quad A(x, D) R_0(x, D) = I + T'_3(x, D), \quad R_0(x, D) A(x, D) = I + T'_4(x, D),$$

where

$$T'_i(x, \xi) \in S^0_{-\delta_2}, \quad \delta_2 = \min(\delta_1, t_0 - \delta), \quad i = 3, 4.$$

In this subsection we construct a more exact regularizer of the operator $A(x,D)$, namely an operator $R(x,D)$ such that

$$(5.6) \qquad \begin{aligned} A(x, D) R(x, D) &= I + T^{(1)}_\infty(x, D), \\ R(x, D) A(x, D) &= I + T^{(2)}_\infty(x, D), \end{aligned}$$

where

$$T^{(i)}_\infty(x, \xi) \in S^0_{-\infty}, \quad T^{(i)}_\infty(\infty, \xi) = 0.$$

LEMMA 5.2. *Let* $a'(x,\xi) \in S^0_{-\gamma}$ *with* $\gamma \geq 0$, $a'(\infty,\xi) = 0$ *and*

$$(1 + |\eta|)^{n+1} |a'(\eta, \xi)| \leqslant \varepsilon (1 + |\xi|)^{-\gamma},$$

where ϵ *is sufficiently small. Then there is a p.d.o.* $r'(x,D)$ *with the symbol*

$$r'(x, \xi) \in S^0_{-\gamma}, \quad r'(\infty, \xi) = 0,$$

such that

$$(5.8)\qquad \begin{aligned} &(I + r'(x, D))\,(I + a'(x, D)) = I,\\ &(I + a'(x, D))\,(I + r'(x, D)) = I. \end{aligned}$$

Proof. By (5.8) we have

$$(5.9)\quad a'(x, D)\,u + r'(x, D)\,u + a'(x, D)\,r'(x, D)\,u = 0, \quad \forall\, u(x) \in S(\mathbf{R}^n).$$

Hence by the definition of a p.d.o. we have

$$(5.10)\qquad \begin{aligned} &\frac{1}{(2\pi)^n}\int_{-\infty}^{\infty} (a'(x, \xi) + r'(x, \xi))\,\tilde{u}(\xi)\,e^{-i(x,\xi)}\,d\xi\\ &+ \frac{1}{(2\pi)^{2n}}\int_{-\infty}^{\infty} a'(x, \eta)\,e^{-i(x,\eta)}\,\tilde{r}'(\eta - \xi, \xi)\,\tilde{u}(\xi)\,d\xi\,d\eta = 0, \end{aligned}$$

where $\tilde{r}'(\eta, \xi) = F_s r'(x, \xi)$. Because $\tilde{u}(\xi)$ is any function from $S(\mathbf{R}^n)$, it follows from (5.10) that

$$(5.11)\qquad \begin{aligned} &r'(x, \xi)\,e^{-i(x,\xi)} + a'(x, \xi)\,e^{-i(x,\xi)}\\ &+ \frac{1}{(2\pi)^n}\int_{-\infty}^{\infty} a'(x, \eta)\,e^{-i(x,\eta)}\,\tilde{r}'(\eta - \xi, \xi)\,d\eta = 0. \end{aligned}$$

When we multiply (5.11) by $e^{i(x,\xi)}$ and make the change of variables $\eta - \xi = \zeta$, we obtain

$$(5.12)\quad r'(x, \xi) + \frac{1}{(2\pi)^n}\int_{-\infty}^{\infty} a'(x, \xi + \zeta)\,e^{-i(x,\zeta)}\,\tilde{r}'(\zeta, \xi)\,d\zeta = -a'(x, \xi).$$

We now prove that the equation (5.12) is solvable. We take the Fourier transformation of (5.12) with respect to x. We obtain

$$(5.13)\qquad \tilde{r}'(\eta, \xi) + \tilde{A}'\,r'(\eta, \xi) = -\tilde{a}'(\eta, \xi),$$

where

$$(5.14)\qquad \tilde{A}\tilde{r}'(\eta, \xi) = \frac{1}{(2\pi)^n}\int_{-\infty}^{\infty} \tilde{a}'(\eta - \zeta, \zeta + \xi)\,\tilde{r}'(\zeta, \xi)\,d\zeta.$$

By (5.7) we have

$$(5.15)\qquad \sup_{\eta,\xi}\,(1 + |\xi|)^{\nu}\,|\tilde{A}'\,\tilde{r}'(\eta, \xi)| \leqslant C\varepsilon \sup_{\eta,\xi}\,(1 + |\xi|)^{\nu}\,|\tilde{r}'(\eta, \xi)|.$$

Thus, if ϵ is sufficiently small, then the operator $\tilde{A}'$ in the space of functions bounded with respect to η and ξ with the weight $(1 + |\xi|)^{\gamma}$ has a norm less than unity. Hence the equation (5.13) has a unique solution $\tilde{r}'(\eta, \xi)$ satisfying the estimate

$$(5.16)\qquad |\tilde{r}'(\eta, \xi)| \leqslant C\,(1 + |\xi|)^{-\nu}, \quad \gamma \geqslant 0,$$

and this solution can be found in the form of a series

$$(5.17)\qquad \widetilde{r}' = -\widetilde{a}' + \widetilde{A}'\widetilde{a}' - \widetilde{A}'^2\widetilde{a}' + \dots + (-1)^n \widetilde{A}'^n\widetilde{a}' + \dots .$$

We multiply (5.13) by η_i, $1 \leq i \leq n$, and write $\widetilde{r}_i'(\eta,\xi) = \eta_i\widetilde{r}'(\eta,\xi)$; then $\widetilde{r}_i'(\eta,\xi)$ satisfies the equation

$$(5.18)\qquad \begin{aligned} \widetilde{r}_i'(\eta,\xi) + \widetilde{A}'\widetilde{r}_i'(\eta,\xi) &= -\eta_i\widetilde{a}'(\eta,\xi) \\ &- \frac{1}{(2\pi)^n}\int_{-\infty}^{\infty}(\eta_i - \zeta_i)\,\widetilde{a}'(\eta-\zeta,\zeta+\xi)\,\widetilde{r}'(\eta,\xi)\,d\zeta. \end{aligned}$$

Because

$$|\widetilde{a}'(\eta,\xi)| \leqslant C_N(1+|\eta|)^{-N}(1+|\xi|)^{-\gamma},\quad \forall N,$$

the right-hand side of (5.18) remains bounded with respect to η and ξ after multiplication by $(1+|\xi|)^{\gamma}$. Hence as in the proof that (5.13) is solvable we obtain

$$|\widetilde{r}_i'(\eta,\xi)| \leqslant C(1+|\xi|)^{-\gamma},\quad 1 \leqslant i \leqslant n.$$

Thus $|r'(\eta,\xi)| \leqq C(1+|\eta|)^{-1}(1+|\xi|)^{-\gamma}$. It can be proved similarly that for any N

$$|\widetilde{r}'(\eta,\xi)| \leqslant C_N(1+|\eta|)^{-N}(1+|\xi|)^{-\gamma};$$

that is,

$$r'(x,\xi) = F_\eta^{-1}\widetilde{r}'(\eta,\xi) \in S^0_{-\gamma}.$$

Thus $(I + a'(x,D))(I + r'(x,D)) = I$. By (5.7) the operator $I + a'(x,D)$ is uniquely invertible in $H_0(\mathbf{R}^n)$. Hence the operator $I + r'(x,D)$ is not only a right but also a left inverse of $I + a'(x,D)$.

We apply Lemma 5.2 to the construction of an operator $R(x,D)$ satisfying (5.6). As in §3 we let $\chi(x) \in C_0^\infty(\mathbf{R}^n)$, $\chi(\xi) = 1$ for $|\xi| \leqq 1$. We put

$$V_1'(x,\xi) = T_3'(x,\xi)(1-\chi(\delta\xi)),\quad V_2'(x,\xi) = T_3'(x,\xi)\chi(\delta\xi).$$

Then $V_2'(x,\xi) \in S^0_{-\infty}$, and for δ sufficiently small $V_1'(x,\xi)$ satisfies (5.7) with $\gamma = \delta_2 - \epsilon$. By Lemma 5.2 there is an operator $I + r'(x,D)$, inverse to $I + V_1'(x,D)$, with $r'(x,\xi) \in S^0_{-\delta_2+\epsilon}$. Hence the operator

$$R(x,D) = R_0(x,D)(I + r'(x,D))$$

satisfies the relation

$$A(x,D)\,R(x,D) = I + T_\infty^{(1)}(x,D).$$

It can be proved similarly that $R(x,D)$ satisfies the second relation in (5.6). For brevity we call a p.d.o. $R(x,D)$ that satisfies (5.6) an *almost inverse* of $A(x,D)$, and denote it by $A^{(-1)}(x,D)$.

3. *"Plus" and "minus" operators*. The symbol $A_+(x,\xi) = A_+(\infty,\xi) + A_+'(x,\xi) \in S_\alpha^t$ will be called *"plus"* if for almost all $\xi' \in \mathbf{R}^{n-1}$ and $\eta \in \mathbf{R}^n$ the operators

$$A_+(\infty,\xi) \quad\text{and}\quad \widetilde{A}_+'(\eta,\xi) = F_x A_+'(x,\xi)$$

have analytic continuations with respect to $\xi_n + i\tau$ to the half-plane $\tau > 0$ that are continuous for $\tau \geqq 0$, and if for $\tau \geqq 0$ they admit the estimates

$$|A_+(\infty, \xi', \xi_n + i\tau)| \leqslant C(1 + |\xi'| + |\xi_n| + |\tau|)^{\alpha}, \tag{5.19}$$

$$|\widetilde{A}'_+(\eta, \xi', \xi_n + i\tau)| \leqslant C_N(1 + |\eta|)^{-N}(1 + |\xi'| + |\xi_n| + |\tau|)^{\alpha}. \tag{5.20}$$

Similarly the symbol

$$A_-(x, \xi) = A_-(\infty, \xi) + A'_-(x, \xi) \in S^t_{\alpha}$$

is called "*minus*" if for almost all η and ξ' the operators $A_-(\infty, \xi)$ and $A'_-(\eta, \xi) = F_x A'(x, \xi)$ have continuations to the half-plane $\tau < 0$ that are analytic in $\xi_n + i\tau$, continuous for $\tau \leqq 0$ and satisfy (5.19) and (5.20) for $\tau \leqq 0$.

Lemma 5.3. *The p.d.o. $A_+(x, D)$ (respectively, $A_-(x, D)$) with the "plus" symbol $A_+(x, \xi) \in S^0_\alpha$ ("minus" symbol $A_-(x, \xi) \in S^0_\alpha$) is bounded from $H^+_s(\mathbf{R}^n)$ into $H_{s-\alpha}(\mathbf{R}^n)$ (from $H^-_s(\mathbf{R}^n)$ into $H^-_{s-\alpha}(\mathbf{R}^n)$) for any s.*

Lemma 5.4. *The symbol of the product of two "plus" ("minus") p.d.o. is also a "plus" ("minus") symbol.*

Lemma 5.5. *Let $\tilde{a}'_+(x, \xi) \in S^0_{-\epsilon}$ be a "plus" symbol, $\epsilon > 0$. There is a "plus" symbol $r'_+(x, \xi) \in S^0_{-\epsilon/2}$ such that*

$$\begin{aligned} (I + a'_+(x, D))\ (I + r'_+(x, D)) &= I + T^+_\infty(x, D), \\ (I + r_+(x, D))\ (I + a'_+(x, D)) &= I + V^+_\infty(x, D), \end{aligned} \tag{5.21}$$

where $T^+_\infty(x, \xi) \in S^0_{-\infty}$ and $V^+_\infty(x, \xi) \in S^0_{-\infty}$ are "plus" symbols.

Proof. Let

$$\alpha_+(x_n) \in C^\infty_0(\mathbf{R}^1_+),\ \tilde{\alpha}_+(\xi_n) = F_{x_n}\alpha_+(x_n),$$

with $\tilde{a}_+(0) = 1$. We put

$$a^{(1)}_+(x, \xi) = a_+(x, \xi)\ (1 - \chi(\delta\xi')\,\tilde{\alpha}_+(\delta\xi_n)),$$

$$\alpha^{(2)}_+(x, \xi) = a_+(x, \xi)\,\chi(\delta\xi')\,\tilde{\alpha}_+(\delta\xi_n).$$

Then $a^{(2)}_+(x, \xi)$ is a "plus" symbol belonging to $S^0_{-\infty}$, and for δ sufficiently small the symbol $a^{(1)}_+(x, \xi)$ satisfies the estimates

$$|\widetilde{a}^{(1)}_+(x, \xi', \xi_n + i\tau)| \leqslant \varepsilon_1(1 + |\eta|)^{-n-1}(1 + |\xi'| + |\xi_n| + |\tau|)^{-\frac{\varepsilon}{2}}, \quad \tau \geqslant 0, \tag{5.22}$$

where ϵ_1 is small. As in the proof of Lemma 5.2, we can show that there is a "plus" p.d.o. $I + r'_+(x, D)$ inverse to $I + a^{(1)}_+(x, D)$. Hence $I + r'(x, D)$ is an almost inverse of $I + a'(x, D)$. An analogous lemma is valid for "minus" pseudodifferential operators.

4. *Factorization of elliptic pseudodifferential operators.*

Lemma 5.6. *Let $q'(x, \xi) \in S^0_{-\gamma}$ and $\partial q'(x, \xi)/\partial\xi_n \in S^0_{-\gamma-1}$, where $0 < \gamma < 1$.*

Then there are a "plus" symbol $q'_+(x,\xi)\in S^0_{-\gamma+\epsilon}$ *and a "minus" symbol* $q'_-(x,\xi)\in S^0_{-\gamma+\epsilon}$ *such that*

$$(I+q'(x,D))(I+q'_+(x,D))=I+q_-(x,D)+T'_\infty(x,D), \tag{5.23}$$

where $T'_\infty(x,\xi)\in S^0_{-\infty}$, *and* $\epsilon>0$ *can be chosen arbitrarily small.*

PROOF. We put

$$q'_1(x,\xi)=q'(x,\xi)(1-\chi(\delta\xi)),\quad q'_2(x,\xi)=q'(x,\xi)\chi(\delta\xi).$$

Obviously $q'_2(x,\xi)\in S^0_{-\infty}$. For any ϵ and ϵ_1 we can find a δ such that

$$\begin{aligned}|\tilde q'_1(\eta,\xi)|\cdot(1+|\eta|)+\left|\frac{\partial\tilde q'_1(\eta,\xi)}{\partial\xi_n}\right|(1+|\xi|)\\ \ll\epsilon_1(1+|\eta|)^{-(n+1)}(1+|\xi|)^{-\gamma+\frac{\varepsilon}{2}}.\end{aligned} \tag{5.24}$$

We choose "plus" and "minus" symbols so that

$$(I+q'_1(x,D))(I+q'_+(x,D))=I+q'_-(x,D). \tag{5.25}$$

As in the proof of Lemma 5.2, it follows from (5.25) that

$$\begin{aligned}q'_+(x,\xi)+\frac{1}{(2\pi)^n}\int_{-\infty}^{\infty}q'_1(x,\zeta+\xi)e^{-i(x,\zeta)}\tilde q'_+(\zeta,\xi)\,d\zeta\\ =-q'_1(x,\xi)+q'_-(x,\xi).\end{aligned} \tag{5.26}$$

We take the Fourier transformation with respect to x in (5.26) and apply the operator $\Pi^+_{\xi_n}$. Because

$$\Pi^+_{\xi_n}\tilde q'_+(\eta,\xi)=\tilde q'_+(\eta,\xi)\ \text{ and }\ \Pi^+_{\xi_n}\tilde q'_-(\eta,\xi)=0,$$

we have

$$\tilde q'_+(\eta,\xi)+\tilde Q\,\tilde q'_+(\eta,\xi)=-\Pi^+_{\xi_n}\tilde q'_1(\eta,\xi), \tag{5.27}$$

where

$$\tilde Q\,\tilde q'_+(\eta,\xi)=\frac{1}{(2\pi)^n}\int_{-\infty}^{\infty}\Pi^+_{\xi_n}\tilde q'_1(\eta-\zeta,\zeta+\xi)\tilde q'_+(\zeta,\xi)\,d\zeta. \tag{5.28}$$

Let H denote the space of functions $\tilde a(\eta,\xi)$ in $\mathbf{R}^n\times\mathbf{R}^n$ with the norm

$$\begin{aligned}\|\tilde a(\eta,\xi)\|^2=\sup_{\eta,\xi}\int_{-\infty}^{\infty}\Big[|\tilde a(\eta,\xi)^2|(1+|\xi|)^{-1+2\gamma-2\varepsilon}(1+|\eta|)^2\\ +\left|\frac{\partial\tilde a(\eta,\xi)}{\partial\xi_n}\right|^2(1+|\xi|)^{1+2\gamma-2\varepsilon}\Big]d\xi_n.\end{aligned} \tag{5.29}$$

It is not hard to show that the operator $\Pi^+_{\xi_n}$ is bounded in the space H and that the operator $\tilde Q$ is also bounded and has a norm less than unity if ϵ_1 is sufficiently small. Hence the equation (5.27) has a unique solution in H.

Hence as in the proof of Lemma 5.2 we find that $q'_+(x,\xi) \in S^0_{-\gamma+\epsilon}$ is a "plus" symbol, and $q'_-(x,\xi) \in S^0_{-\gamma+\epsilon}$ is a "minus" symbol.

REMARK 5.1. We have defined the classes of "plus" and "minus" symbols for all $x \in \mathbf{R}^n$. We can introduce wider classes of symbols that are "plus" for $x \in \mathbf{R}^n_-$ or "minus" for $\mathbf{R}^n_+$ for which the assertions of this section remain valid. We shall not go into this question here.

Let $A(x,\xi) \in \hat{O}^\infty_{\alpha+i\beta}$ be a scalar elliptic symbol, and let $A_0(x,\xi)$ be the principal homogeneous symbol of the p.d.o. $A(x,D)$. For each $x \in \mathbf{R}^n$ we factorize $A_0(x,\xi)$ with respect to ξ_n: $A_0(x,\xi) = A_0^-(x,\xi)A_0^+(x,\xi)$. We suppose that the oscillation of the real part of the index of the factorization of $A_0(x,\xi)$ is less than unity; that is, there are numbers κ_+ and κ_- such that

$$\kappa_- < \operatorname{Re}\kappa(x) < \kappa_+, \quad \kappa_+ - \kappa_- < 1. \tag{5.30}$$

The following assertion is an easy consequence of §5.1 and Lemma 5.6: *there are a "plus" p.d.o. $A_+(x,D)$ and a "minus" p.d.o. $A_-(x,D)$ such that*

$$A(x,D) = A_-(x,D)A_+(x,D) + T'_\infty(x,D), \tag{5.31}$$

where $T'_\infty(x,\xi) \in S^0_{-\infty}$.

§6. The parametrix of boundary value problems

1. By the right parametrix of the boundary value problem (3.17), (3.18) we mean the operator $\mathscr{R}_{\mathrm{R}}$ that satisfies the relation

$$\mathfrak{A}\,\mathscr{R}_{\mathrm{R}} = I_2 + T^\infty_{\mathrm{R}}, \tag{6.1}$$

where T^∞_{R} is a bounded operator from $\mathscr{H}^{(2)}_s(G)$ into $\mathscr{H}^{(2)}_\infty(G)$. Similarly an operator $\mathscr{R}_{\mathrm{L}}$ is called a left parametrix if

$$\mathscr{R}_{\mathrm{L}}\mathfrak{A} = I_1 + T^\infty_{\mathrm{L}}, \tag{6.2}$$

$$T^\infty_{\mathrm{L}} \text{ is a bounded operator from } \mathscr{H}^{(1)}_{-\infty}(G) \text{ into } \mathscr{H}^{(1)}_s(G). \tag{6.3}$$

Here we shall construct parametrices for a boundary value problem of the form (3.17), (3.18) in a half-space. Parametrices for bounded regions and also applications to the case of p.d.o.'s of the class $D^{(0)}_{\alpha+i\beta}$ will be considered elsewhere.

We note that in the general case the parametrix can be constructed at once in the neighborhood $\Gamma \times (0,\delta)$ of the entire boundary of the domain G.

2. We consider a boundary value problem of the form (3.17), (3.18) in the half-space $\mathbf{R}^n_+$

$$pA(x,D)u_+ + \sum_{k=1}^{m_-} pC_k(x,D)(v_k(x') \times \delta(x_n)) = f, \tag{6.4}$$

$$p'B_j(x,D)u_+ + \sum_{k=1}^{m_-} E_{jk}(x',D')v_k(x') = g_j(x'), \quad 1 \leqslant j \leqslant m_+. \tag{6.5}$$

We construct the left parametrix of the boundary value problem (6.4), (6.5).

Lemma 6.1. *Let $A(x,D)$ be an operator of the class $\hat{O}^{\infty}_{\alpha+i\beta}$, let $A_0(x,\xi)$ be the principal homogeneous symbol of the operator $A(x,D)$, and $\det A_0(x,\xi)\neq 0$ for $\xi\neq 0$. Let s satisfy the inequality* (1.38). *Then there are a "plus" p.d.o. $Q_+(x,D)$ and a "minus" p.d.o. $Q_-(x,D)$ such that*

$$A(x,D)\hat{\Lambda}_-^{-N}\hat{\Lambda}_+^{N-s}P_{2N}\hat{\Lambda}_+^{-2N}Q_+=Q_-(x,D)\hat{\Lambda}_-^{\alpha+i\beta-s}(D)+T'_\infty(x,D), \tag{6.6}$$

where $T'_\infty(x,\xi)\in S^0_{-\infty}$. In addition $Q_+(x,D)$ has a "plus" almost inverse $Q_+^{(-1)}$, and $Q_-(x,D)$ has a "minus" almost inverse $Q_-^{(-1)}$.

Proof. By Theorem 4.2 and Lemma 5.1 we have

$$A(x,D)\hat{\Lambda}_-^{-N}\hat{\Lambda}_+^{N-s}P_{2N}\hat{\Lambda}_+^{-2N}Q_+^{(0)}=Q_-^{(0)}(x,D)\hat{\Lambda}_-^{\alpha+i\beta-s}+A_1(x,D)\hat{\Lambda}_-^{\alpha+i\beta-s}, \tag{6.7}$$

where

$$A_1(x,\xi)\in S^0_{-\frac{1}{2}-\delta_0},\quad Q_+^{(0)}=(I+\overset{+}{q}_N(x,\xi))^{-1}b_+^{-1}(x,\xi).$$

$$Q_-^{(0)}(x,\xi)=(I+q_N^-)\,a_+\,\hat{b}_-^{-1}.$$

When we use Lemmas 5.1 and 5.5 and a property of the symbol $Q_-^{(0)}(x,\xi)$ (see Theorem 4.2) we can prove that there is "minus" p.d.o. $Q_-^{(1)}(x,D)$ almost inverse to $Q_-^{(0)}(x,D)$. When we apply it to the left in (6.7) we obtain

$$\begin{aligned} Q_-^{(1)}A\hat{\Lambda}_-^{-N}\hat{\Lambda}_+^{N-s}\;P_{2N}\hat{\Lambda}_+^{-2N}Q_+^{(0)}&=(I+a_1(x,D))\hat{\Lambda}_-^{\alpha+i\beta-s}\\ &=\hat{\Lambda}_-^{\alpha+i\beta-s}(I+a_2(x,D)),\end{aligned} \tag{6.8}$$

where

$$a_1(x,\xi)\in S^0_{-2\delta_0},\quad a_2(x,D)=\hat{\Lambda}_-^{-\alpha-i\beta+s}a_1\hat{\Lambda}_-^{\alpha+i\beta-s},\quad a_2(x,\xi)\in S^0_{-2\delta_0}.$$

By Lemma 5.6 there is a "plus" symbol $a_2^+(x,\xi)\in S^0_{-\delta_0}$ and a "minus" symbol $a_2^-(x,\xi)\in S^0_{-\delta_0}$ such that

$$(I+a_2(x,D))(I+a_2^+(x,D))=I+a_2^-(x,D)+T'_\infty(x,D).$$

When we apply $I+a_2^+(x,D)$ to the right and $Q_-^{(0)}(x,D)$ to the left in (6.8), we obtain (6.6).

We note without proof the following two simple lemmas.

Lemma 6.2. *Let $B(x,\xi)\in S^t_\beta$, $\beta<-1$. Then the operator $p'B(x,D)\times(v(x')\times\delta(x_n))$ is the p.d.o. in $\mathbf{R}^{n-1}$ with the symbol $b(x',\xi')\in S^t_{\beta+1}$ defined by the formula*

$$b(x',\xi')=\frac{1}{2\pi}\int_{-\infty}^{\infty}B(x',0,\xi',\xi_n)\,d\xi_n. \tag{6.9}$$

Lemma 6.3. *Let*

$$B_1(x,\xi)\in S^t_{\beta_1},\quad B_2(x,\xi)\in S^t_{\beta_2},\quad \beta_1<0,\ \beta_2<0,\ \beta_1+\beta_2<-1.$$

Then

$$p'B_1(x,D)\Theta^+B_2(x,D)(v(x')\times\delta(x_n))=b_3(x',D')v(x'),$$

where

$$b_3(x',\xi')\in S^t_{\beta_1+\beta_2+1}(\mathbf{R}^{n-1}),$$

$$b_3(x',\xi')=\frac{1}{2\pi}\int_{-\infty}^{\infty}B_1(x',0,\xi)\Pi^+_{\xi_n}B_2(\infty,\xi)\,d\xi_n \tag{6.10}$$
$$+\frac{1}{(2\pi)^{n+1}}\int_{-\infty}^{\infty}B_1(x',0,\xi+\zeta)e^{-i(x',\zeta')}\Pi^+_{\xi_n}B_2'(\zeta,\xi)\,d\zeta\,d\xi_n.$$

Let R_L denote the operator defined by

$$R_Lg=\hat\Lambda_+^{-s}(D)P_{2N}(x,D)\hat\Lambda_+^{-2N}(D)Q_+(x,D)\Theta^+\hat\Lambda_-^{-\alpha-i\beta+s}Q_-^{(-1)}lg. \tag{6.11}$$

We apply R_L to the left in the system (6.4). We obtain

$$\hat\Lambda_+^{-s}P_{2N}\hat\Lambda_+^{-2N}Q_+\Theta^+(Q_+^{(-1)}\hat\Lambda_+^{2N}P_{2N}^{(-1)}\hat\Lambda_-^{N}\hat\Lambda_+^{s-N}+T'_\infty)u_+ \tag{6.12}$$
$$+\sum_{k=1}^{m}R_LpC_k(v_k\times\delta)=R_Lf.$$

We have $\theta^+w=w-\theta^-w$, where $w=Q_+^{(-1)}\hat\Lambda_+^{2N}P_{2N}^{(-1)}\hat\Lambda_-^{N}\hat\Lambda_+^{s-N}u_+$. We show that

$$\Theta^-w=\sum_{k=1}^{2N}\Theta^-Q_{k1}(x,D)(p'Q_{k2}(x,D)\hat\Lambda_+^{s}u_+\times\delta)+T^\infty u_+, \tag{6.13}$$

where $T^\infty u_+$ is extended by continuity to a bounded operator from $H^+_{-\infty}(\mathbf{R}^n)$ into $H^-_{-1/2+\delta_0}(\mathbf{R}^n)$.

For,

$$\Theta^-w=\Theta^-Q_+^{(-1)}\Theta^-\hat\Lambda_+^{2N}P_{2N}^{(-1)}\hat\Lambda_-^{N}\hat\Lambda_+^{s-N}u_+, \tag{6.14}$$

because $Q_+^{(-1)}$ is a "plus" operator.

The operator $\hat\Lambda_+^{2N}$ is a differential operator with respect to x_n. Hence by Leibniz' formula (compare this with the formula in [20] for the expansion of a Cauchy type integral) we obtain

$$\hat\Lambda_+^{2N}\Theta^-P_{2N}^{(-1)}\hat\Lambda_-^{N}\hat\Lambda_+^{s-N}u_+=\Theta^-\hat\Lambda_+^{2N}P_{2N}^{(-1)}\hat\Lambda_-^{N}\hat\Lambda_+^{s-N}u_+ \tag{6.15}$$
$$+\sum_{k=1}^{2N}\hat q_{k1}(D)(p'D_n^{k-1}P_{2N}^{(-1)}\hat\Lambda_-^{N}\hat\Lambda_+^{s-N}u_+\times\delta(x_n)),$$

where $\hat q_{k1}(D)$ are differential operators with respect to x_n. Similarly we have

$$P_{2N}(x,D)\Theta^-P_{2N}^{(-1)}\hat\Lambda_-^{N}\hat\Lambda_+^{s-N}u_+=\Theta^-(I+T_\infty^{(1)})\hat\Lambda_-^{N}\hat\Lambda_+^{s-N}u_++$$

(6.16)
$$+\sum_{k=1}^{2N} q_{k2}(x, D)\,(p'\,D_n^{k-1}\,P_{2N}^{(-1)}\,\hat{\Lambda}_-^N\,\hat{\Lambda}_+^{s-N}\,u_+\times\delta(x_n)).$$

Because

(6.17)
$$\theta^-\hat{\Lambda}_-^N\hat{\Lambda}_+^{s-N}u_+=0,$$

by (6.15) and (6.16) we have

(6.18)
$$\begin{aligned}\Theta^-\hat{\Lambda}_+^{2N}P_{2N}^{(-1)}\hat{\Lambda}_-^N\hat{\Lambda}_+^{s-N}u_+ &= \sum_{k=1}^{2N}\hat{\Lambda}_+^{2N}P_{2N}^{(-1)}q_{k2}(p'\,D_n^{k-1}\,P_{2N}^{(-1)}\hat{\Lambda}_-^N\hat{\Lambda}_+^{s-N}u_+\times\delta)\\ &-\sum_{k=1}^{2N}\hat{q}_{k1}(D)\,(p'\,D_n^{k-1}\,P_{2N}^{(-1)}\hat{\Lambda}_-^N\hat{\Lambda}_+^{s-N}u_+\times\delta)+T_\infty^{(2)}u_+.\end{aligned}$$

We note that the left-hand member of (6.18) belongs to $H_0^-(\mathbf{R}^n)$. Hence if we extract the differential operators with respect to x_n from the first sum on the right, then it contracts with the second sum in (6.18). As a result we obtain (6.13) from (6.18) and (6.14).

Thus by (6.12) and (6.13) we have

(6.19)
$$\begin{aligned}\hat{\Lambda}_-^N(D)\,\hat{\Lambda}_+^{-N}(D)\,u_+ &= \sum_{k=1}^{2N}\hat{\Lambda}_+^{-s}P_{2N}(x, D)\,\hat{\Lambda}_+^{-2N}Q_+\Theta^-Q_{k1}(w_k\times\delta)\\ &-R_{\mathrm{L}}\sum_{k=1}^{m_-}pC_k(x, D)\,(v_k\times\delta)+R_{\mathrm{L}}f+T_{\mathrm{L}}^\infty u_+,\end{aligned}$$

where we have written

(6.20)
$$w_k(x')=p'Q_{k2}(x, D)\,\hat{\Lambda}_+^s u_+,\quad 1\leqslant k\leqslant 2N.$$

In (6.19) and below we write T_{L}^∞ with various indices for bounded operators from $H_{-\infty}^+(\mathbf{R}^n)$ into $\mathscr{A}_s^{(1)}(\mathbf{R}^n)$.

Let $\Phi_+(x)$ denote the right-hand member of (6.19). For the equation $\hat{\Lambda}_-^N\hat{\Lambda}_+^{-N}u_+=\Phi_+$, $\Phi_+\in H_s^+$, to have a solution $u_+\in H_s^+$ it is necessary and sufficient that $p'\hat{\Lambda}^{-k}\hat{\Lambda}_+^s\Phi_+=0$. Then the solution u_+ has the form

$$u_+=\hat{\Lambda}_+^N\hat{\Lambda}_-^{-N}\Phi_+=\hat{\Lambda}_+^{-s}\Theta^+\hat{\Lambda}_+^{N+s}\hat{\Lambda}_-^{-N}\Phi_+.$$

Thus the system (6.19) is equivalent to the system

(6.21)
$$\begin{aligned}u_+ &= \hat{\Lambda}_+^{N-s}\hat{\Lambda}_-^{-N}P_{2N}\hat{\Lambda}_+^{-2N}Q_+\Theta^-Q_{k1}(w_k\times\delta)\\ &-\sum_{k=1}^{m_-}\hat{\Lambda}_+^N\hat{\Lambda}_-^{-N}R_{\mathrm{L}}pC_k(v_k\times\delta)+\hat{\Lambda}_+^N\hat{\Lambda}_-^{-N}R_{\mathrm{L}}f+T_{\mathrm{L},1}^\infty u_+,\end{aligned}$$

and the following conditions hold:

(6.22)
$$\begin{aligned}&\sum_{k=1}^{2N} b_{jk}^{(1)}(x', D')\,w_k(x')-\sum_{k=1}^{m_-}b_{jk}^{(2)}(x', D')\,v_k(x')\\ &=p'\,\hat{\Lambda}_-^{-j}\hat{\Lambda}_+^s R_{\mathrm{L}}f+T_{\mathrm{L}2}^\infty u_+,\ 1\leqslant j\leqslant N,\end{aligned}$$

where $b_{jk}^{(1)}(x',D')$ and $b_{jk}^{(2)}(x',D')$ are pseudodifferential operators in $\mathbf{R}^{n-1}$ that by Lemmas 5.1, 6.2 and 6.3 are defined by the equalities

$$(6.23)\qquad b_{jk}^{(1)}(x',D')w_k = p'\hat{\Lambda}_-^{-j}P_{2N}\hat{\Lambda}_+^{-2N}Q_+\Theta^-Q_{k1}(w_k\times\delta),$$

$$(6.24)\qquad b_{jk}^{(2)}(x',D')v_k(x') = p'\hat{\Lambda}_-^{-j}P_{2N}\hat{\Lambda}_+^{-2N}Q_+\Theta^+\hat{\Lambda}_-^{-\alpha-i\beta+s}Q_-^{(-1)}C_k(v_k\times\delta).$$

We substitute (6.21) in the boundary conditions (6.5). We obtain

$$(6.25)\qquad \begin{gathered}\sum_{k=1}^{2N} b_{jk}^{(3)}(x',D')w_k(x') + \sum_{k=1}^{m_-} b_{jk}^{(4)}(x',D')v_k(x')\\ = g_j(x') - p'B_j\hat{\Lambda}_+^N\hat{\Lambda}_-^{-N}R_{\mathrm{L}}f + T_{\mathrm{L}3}^\infty u_+,\quad 1\leqslant j\leqslant m_+,\end{gathered}$$

where the following notation is used:

$$(6.26)\qquad b_{jk}^{(3)}(x',D')w_k(x') = p'B_j\hat{\Lambda}_+^{N-s}\hat{\Lambda}_-^{-N}P_{2N}\hat{\Lambda}_+^{-2N}\Theta^-Q_{k1}(w_k\times\delta),$$

$$(6.27)\qquad \begin{gathered}b_{jk}^{(4)}(x',D')v_k(x') = E_{jk}(x',D')v_k(x')\\ - p'B_j\hat{\Lambda}_+^{N-s}\hat{\Lambda}_-^{-N}P_{2N}\hat{\Lambda}_+^{-2N}Q_+\Theta^+\hat{\Lambda}_-^{-\alpha-i\beta+s}Q_-^{(-1)}C_k(v_k\times\delta).\end{gathered}$$

We now apply the operator $p'Q_{j2}(x,D)\hat{\Lambda}_+^s$ to (6.21). We obtain

$$(6.28)\qquad \begin{gathered}w_j(x') = \sum_{k=1}^{2N} b_{jk}^{(5)}(x',D')w_k(x') - \sum_{k=1}^{m_-} b_{jk}^{(6)}(x',D')v_k(x')\\ + p'Q_{j2}\hat{\Lambda}_+^{N+s}\hat{\Lambda}_-^{-N}R_{\mathrm{L}}f + T_{\mathrm{L},4}^\infty u_+,\quad 1\leqslant j\leqslant 2N,\end{gathered}$$

where we have written

$$(6.29)\qquad b_{jk}^{(5)}(x',D')w_k(x') = p'Q_{j2}\hat{\Lambda}_+^N\hat{\Lambda}_-^{-N}P_{2N}\hat{\Lambda}_+^{-2N}Q_+\Theta^-Q_{k1}(w_k\times\delta),$$

$$(6.30)\qquad \begin{gathered}b_{jk}^{(6)}(x',D')v_k(x')\\ = p'Q_{j2}\hat{\Lambda}_+^N\hat{\Lambda}_-^{-N}P_{2N}\hat{\Lambda}_+^{-2N}Q_+\Theta^+\hat{\Lambda}_-^{-\alpha-i\beta+s}Q_-^{(-1)}C_k(v_k\times\delta).\end{gathered}$$

Thus the solution of the boundary value problem (6.4), (6.5) is reduced to the solution of the system of pseudodifferential equations (6.22), (6.25), (6.28) with respect to $w_k(x')$, $1\leqq k\leqq 2N$, and $v_k(x')$, $1\leqq k\leqq m_-$.

Let the assumptions of Theorem 3.2 hold. In (6.4), (6.5) we fix the principal homogeneous symbols of the operators $A(x,D)$, $B_j(x,D)$, $C_k(x,D)$ and $E_{jk}(x',D')$, and also $x_0=(x_0',0)$ and ξ_0', and consider a boundary value problem of the form (3.5), (3.6) on the half-line. Because the operator defined by

$$\begin{gathered}R_{\mathrm{L}}^0 g(x_n) = \Lambda_+^{-2N-s}(\xi_0',D_n)P_{2N}(x_0,\xi_0',D_n)Q_+^{(0)}(x_0,\xi_0',D_n)\\ \times\Theta^+\Lambda_-^{-\alpha-i\beta+s}(\xi_0',D_n)(Q_-^{(0)}(x_0,\xi_0',D_n))^{-1}lg(x_n)\end{gathered}$$

has no kernel in $H_{s-\alpha}(\mathbf{R}_+^1)$, when we apply the operator R_{L}^0 to the left in

(3.5) and repeat the transformations that led to the formulas (6.21), (6.22), (6.25) and (6.28) we obtain an algebraic system of equations, which together with an equation of the form (6.21) is equivalent to the original system (3.5), (3.6). We denote the matrix of this system by $L(x_0', \xi_0')$. Because by hypothesis the system (3.5), (3.6) has a unique solution, $L(x_0', \xi_0')$ has maximal rank; hence it has a left inverse. Because the principal part of the symbol of the system (6.22), (6.25), (6.28) coincides with the matrix $L(x_0', \xi_0')$, the p.d.o. defined by the system (6.22), (6.25), (6.28) has a left almost inverse (see Lemmas 5.1 and 5.2). When we apply this to the left in (6.22), (6.25), (6.28) we obtain together with (6.21) the expression for the left parametrix.

The right parametrix is constructed similarly; it can also be obtained from the left parametrix for the operator $\mathfrak{A}^*$ by going over to the adjoint operator.

Received 17/OCT/71

Bibliography

[1] L. Boutet de Monvel, *Opérateurs pseudo-differéntiels analytiques et problèmes aux limites elliptiques,* Ann. Inst. Fourier (Grenoble) **19** (1970), fasc. 2, 169-268.

[2] ———, *Boundary problems for pseudo-differential operators,* Acta Math. **126** (1971), 11-51.

[3] N. P. Vekua, *Systems of singular integral equations and certain boundary value problems,* 2nd rev. ed., "Nauka", Moscow, 1970. (Russian) MR **42** # 4979.

[4] M. I. Višik and G. I. Èskin, *Elliptic equations in convolution in a bounded domain and their applications,* Uspehi Mat. Nauk **22** (1967), no. 1 (133), 15-76 = Russian Math. Surveys **22** (1967), no. 1, 13-75. MR **35** # 5757.

[5] ———, *Normally solvable problems for elliptic systems of equations in convolutions,* Mat. Sb. **74** (**116**) (1967), 326-356 = Math. USSR Sb. **3** (1967), 303-332. MR **36** # 4125.

[6] F. R. Gantmaher, *The theory of matrices,* GITTL, Moscow, 1953; English transl., Vols. 1, 2, Chelsea, New York, 1959. MR **16**, 438; **21** # 6372c.

[7] I. C. Gohberg and M. G. Kreĭn, *Systems of integral equations on a half line with kernels depending on the difference of arguments,* Uspehi Mat. Nauk **13** (1958), no. 2 (80), 3-72; English transl., Amer. Math. Soc. Transl. (2) **14** (1960), 217-287. MR **21** # 1506; **22** # 3954.

[8] I. C. Gohberg and I. A. Fel′dman, *Convolution equations and projection methods for their solution,* "Nauka", Moscow, 1971; English transl., Transl. Math. Monographs, vol. 41, Amer. Math. Soc., Providence, R. I., 1974.

[9] I. C. Gohberg and N. Ja. Krupnik, *The algebra generated by the one-dimensional singular integral operators with piecewise continuous coefficients,* Funkcional. Anal. i Priložen. **4** (1970), no. 3, 26-36 = Functional Anal. Appl. **4** (1970), 193-201. MR **42** # 5057.

[10] A. S. Dynin, *On the theory of pseudodifferential operators on a manifold with a boundary,* Dokl. Akad. Nauk SSSR **186** (1969), 251-253 = Soviet Math. Dokl. **10** (1969), 575-578. MR **40** # 6307.

[11] H. O. Cordes and E. A. Hermann, *Pseudo-differential operators on a halfline,* J. Math. Mech. **18** (1969), 893-908.

[12] I. B. Simonenko, *A new general method of investigating linear operator equations of singular integral equation type.* I, II, Izv. Akad. Nauk SSSR Ser. Mat. **29** (1965), 567-586, 757-782. (Russian) MR **31** # 3876; **32** # 6174.

[13] E. C. Titchmarsh, *Introduction to the theory of Fourier integrals,* Clarendon Press, Oxford, 1937; Russian transl., GITTL, Moscow, 1948.

[14] K. O. Friedrichs, *Pseudo-differential operators. An introduction,* Courant Inst. Math. Sci., New York University, New York, 1970. MR **44** # 859.

[15] Čan Zuĭ Ho and G. I. Èskin, *Boundary value problems for parabolic systems of pseudo-differential equations,* Dokl. Akad. Nauk SSSR **198** (1971), 50-53 = Soviet Math. Dokl. **12** (1971), 739-743. MR **45** # 7533.

[16] E. Shamir, *Elliptic systems of singular integral operators*. I. *The half-space case,* Trans. Amer. Math. Soc. **127** (1967), 107-124. MR **35** #5981.

[17] M. A. Šubin, *Factorization of matrices depending on a parameter, and elliptic equations in a halfspace,* Mat. Sb. **85** (**127**) (1971), 65-84 = Math. USSR Sb. **14** (1971), 65-84.

[18] G. I. Èskin, *A general boundary value problem for equations of principal type in a plane domain with corner points,* Uspehi Mat. Nauk **18** (1963), no. 3 (117), 241-242. (Russian)

[19] ———, *The conjugacy problem for equations of principal type with two independent variables,* Trudy Moskov. Mat. Obšč. **21** (1970), 245-292 = Trans. Moscow Math. Soc. **21** (1970), 263-316. MR **42** #3429.

[20] ———, *Generalized functions and pseudodifferential operators,* Preprint no. 5, Institute of Problems of Mechanics, Moscow, 1971. (Russian)

[21] ———, *Boundary value problems in a half-space for elliptic pseudodifferential equations,* Preprint no. 7, Institute of Problems of Mechanics, Moscow, 1971. (Russian)

[22] ———, *Boundary problems for elliptic pseudodifferential operators,* "Nauka", Moscow, 1973. (Russian)

Translated by
J. L. B. COOPER

Trudy Moskov. Mat. Obšč.
Tom 28 (1973)

Trans. Moscow Math. Soc.
Vol. 28 (1973)

DEVELOPMENT OF A PLANE-PARALLEL SYMMETRIC BOUNDARY LAYER IN A SUDDEN GENERATION OF MOTION

V. N. SAMOHIN

In this article we investigate the formation of the boundary layer on a body which is instantly set in motion in a viscous incompressible fluid that is at rest. Under certain conditions we prove the existence and uniqueness of the solution of the appropriate boundary value problem for Prandtl's system of equations on some time interval $0 \leqq t \leqq T$ along the whole of a streamlined body. We give a method of constructing approximate solutions of the problem, and prove that they are convergent.

The problem of the generation of the boundary layer on a body which is set in motion in a fluid that is at rest has also been considered by Blasius [1]. The method he suggested for solving the problem consisted of finding the stream function in the form of an asymptotic series in powers of the time variable; he gave the explicit form of the first two terms of the expansion. A short account of these results and a mathematical statement of the problem are given in the books [2] and [3]. Oleĭnik [4], [5], [6] has investigated the problem of the development of the boundary layer under a gradual acceleration.

For the plane-parallel symmetric flow around a body, the problem of the formation of the boundary layer when motion is suddenly started leads one to consider the following system of equations:

$$u_t + uu_x + vu_y = -p_x + \nu u_{yy}, \qquad u_x + v_y = 0 \tag{1}$$

in the domain $D_T\{0 \leqq t \leqq T, 0 \leqq x \leqq X, 0 \leqq y < \infty\}$, with the conditions

$$\begin{gathered} u|_{t=0} = U(0, x), \quad u|_{y=0} = 0 \quad \text{for } t > 0, \quad u|_{x=0} = 0, \\ v|_{y=0} = v_0(t, x), \quad u \to U(t, x) \quad \text{as } y \to \infty. \end{gathered} \tag{2}$$

Here u and v are the velocity components, $U(t,x)$ is the longitudinal component of the velocity of the external flow, $U(t,0) = 0$, $U(t,x) > 0$ for $x > 0$, and $-p_x = U_t + UU_x$.

In this article we prove the existence and uniqueness of the solution of the problem (1), (2) for an arbitrary X and for $0 \leqq t \leqq T$, where T depends on the functions U and v and the coefficient of viscosity ν. We assume that $v_0(t,x)$, $U(t,x)$, U_x, and U_t/U are bounded, and $U_x + U_t/U > 0$.

Analogously one can consider the problem of the development of the

---

AMS (MOS) subject classifications (1970). Primary 76D10, 35Q99; Secondary 35A35.

boundary layer when $U(t,x) > 0$ for $0 \leqq x \leqq X$, and with the initial velocity profile $u|_{x=0} = u_0(t,y)$, where $u_0 > 0$ for $y > 0$, instead of the condition $u|_{x=0} = 0$.

In the system (1) we make the following change of independent variables

(3) $$\tau = \sqrt{t}, \quad \xi = x, \quad \eta = \frac{u(t, x, y)}{U(t, x)}$$

and introduce the new unknown function

(4) $$w(\tau, \xi, \eta) = \frac{\sqrt{t}\, u_y(t, x, y)}{U(t, x)}.$$

Then the problem (1), (2) is reduced to one for the single equation

(5) $$\nu w^2 w_{\eta\eta} - \frac{\tau}{2} w_\tau - \tau^2 \eta U w_\xi + A w_\eta + B w = 0$$

in the domain $\Omega\{0 \leqq \tau \leqq \sqrt{T}, 0 \leqq \xi \leqq X, 0 \leqq \eta \leqq 1\}$, with the boundary conditions

(6) $$w|_{\eta=1} = 0, \quad (\nu w w_\eta - \tau v_0 w + C)|_{\eta=0} = 0,$$

where

$$A = \tau^2(\eta^2 - 1) U_x + \tau^2(\eta - 1)\frac{U_t}{U}, \quad B = -\eta\tau^2 U_x - \tau^2 \frac{U_t}{U} + \frac{1}{2},$$

$$C = \tau^2 U_x + \tau^2 \frac{U_t}{U}.$$

We use the methods developed in the article [4] to construct a solution of the problem (5), (6) for an arbitrary X when T is sufficiently small. Hence from the transformations (3) and (4) we obtain the corresponding results for the solution of the problem (1), (2). We obtain estimates that characterize the behavior of $u(t,x,y)$ as $y \to \infty$, and as $t \to 0$.

For an arbitrary function $f(\tau, \xi, \eta)$ we denote $f(kh, lh, \eta)$, where h is a positive constant, by $f^{k,l}(\eta)$. We replace the equation (5) with the conditions (6) by the system of ordinary differential equations

(7) $$L^{k,l}(w) \equiv \nu (w^{k,l})^2 w^{k,l}_{\eta\eta} - \frac{kh}{2}\frac{w^{k,l} - w^{k-1,l}}{h} - (\eta + \lambda h)(kh)^2 U^{k,l} \frac{w^{k,l} - w^{k,l-1}}{h} + A^{k,l} w^{k,l}_\eta + B^{k,l} w^{k,l} = 0,$$

on the interval $0 \leqq \eta \leqq 1$, where $k = 0, 1, \cdots, [\sqrt{T}/h]$ and $l = 0, 1, \cdots, [X/h]$, with the conditions

(8) $$w^{k,l}|_{\eta=1} = 0, \quad l^{k,l}(w^{k,l}) \equiv (\nu w^{k,l} w^{k,l}_\eta - kh\, v_0^{k,l} w^{k,l} + C^{k,l})|_{\eta=0} = 0.$$

In the equations (7) λ is a positive constant, which we determine below.

From the system (7) and the conditions (8) with $k = 0$ we obtain the equations

(9) $$\nu w^{0,l} w^{0,l}_{\eta\eta} + \frac{1}{2} = 0, \quad l = 0, 1, \ldots, \left[\frac{X}{h}\right]$$

with the conditions

$$w^{0,l}(1) = 0, \qquad w^{0,l}_\eta(0) = 0. \tag{10}$$

It is not difficult to show that the solutions of these equations are equal to $(\pi\nu)^{-1/2}$ when $\eta = 0$, and that they are given by the formula

$$\operatorname{erf}(x)\left(\sqrt{-\log\sqrt{\pi\nu}\, w^{0,l}(\eta)}\right) = \eta, \tag{11}$$

where

$$\operatorname{erf}(x) = \frac{2}{\sqrt{\pi}}\int_0^x e^{-S^2}\, ds.$$

Everywhere in what follows C_i and M_i denote positive constants that are independent of h.

Lemma 1. *The solutions of the problem* (9), (10) *have the following properties*:

$$C_0(1-\eta)\sigma_\mu \leqq w^{0,l} \leqq C_1(1-\eta)\sigma_{\mu_1} \tag{12}$$

where

$$\sigma_\gamma = \sqrt{-\log\gamma(1-\eta)}, \quad 0 < \gamma < 1, \quad 0 \leqslant \eta < 1, \quad 0 < \mu_1 < \frac{1}{\sqrt{e}} < \mu < 1,$$

$$-C_2\sigma_\mu + C_2\sqrt{-\log\mu} \leqslant w^{0,l}_\eta \leqslant -C_3\sigma_{\mu_1} + C_3\sqrt{-\log\mu_1}. \tag{13}$$

Proof. It follows from (11) that $w^{0,l} > 0$ for $\eta < 1$, and $w^{0,l}(1) = 0$. We consider the function $\kappa_0 = C_0(1-\eta)\sigma_\mu$. We have

$$\begin{aligned} \nu\kappa_0\kappa_{0\eta\eta} + \frac{1}{2} &= \nu C_0^2(1-\eta)\sigma_\mu\left(-\frac{1}{2\sigma_\mu(1-\eta)} - \frac{1}{4\sigma_\mu^3(1-\eta)}\right) \\ &+ \frac{1}{2} = -\frac{\nu C_0^2}{2} - \frac{\nu C_0^2}{4\sigma_\mu^2} + \frac{1}{2} > 0, \end{aligned} \tag{14}$$

if C_0 satisfies the condition $1/C_0^2 > \nu - \nu/2\log\mu$.

In addition we have

$$\kappa_{0\eta}\big|_{\eta=0} = C_0\left(-\sigma_\mu + \frac{1}{2\sigma_\mu}\right) > 0 \tag{15}$$

by the choice of μ. It follows from (9), (10), (14) and (15) that

$$w^{0,l} z^{0,l}_{\eta\eta} + \kappa_{0\eta\eta} z^{0,l} > 0, \quad z^{0,l}(1) = 0, \quad z^{0,l}_\eta(0) > 0, \tag{16}$$

where $z^{0,l} = \kappa_0 - w^{0,l}$, $l = 0, 1, \cdots, [X/h]$. It follows from (16) that $z^{0,l}$ cannot have a positive maximum for $0 \leqq \eta < 1$, and $z^{0,l}(1) = 0$. Therefore $z^{0,l} \leqq 0$ for $0 \leqq \eta < 1$; that is, $w^{0,l} \geqq \kappa_0$. Similarly one can prove the estimate from above for $w^{0,l}$.

We now prove the estimates (13). For this we note that

$$(w^{0,l}_\eta)_\eta = -\frac{1}{2\nu w^{0,l}} \text{ and } w^{0,l}_\eta\Big|_{\eta=0} = 0.$$

Therefore when we use (12) we find that

$$-\frac{1}{2\nu}\int_0^\eta \frac{d\eta}{C_0(1-\eta)\sigma_\mu} \leqslant w_\eta^{0,l} \leqslant -\frac{1}{2\nu}\int_0^\eta \frac{d\eta}{C_1(1-\eta)\sigma_{\mu_1}};$$

hence we obtain the inequalities (13) with the constants $C_2 = 1/\nu C_0$ and $C_3 = 1/\nu C_1$. It follows from (13) that in a neighborhood of $\eta = 1$ we have

$$-C_4\sigma_\mu \leqq w_\eta^{0,l} \leqq -C_5\sigma_\mu, \tag{17}$$

where $C_4 > C_2$ and $C_5 < C_3$. This proves the lemma.

Lemma 2. *Assume that U_x is continuous, $U_x(0,0) > 0$, and $U_t/U > -M_1U$ in some neighborhood of $x = 0$. Then for $0 \leqq lh \leqq X$ and $0 \leqq kh \leqq \sqrt{T}$, where T depends on U, v_0 and ν, the system (7) with the conditions (8) has the solution $w^{k,l}(\eta)$, which is continuous for $0 \leqq \eta \leqq 1$, and has a continuous third order derivative for $0 \leqq \eta < 1$. The following estimates hold for this solution:*

$$C_6\, kh\,(1-\eta)\,e^{-\alpha lh} \leqslant w^{k,l} \leqslant C_7\,(1-\eta)\,\sigma_{\mu_1}, \tag{18}$$

where $\alpha = \alpha(h) > 0$.

Proof. The existence of $w^{0,l}(\eta)$ has been proved above. For $k \geqq 1$ we obtain the solution $w^{k,l}(\eta)$ of the problem (7), (8) as the limit as $\epsilon \to 0$ of the solution of the system

$$L_\varepsilon^{k,l}(w) \equiv (\nu\,(w^{k,l})^2 + \varepsilon)\,w_{\eta\eta}^{k,l} - \frac{kh}{2}\,\frac{w^{k,l} - w_\varepsilon^{k-1,l}}{h} - (\eta + \lambda h)\,(kh)^2\,U^{k,l}$$

$$\times\,\frac{w^{k,l} - w^{k,l-1}}{h} + A^{k,l}w_\eta^{k,l} + B^{k,l}w^{k,l} = 0, \quad \varepsilon > 0, \tag{19}$$

$$k = 1,\ 2,\ \ldots,\ \left[\frac{\sqrt{T}}{h}\right], \quad l = 0,\ 1,\ \ldots,\ \left[\frac{X}{h}\right]$$

with the conditions

$$w^{k,l}(1) = 0, \quad l^{k,l}(w) \equiv (\nu w^{k,l}w_\eta^{k,l} - kh\,v_0^{k,l}\,w^{k,l} + C^{k,l})\,|_{\eta=0} = 0. \tag{20}$$

In the equations (19) $w_\epsilon^{k,l} = w^{k,l}$ for $k \geqq 1$, and for $k = 0$ we put $w_\epsilon^{k,l} = w^{0,l}$ for $\eta \leqq \frac{1}{2}$ and $w_\epsilon^{k,l} = w^{0,l}(\eta - \epsilon) - w^{0,l}(1-\epsilon)$ for $\frac{1}{2} + \epsilon \leqq \eta \leqq 1$; we extend $w_\epsilon^{k,l}$ smoothly to the segment $[\frac{1}{2}, \frac{1}{2} + \epsilon]$ so that $w_\epsilon^{k,l} \geqq 0$.

We first assume that $w^{k,l}(0) > 0$; we obtain an a priori estimate from below for the solution of the problem (19), (20). For this we consider the function $\kappa_6 = C_6kh(1-\eta)e^{-\alpha lh}$. We have

$$L_\varepsilon^{k,l}(\varkappa_6) = C_6\,(kh)^3\,(1-\eta)\,e^{-\alpha lh}\,[U_x^{k,l} + (\eta + \lambda h)\,\alpha e^{\alpha h'}\,U^{k,l}], \tag{21}$$

where $0 < h' < h$. The expression (21) is positive for small values of lh and kh because $U_x(0,0) > 0$. For $lh \geqq \delta > 0$ this expression is positive for $\alpha(h)$ sufficiently large.

For $\eta = 0$ we have

$$l^{k,l}(\varkappa_6) = -\nu C_6^2 (kh)^2 e^{-2\alpha lh} - C_6 (kh)^2 v_0^{k,l} e^{-\alpha lh} + (kh)^2 \left(U_x^{k,l} + \frac{U_t^{k,l}}{U^{k,l}} \right) > 0,$$

if C_6 is taken sufficiently small. We consider the difference $\kappa_6^{k,l} - w^{k,l} = z^{k,l}$. For $k = 0$ the function $z^{k,l}$ is equal to $\kappa_6^{0,l} - w_\epsilon^{0,l}$. In addition, $z^{k,l}(1) = 0$ for $k \geqq 0$. In the inequalities

$$L_\varepsilon^{k,l}(x_6) - L_\varepsilon^{k,l}(w) > 0 \quad \text{and} \quad \frac{l^{k,l}(\varkappa_6)}{\varkappa_6(0)} - \frac{l^{k,l}(w)}{w(0)} > 0$$

we go from $z^{k,l}$ to $\bar{z}^{k,l} = z^{k,l} e^{-\beta lh}$, where $\beta(h) = \text{const} > 0$. For $\bar{z}^{k,l}$ we have the inequalities

$$(\nu (w^{k,l})^2 + \varepsilon) \bar{z}_{\eta\eta}^{k,l} - (\eta + \lambda h)(kh)^2 U^{k,l} e^{-\beta h} \frac{\bar{z}^{k,l} - \bar{z}^{k,l-1}}{h} + A^{k,l} \bar{z}_\eta^{k,l}$$

(22) $$+ \left[B^{k,l} - \frac{k}{2} - (\eta + \lambda h) \beta (kh)^2 U^{k,l} e^{-\beta h'} \right] \bar{z}^{k,l} > -\frac{k}{2} \bar{z}^{k-1,l},$$

$$1 \leqslant k \leqslant \left[\frac{\sqrt{T}}{h} \right], \quad l \geqslant 0$$

and

(23) $$\left(\nu \bar{z}_\eta^{k,l} - \frac{C^{k,l} \bar{z}^{k,l}}{\varkappa_6^{k,l} w^{k,l}} \right) \bigg|_{\eta=0} > 0, \quad \bar{z}^{k,l}(1) = 0, \quad 0 < h' < h.$$

The coefficient of $\bar{z}^{k,l}$ in (22) is

(24) $$-\eta (kh)^2 U_x^{k,l} - (kh)^2 \frac{U_t^{k,l}}{U^{k,l}} + \frac{1}{2} - \frac{k}{2} - (\eta + \lambda h) \beta (kh)^2 U^{k,l} e^{-\beta h'}.$$

Let $\beta = 1/h$. If lh and kh are sufficiently small, then $U_x^{k,l} > 0$, and the expression (24) does not exceed $(kh)^2(-U_t^{k,l}/U^{k,l} - \lambda U^{k,l} e^{-1})$. If we take $\lambda \geqq M_1 e$, then it follows from the assumptions of Lemma 2 that $(-U_t^{k,l}/U^{k,l} - \lambda U^{k,l} e^{-1}) < 0$. If $lh \geqq \delta > 0$, then the expression (24) is negative for $\lambda > e \max_{t,s} |U_x| / \min_{t,x \geqq \delta} U$.

Because $w^{k,l}(0) > 0$ by hypothesis, it is obvious that the coefficient of $\bar{z}^{k,l}$ in the first of the conditions (23) is negative. It follows from the definition of $w_\epsilon^{k,l}$ that $\bar{z}^{0,l} \leqq 0$.

We use the maximum principle and induction on k to show that $\bar{z}^{k,l} \leqq 0$ for $1 \leqq k \leqq [\sqrt{T}/h]$ and $l \geqq 0$. In fact, we assume that $\bar{z}^{k,l}$ takes positive values; then at some point $\bar{z}^{k,l}$ has a greatest positive value. The point at which $\bar{z}^{k,l}$ has a positive maximum cannot lie in the interior of $(0, 1)$ because at this point we have $\bar{z}^{k,l} > 0$, $\bar{z}_\eta^{k,l} = 0$, and $\bar{z}_{\eta\eta}^{k,l} \leqq 0$, which is inconsistent with (22). If $\bar{z}^{k,l}$ has a positive maximum at $\eta = 0$, then $\bar{z}_\eta^{k,l}|_{\eta=0} \leqq 0$; this is impossible by (23). For $\eta = 1$ we have $\bar{z}^{k,l} = 0$. Hence $\bar{z}^{k,l} \leqq 0$, and so $z^{k,l} \leqq 0$ also. Hence it follows that $\kappa_6^{k,l} \leqq w^{k,l}$, and the estimate from below for $w^{k,l}$ is proved.

We now estimate $w^{k,l}$ from above uniformly with respect to ϵ and h. Let $\kappa_7 = C_7(1 - \eta)\sigma_{\mu_1}$, $0 < \mu_1 < e^{-1/2}$. We have

$$L_\varepsilon^{k,l}(\varkappa_7) = (\nu(C_7(1-\eta)\sigma_{\mu_1})^2 + \varepsilon)\left(-\frac{1}{2(1-\eta)\sigma_{\mu_1}} - \frac{1}{4\sigma_{\mu_1}^3(1-\eta)}\right)C_7$$
$$+ A^{k,l}\left(-\sigma_{\mu_1} + \frac{1}{2\sigma_{\mu_1}}\right)C_7 + B^{k,l}C_7(1-\eta)\sigma_{\mu_1} < 0$$

for a sufficiently large C_7 and for $kh \leqq \sqrt{T}$. In addition we have

$$l^{k,l}(\varkappa_7) = \nu C_7^2\left(\log\mu_1 + \frac{1}{2}\right) - kh\, v_0^{k,l} C_7 \sqrt{-\log\mu_1}$$
$$+ (kh)^2\left(U_x^{k,l} + \frac{U_t^{k,l}}{U^{k,l}}\right) < 0,$$

if C_7 is sufficiently large and $kh \leqq \sqrt{T}$, because $\log\mu_1 + \frac{1}{2} < 0$ by the choice of μ_1. For the difference $\kappa_7^{k,l} - w^{k,l} = z^{k,l}$ when $k \geqq 1$ we have the following inequalities:

$$(\nu(w^{k,l})^2 + \varepsilon) z_{\eta\eta}^{k,l} - \frac{kh}{2}\,\frac{z^{k,l} - z^{k-1,l}}{h} - (\eta + \lambda h)(kh)^2 U^{k,l}\,\frac{z^{k,l} - z^{k,l-1}}{h}$$

(25) $$+ A^{k,l} z_\eta^{k,l} + B^{k,l} z^{k,l} + \nu\varkappa_{7\eta\eta}^{k,l}(\varkappa_7^{k,l} + w^{k,l}) z^{k,l} < 0$$

for $l \geqq 0$, where $z^{0,l} = \kappa_7^{0,l} - w_\epsilon^{0,l}$, and the following conditions:

(26) $$\left.\left(\nu z_\eta^{k,l} - \frac{C^{k,l} z^{k,l}}{\varkappa_7^{k,l} w^{k,l}}\right)\right|_{\eta=0} < 0, \quad z^{k,l}(1) = 0.$$

The coefficient of $z^{k,l}$ in (25) is

$$B^{k,l} - \frac{k}{2} + \nu\varkappa_{7\eta\eta}^{k,l}(\varkappa_7^{k,l} + w^{k,l}) - \left(\frac{\eta + \lambda h}{h}\right)(kh)^2 U^{k,l}.$$

Because $w^{k,l} \geqq 0$ but $\kappa_{7\eta\eta}^{k,l}\kappa_7^{k,l} \leqq -C_7^2/2$, this coefficient is negative for a sufficiently large C_7 and for $kh \leqq \sqrt{T}$. If we choose C_7 so that $\kappa_7 \geqq w^{0,l}$ (this can be done by (12)), then when we use the maximum principle we obtain the inequality

$$w^{k,l} \leqq C_7(1-\eta)\sigma_{\mu_1}, \qquad k \geqq 0, \quad l \geqq 0.$$

We now prove the existence of a solution $w^{k,l}$ of the problem (7), (8) that is positive for $\eta = 0$. With this objective we construct a smooth function $\Psi_{kh}(w)$, which we define so that

$$\Psi_{kh}(w) = \begin{cases} w & \text{for } w \geqslant C_6 kh\, e^{-\alpha X}, \\ \dfrac{C_6 kh\, e^{-\alpha X}}{2} & \text{for } w \leqslant \dfrac{C_6\, kh\, e^{-\alpha X}}{4} \end{cases}$$

and $0 \leqq (\Psi_{kh})' \leqq 1$ for $k \geqq 1$. We consider the system (19) with the boundary conditions

$$w^{k,l}(1) = 0, \quad l_\Psi^{k,l}(w) \equiv (\nu\Psi_{kh}(w^{k,l}) w_\eta^{k,l} - kh\, v_0^{k,l}\,\Psi_{kh}(w^{k,l}) + C^{k,l})|_{\eta=0} = 0.$$

(27)

We show that every solution of the problem (19), (27) is a solution of the problem (19), (20) that is positive for $\eta = 0$. We take $\kappa_6 = C_6 kh(1-\eta)e^{-\alpha lh}$, where C_6 and α are the constants defined in the a priori estimate (18), and which occur in the definition of $\Psi_{kh}(w)$. We have shown earlier that $L_\epsilon^{k,l}(\kappa_6) \geqq 0$. We consider the expression

$$l_\Psi^{k,l}(\varkappa_6) = -\nu \frac{C_6^2}{2}(kh)^2 e^{-\alpha(lh+X)} - \frac{C_6}{2}(kh)^2 v_0^{k,l} e^{-\alpha X} + (kh)^2 \left(U_x^{k,l} + \frac{U_t^{k,l}}{U^{k,l}}\right).$$

It is obvious that by the choice of α and C_6 we have $l_\Psi^{k,l}(\kappa_6) > 0$.

When we consider the inequalities

$$L_\varepsilon^{k,l}(\varkappa_6) - L_\varepsilon^{k,l}(w) \geqslant 0 \quad \text{and} \quad \left(\nu S_\eta^{k,l} - \frac{C^{k,l} S^{k,l} \Psi'_{kh}(\vartheta)}{\Psi(\varkappa_6)\Psi(w^{k,l})}\right)\Bigg|_{\eta=0} > 0,$$

where $S^{k,l} = \kappa_6^{k,l} - w^{k,l}$ and $\vartheta \in (\kappa_6(0), w^{k,l}(0))$, we find that $S^{k,l} \leqq 0$. Hence $w^{k,l} \geqq \kappa_6$, where $w^{k,l}$ is a solution of the problem (19), (27). Hence it follows that $w^{k,l}|_{\eta=0} \geqq \kappa_6|_{\eta=0} \geqq C_6 e^{-\alpha X} kh$. This means that $\Psi_{kh}(w^{k,l}) = w^{k,l}$ for every solution of the problem (19), (27). Hence every solution of the problem (19), (27) is a solution of the problem (19), (20) that is positive for $\eta = 0$.

The existence of a solution of the problem (19), (27) is proved by using the Leray-Schauder theorem in complete analogy to the way this was done in the article [4]. In exactly the same way as in [4] it is easily proved that, as $\epsilon \to 0$, a solution of the problem (19), (20) is convergent to a solution of the problem (7), (8) with the properties stated in Lemma 2. This proves the lemma.

We now derive an estimate from below, uniform with respect to h, for the solutions of the problem (7), (8).

LEMMA 3. *The following equality holds for* $w^{k,l}$:

(28) $$C_8(1-\eta)\sigma_\mu \leqq w^{k,l},$$

where $l \geqq 0$, $0 \leqq kh \leqq \sqrt{T}$, *$T$ depends on U, v_0 and ν, and* $e^{-1/2} < \mu < 1$.

PROOF. Let $\kappa_8 = C_8(1-\eta)\sigma_\mu$. Then we have

(29) $$\begin{aligned} L^{k,l}(\varkappa_8) &= \nu(C_8(1-\eta)\sigma_\mu)^2\left(-\frac{1}{2\sigma_\mu(1-\eta)} - \frac{1}{4\sigma_\mu^3(1-\eta)}\right)C_8 \\ &\quad + C_8\left(-\sigma_\mu + \frac{1}{2\sigma_\mu}\right)A^{k,l} + B^{k,l} C_8(1-\eta)\sigma_\mu \geqslant 0, \end{aligned}$$

if C_8 and T are chosen sufficiently small.

For $\eta = 0$ we have

(30) $$\begin{aligned} l^{k,l}(\varkappa_8) &= \nu C_8^2\left(\log\mu + \frac{1}{2}\right) - kh\, v_0^{k,l} C_8 \sqrt{-\log\mu} + (kh)^2 \\ &\quad \times \left(U_x^{k,l} + \frac{U_t^{k,l}}{U^{k,l}}\right) > 0, \end{aligned}$$

because $\log\mu + \frac{1}{2} > 0$ by our choice of μ, and T has been chosen sufficiently small.

By using the maximum principle it is not difficult to prove that (29) and (30) imply (28).

Thus we have constructed approximate solutions of the problem (7), (8), and for them we have proved estimates that are uniform with respect to h. To prove that the solutions of the problem (7), (8) are convergent to a solution of the problem (5), (6), we use the method given in [4] to obtain estimates, uniform with respect to h, for the quantities

$$(31) \qquad r^{k,l} = \frac{w^{k,l} - w^{k,l-1}}{h}, \qquad \rho^{k,l} = \frac{w^{k,l} - w^{k-1,l}}{h}, \qquad z^{k,l} = w^{k,l}_{\eta}.$$

Lemma 4. *Assume that the conditions of Lemma 2 are fulfilled and, in addition, that* $v_0 \leqq M_2\tau^{1+\epsilon}$, $\epsilon > 0$, $|v_{0x}| \leqq M_3\tau$, *and the functions* $v_{0\tau}$, U_{xx}, $U_{x\tau}$, $(U_t/U)_x$ *and* $(U_t/U)_\tau$ *are bounded. Then for* $0 \leqq lh \leqq X$ *and* $0 \leqq kh \leqq \sqrt{T}$, *where* T *depends on* U, v_0 *and* ν, *the following inequalities hold*:

$$(32) \qquad \left| \frac{w^{k,l} - w^{k,l-1}}{h} \right| \leqslant C_9 (1 - \eta)\, \sigma_{\mu_1},$$

$$(33) \qquad \left| \frac{w^{k,l} - w^{k-1,l}}{h} \right| \leqslant C_{10} (1 - \eta)\, \sigma_{\mu_1},$$

$$(34) \qquad -C_{11}\, \sigma_{\mu_1} \leqslant w^{k,l}_{\eta} \leqslant C_{12}\, \Phi(\eta);$$

here $\Phi(\eta) = -\eta^2$ *for* $0 \leqq \eta \leqq 1 - \delta$, *and* $\Phi(\eta) = -\sigma_\mu$ *for* $1 - \delta < \eta \leqq 1$, $\delta > 0$. *In addition* $w^{k,l}w^{k,l}_{\eta\eta} < -1/4\nu$ *and*

$$(35) \qquad |w^{k,l}w^{k,l}_{\eta\eta}| \leqq C_{13}.$$

Proof. We prove the inequalities (32) – (35) by induction. For $k = 0$ the function $\rho^{k,l}$ is not defined. It follows from Lemma 1 that (34) holds for $z^{0,l}$, and $r^{0,l} = 0$. For $l = 0$ the function $r^{k,l}$ is not defined. We assume that the inequalities (32) – (35) hold for $w^{k',l'}(\eta)$ for $k' \leqq k - 1$, and for $k' = k$ and $l' \leqq l - 1$. We show that (32) – (35) hold also for $k' = k$ and $l' = l$, with constants that are independent of h.

We form the equation satisfied by $r^{k,l}$. For this we subtract from equation (7) for $w^{k,l}$ the equation for $w^{k,l-1}$ and divide the difference by h. We obtain

$$R^{k,l}(r^{k,l}) \equiv \nu\, (w^{k,l})^2\, r^{k,l}_{\eta\eta} - \frac{kh}{2}\, \frac{r^{k,l} - r^{k-1,l}}{h}$$

$$- (\eta + \lambda h)(kh)^2 U^{k,l-1}\, \frac{r^{k,l} - r^{k,l-1}}{h} + A^{k,l} r^{k,l}_{\eta} + B^{k,l} r^{k,l}$$

$$(36) \qquad + r^{k,l}\left[\frac{w^{k,l} + w^{k,l-1}}{(w^{k,l-1})^2}\left(\frac{kh}{2}\,\rho^{k,l-1} + (kh)^2(\eta + \lambda h)\, U^{k,l-1} r^{k,l-1} - A^{k,l-1} z^{k,l-1} - B^{k,l-1} w^{k,l-1}\right)\right]$$

$$- (\eta + \lambda h)(kh)^2\, \frac{U^{k,l} - U^{k,l-1}}{h}\, r^{k,l} =$$

$$= -\frac{A^{k,l}-A^{k,l-1}}{h} z^{k,l-1} - \frac{B^{k,l}-B^{k,l-1}}{h} w^{k,l-1}, \quad l \geqslant 1,\ 0 \leqslant k \leqslant \left[\frac{\sqrt{T}}{h}\right].$$

From the boundary conditions (8) we obtain

$$\lambda^{k,l}(r^{k,l}) \equiv \left[\nu w^{k,l} r_\eta^{k,l} + \left(kh\, v_0^{k,l-1} - kh\, v_0^{k,l} - \frac{C^{k,l-1}}{w^{k,l-1}} \right) r^{k,l} \right]\Bigg|_{\eta=0}$$

(37)

$$= kh\, w^{k,l-1}(0) \frac{v_0^{k,l} - v^{k,l-1}}{h} - \frac{C^{k,l} - C^{k,l-1}}{h}, \quad r^{k,l}(1) = 0.$$

In order to estimate $r^{k,l}$ we consider the function $\varkappa_{14} = C_{14} w$. By (7) we obtain

$$R^{k,l}(\varkappa_{14}) = C_{14} w^{k,l} \left[\frac{w^{k,l} + w^{k,l-1}}{(w^{k,l-1})^2} \left(\frac{kh}{2} \rho^{k,l-1} + (kh)^2 (\eta + \lambda h) U^{k,l-1} r^{k,l-1} \right.\right.$$

(38)

$$\left.\left. - A^{k,l-1} z^{k,l-1} - B^{k,l-1} w^{k,l-1} \right) \right]$$

$$- (\eta + \lambda h)(kh)^2 \frac{U^{k,l} - U^{k,l-1}}{h} C_{14} w^{k,l-1}.$$

We take T sufficiently small that $khC_9 \leqq 1$ for $kh \leqq \sqrt{T}$. The order of all the terms in (38) apart from $B^{k,l-1} w^{k,l-1}$ is at least kh; therefore we can take T sufficiently small so that

$$R^{k,l}(\varkappa_{14}) < -\frac{C_{14}}{4} \frac{w^{k,l}}{w^{k,l-1}} (w^{k,l} + w^{k,l-1}).$$

Hence it follows that

(39) $$R^{k,l}(\varkappa_{14}) + \left| -\frac{A^{k,l}-A^{k,l-1}}{h} z^{k,l-1} - \frac{B^{k,l}-B^{k,l-1}}{h} w^{k,l-1} \right| < 0,$$

if C_{14} is chosen sufficiently large and $kh \leqq \sqrt{T}$. Here we have used the assumptions of the lemma and the result $\lim_{\eta \to 1} \sigma_\mu / \sigma_{\mu_1} = 1$ for every μ and μ_1 from $(0, 1)$. In addition we consider

$$\lambda^{k,l}(\varkappa_{14}) = -C_{14} C^{k,l} + \left(kh\, v_0^{k,l-1} - \frac{C^{k,l-1}}{w^{k,l-1}(0)} \right) C_{14} w^{k,l}(0).$$

For a sufficiently small $kh \leqq \sqrt{T}$ the expression for $\lambda^{k,l}(\varkappa_{14})$ is negative, and its order is not higher than $(kh)^2$. Hence it follows that for a sufficiently large constant C_{14} we obtain the inequality

(40) $$\lambda^{k,l}(\varkappa_{14}) + \left| kh\, w^{k,l-1}(0) \frac{v_0^{k,l} - v_0^{k,l-1}}{h} - \frac{C^{k,l} - C^{k,l-1}}{h} \right| < 0.$$

It follows from (39) and (40) that

(41) $$R^{k,l}(\varkappa_{14} \pm r) < 0, \quad \lambda^{k,l}(\varkappa_{14} \pm r) < 0, \quad (\varkappa_{14} \pm r)|_{\eta=1} = 0,$$

where C_{14} depends only on the data of the problem (7), (8). The function

$r^{k,0}$ does not enter into the equation (36); hence we can assume that (32) holds for $r^{k,0}$. We put $C_9 = C_{14}C_7$. By the induction hypothesis we obtain $\kappa_{14}^{k,l-1} \pm r^{k,l-1} \geqq 0$. When we use the maximum principle we find from (41) and the last inequality that $\kappa_{14}^{k,l} \pm r^{k,l} \geqq 0$ for $0 \leqq \eta \leqq 1$ and $kh \leqq 1/C_7C_{14}$. This means that

$$|r^{k,l}| \leqq C_{14}w^{k,l} \leqq C_{14}C_7(1-\eta)\sigma_{\mu_1},$$

and the inequality (32) is proved.

From (7) and (8) we see that $\rho^{k,l}$ satisfies the equation

$$P^{k,l}(\rho^{k,l}) \equiv \nu (w^{k,l})^2 \rho_{\eta\eta}^{k,l} - \frac{(k-1)h}{2} \frac{\rho^{k,l} - \rho^{k-1,l}}{h} - (\eta + \lambda h)(kh)^2 U^{k,l}$$

$$\times \frac{\rho^{k,l} - \rho^{k,l-1}}{h} + A^{k,l}\rho_\eta^{k,l} + B^{k,l}\rho^{k,l} + \rho^{k,l}\left[\frac{w^{k,l} + w^{k-1,l}}{(w^{k-1,l})^2}\left(\frac{(k-1)h}{2}\rho^{k-1,l}\right.\right.$$

$$\left.\left. + (\eta + \lambda h)((k-1)h)^2 U^{k-1,l} r^{k-1,l} - A^{k-1,l} z^{k-1,l} - B^{k-1,l} w^{k-1,l}\right) - \frac{1}{2}\right]$$

$$= (\eta + \lambda h)(kh)^2 r^{k-1,l} \frac{U^{k,l} - U^{k-1,l}}{h} - (-2k + 1)h(\eta + \lambda h) r^{k-1,l} U^{k-1,l}$$

$$- \frac{A^{k,l} - A^{k-1,l}}{h} z^{k-1,l} - \frac{B^{k,l} - B^{k-1,l}}{h} w^{k-1,l},$$

$$1 \leqslant k \leqslant \left[\frac{\sqrt{T}}{h}\right], \quad l > 0$$

and the boundary conditions

$$\Gamma^{k,l}(\rho^{k,l}) \equiv \left[\nu w^{k,l} \rho_\eta^{k,l} - \frac{C^{k-1,l}}{w^{k-1,l}} \rho^{k,l}\right]\Bigg|_{\eta=0}$$

$$= w^{k,l}(0)\left(kh \frac{v_0^{k,l} - v_0^{k-1,l}}{h} + v_0^{k-1,l}\right) - \frac{C^{k,l} - C^{k-1,l}}{h}, \quad \rho^{k,l}(1) = 0.$$

Let $\kappa_{15} = C_{15}(1-\eta)\sigma_{\mu_1}$. We have

$$P^{k,l}(\varkappa_{15}) = \nu (w^{k,l})^2 C_{15}\left(-\frac{1}{2\sigma_{\mu_1}(1-\eta)} - \frac{1}{4\sigma_{\mu_1}^3(1-\eta)}\right)$$

$$+ C_{15}\left(-\sigma_{\mu_1} + \frac{1}{2\sigma_{\mu_1}}\right) A^{k,l} + C_{15}(1-\eta)\sigma_{\mu_1} B^{k,l} + C_{15}(1-\eta)\sigma_\mu$$

$$\times \left[\frac{w^{k,l} + w^{k-1,l}}{(w^{k-1,l})^2}\left(\frac{(k-1)h}{2}\rho^{k-1,l} + ((k-1)h)^2(\eta + \lambda h) U^{k-1,l} r^{k-1,l}\right.\right.$$

$$\left.\left. - A^{k-1,l} z^{k-1,l} - B^{k-1,l} w^{k-1,l}\right) - \frac{1}{2}\right].$$

We assume that T and h are sufficiently small that

$$\frac{(k-1)h}{2} C_{10}(1-\eta)\sigma_{\mu_1} \leqslant \frac{C_8(1-\eta)\sigma_\mu}{2}.$$

Hence it is clear that $P^{k,l}(\kappa_{15})$ is negative, and its modulus can be made arbitrarily large if we take T sufficiently small for a sufficiently large constant C_{15}. Hence it follows that

$$P^{k,l}(\varkappa_{15}) + \Big|(\eta+\lambda h)(kh)^2 r^{k-1,l}\frac{U^{k,l}-U^{k-1,l}}{h} - \frac{A^{k,l}-A^{k-1,l}}{h}z^{k-1,l} - (-2k+1)h(\eta+\lambda h)r^{k-1,l}U^{k-1,l} - \frac{B^{k,l}-B^{k-1,l}}{h}w^{k-1,l}\Big| < 0, \tag{42}$$

provided $kh \leqq \sqrt{T}$, and C_{15} is chosen sufficiently large.

For $\eta=1$ we have $\kappa_{15}^{k,l}=0$, and for $\eta=0$ we find that

$$\Gamma^{k,l}(\varkappa_{15}) + \left|w^{k,l}(0)\left(kh\frac{v_0^{k,l}-v_0^{k-1,l}}{h}+v_0^{k-1,l}\right) - \frac{C^{k,l}-C^{k-1,l}}{h}\right|$$

$$= \nu w^{k,l}(0)C_{15}\left(-\sqrt{-\log\mu_1}+\frac{1}{2\sqrt{-\log\mu_1}}\right) - \frac{C^{k-1,l}}{w^{k-1,l}}C_{15}\sqrt{-\log\mu_1}$$

$$+\left|w^{k,l}(0)\left(kh\frac{v_0^{k,l}-v_0^{k-1,l}}{h}+v_0^{k-1,l}\right) - \frac{C^{k,l}-C^{k-1,l}}{h}\right| < 0, \tag{43}$$

if $kh \leqq \sqrt{T}$ and C_{15} is chosen sufficiently large.

When we take $C_{10} \leqq C_{15}$ and consider the function $\kappa_{15}^{k,l} \pm \rho^{k,l}$, from (42) and (43) we deduce the inequality (33) with a constant independent of h; the result is easily proved by using the maximum principle as was done in the proof of (32).

We now estimate $z^{k,l}=w_\eta^{k,l}$. When we differentiate (7) with respect to η we obtain

$$\Lambda^{k,l}(z) \equiv \nu(w^{k,l})^2 z_{\eta\eta}^{k,l} - \frac{kh}{2}\frac{z^{k,l}-z^{k-1,l}}{h} - (\eta+\lambda h)(kh)^2 U^{k,l}\frac{z^{k,l}-z^{k,l-1}}{h}$$
$$+A^{k,l}z_\eta^{k,l} + B^{k,l}z^{k,l} + 2\nu w^{k,l}z^{k,l}z_\eta^{k,l} + A_\eta^{k,l}z^{k,l} - (kh)^2U^{k,l}r^{k,l} + B_\eta^{k,l}w^{k,l} = 0. \tag{44}$$

It follows from (8) that

$$z^{k,l}|_{\eta=0} = \frac{1}{\nu}\left(kh\,v_0^{k,l} - \frac{C^{k,l}}{w^{k,l}(0)}\right). \tag{45}$$

The quantities $r^{k,l}$ and $w^{k,l}$ in (44) have already been estimated uniformly with respect to h. We are going to estimate $z^{k,l}$ from below. With this objective we consider $\kappa_{16} = -C_{16}\sigma_{\mu_1}$, where $0<\mu_1<e^{-1/2}$ and $C_{16}=\text{const}>0$. We have

$$\Lambda^{k,l}(\varkappa_{16}) = -\nu(w^{k,l})^2\left(\frac{1}{2\sigma_{\mu_1}(1-\eta)^2} - \frac{1}{4\sigma_{\mu_1}^3(1-\eta)^2}\right)C_{16}$$
$$+C_{16}\left(-\frac{1}{2\sigma_{\mu_1}(1-\eta)}\right)A^{k,l} + B^{k,l}C_{16}(-\sigma_{\mu_1}) + 2\nu w^{k,l}C_{16}^2\frac{1}{2(1-\eta)}$$
$$-C_{16}\sigma_{\mu_1}A_\eta^{k,l} - (kh)^2U^{k,l}r^{k,l} + B_\eta^{k,l}w^{k,l} > 0,$$

if $kh \leqq \sqrt{T}$ and C_{16} is chosen sufficiently large. Here we have again used the equivalence of σ_μ and σ_{μ_1} in a neighborhood of $\eta = 1$. It follows from (18) that there exists a sequence of points $\eta_n^{k,l} \to 1$ at which $z^{k,l} \geqq -C_{17}\sigma_{\mu_1}$. For $\eta = 0$ it follows from (45) that $z^{k,l}(0)$ is bounded from below. Hence we can choose C_{16} so that $s^{k,l} = \kappa_{16}^{k,l} - z^{k,l} \leqq 0$ at the points $\eta_n^{k,l}$ and for $\eta = 0$.

In addition, for $s^{k,l}$ we have the inequality

$$\begin{aligned}\Lambda^{k,l}(\varkappa_{16}) - \Lambda^{k,l}(z) &\equiv \nu (w^{k,l})^2 s_{\eta\eta}^{k,l} - \frac{kh}{2}\,\frac{s^{k,l} - s^{k-1,l}}{h} \\ &- (\eta + \lambda h)(kh)^2 U^{k,l}\,\frac{s^{k,l} - s^{k,l-1}}{h} + A^{k,l} s_\eta^{k,l} + B^{k,l} s^{k,l} \\ &+ A_\eta^{k,l} s^{k,l} + 2\nu w^{k,l} \varkappa_{16\eta}^{k,l} s^{k,l} + 2\nu w^{k,l} z^{k,l} s_\eta^{k,l} > 0.\end{aligned} \tag{46}$$

The coefficient of $s^{k,l}$ in (46) is $B^{k,l} + A_\eta^{k,l} + 2\nu w^{k,l}\kappa_{16\eta}^{k,l}$, and it is negative if C_{16} is sufficiently large. Therefore by using the maximum principle we deduce from (46) that $s^{k,l} \leqq 0$ for $0 \leqq \eta \leqq \eta_n^{k,l}$; that is, $z^{k,l} \geqq -C_{16}\sigma_{\mu_1}$. In the statement of Lemma 4 we can take $C_{11} \geqq C_{16}$.

We derive the estimate from above for $z^{k,l}$ in two steps. First we estimate $z^{k,l}$ for $1 - \delta_0 \leqq \eta < 1$, where $0 < \delta_0 < 1$, and then for $0 \leqq \eta \leqq 1 - \delta_0$.

It follows from (28) that there exists a sequence $\bar{\eta}_n^{k,l} \to 1$, at the points of which $z^{k,l} \leqq -C_{18}\sigma_\mu$ for $k \geqq 1$. We consider the function $\kappa_{19} = -C_{19}\sigma_\mu$ on the interval $\bar{\eta}_{n_0}^{k,l} \leqq \eta < 1$, where $\bar{\eta}_{n_0}^{k,l}$ is some fixed point of $\{\bar{\eta}_n^{k,l}\}$. We obtain

$$\begin{aligned}\Lambda^{k,l}(\varkappa_{19}) \leqslant &\Bigg[-\frac{\nu C_8^2}{2} C_{19}\sigma_\mu + \frac{\nu C_7^2 \sigma_{\mu_1}^2 C_{19}}{4\sigma_\mu^3} + A^{k,l}\left(-\frac{1}{2\sigma_\mu(1-\eta)}\right) C_{19} \\ &\qquad + B_\eta^{k,l} w^{k,l} - (kh)^2 r^{k,l} U^{k,l}\Bigg] \\ &+ \Bigg[-\frac{C_{19}\sigma_\mu}{2} \\ &\qquad + \left(\frac{1}{2} - B^{k,l}\right) C_{19}\sigma_\mu + \nu C_{19}^2 C_7 \sigma_{\mu_1} - A_\eta^{k,l} C_{19}\sigma_\mu\Bigg].\end{aligned} \tag{47}$$

If $\bar{\eta}_{n_0}^{k,l}$ is sufficiently close to 1, then the sum

$$\begin{aligned}-\frac{\nu C_8^2 C_{19}\sigma_\mu}{2} + \frac{\nu C_7^2 C_{19}\sigma_{\mu_1}^2}{4\sigma_\mu^3} + A^{k,l}\left(-\frac{1}{2\sigma_\mu(1-\eta)}\right) C_{19} \\ + B_\eta^{k,l} w^{k,l} - (kh)^2 r^{k,l} U^{k,l}\end{aligned}$$

is nonpositive. The second group of terms in (47) is negative if T and C_{19} are sufficiently small that for $kh \leqq \sqrt{T}$ we have

$$\nu C_{19} C_7 \frac{\sigma_{\mu_1}}{\sigma_\mu}\Bigg|_{\eta = \bar{\eta}_{n_0}^{k,l}} - U_x^{k,l}(kh)^2 < \frac{1}{2}.$$

Thus the point $\bar{\eta}_{n_0}^{kl}$, the time T and the constant C_{19} can be chosen so that for $kh \leqq \sqrt{T}$ we have $\Lambda^{k,l}(\kappa_{19}) < 0$ on the interval $\bar{\eta}^{k,l}_{n_0} \leqq \eta < 1$. We suppose, in addition, that $C_{19} \leqq C_{18}$. Then at the points $\bar{\eta}^{k,l}_{n}$ the difference $\kappa^{k,l}_{19} - z^{k,l} = s^{k,l}$ is nonnegative. For $s^{k,l}$ we have the inequality

$$(48)\quad \begin{aligned} &\nu\,(w^{k,l})^2 s^{k,l}_{\eta\eta} + A^{k,l} s^{k,l}_{\eta} + B^{k,l} s^{k,l} + A^{k,l}_{\eta} s^{k,l} + 2\nu w^{k,l} \varkappa^{k,l}_{19\eta} s^{k,l} \\ &+ 2\nu w^{k,l} z^{k,l} s^{k,l}_{\eta} - \frac{k}{2} s^{k,l} - (\eta + \lambda h)(kh)^2 U^{k,l} \frac{s^{k,l} - s^{k,l-1}}{h} \\ &\leqq -\frac{k}{2} s^{k-1,l}, \quad 1 \leqslant k \leqslant \left[\frac{\sqrt{T}}{h}\right]. \end{aligned}$$

The coefficient of $s^{k,l}$ in (48) is $(kh)^2\eta U^{k,l}_x - C_{19}\nu - k/2 + \frac{1}{2}$; this sum is negative if $kh \leqq \sqrt{T}$ and T is sufficiently small. It follows from (17) that we can choose C_{19} so that $s^{0,l} \geqq 0$. When we take this into account we obtain from (51), by induction on k and by using the maximum principle, that $s^{k,l} \geqq 0$. Therefore $\kappa^{k,l}_{19} \geqq z^{k,l}$ for $\bar{\eta}^{k,l}_{n_0} \leqq \eta < 1$.

To estimate $z^{k,l}$ from above for $0 \leqq \eta \leqq \bar{\eta}^{k,l}_{n_0}$ we take the function $\kappa_{20} = -C_{20}\eta^2$. We have

$$(49)\quad \begin{aligned} \Lambda^{k,l}(\varkappa_{20}) = &-2C_{20}\nu(w^{k,l})^2 - 2C_{20}\eta A^{k,l} - \frac{C_{20}\eta^2}{2} - C_{20}\eta^2 \\ &\times \left(-\eta(kh)^2 U^{k,l}_x - (kh)^2 \frac{U^{k,l}_t}{U^{k,l}}\right) + 4\nu w^{k,l} C^2_{20}\eta^3 - C_{20}\eta^2 \cdot \\ &\times \left(2\eta(kh)^2 U^{k,l}_x + (kh)^2 \frac{U^{k,l}_t}{U^{k,l}}\right) - (kh)^2 U^{k,l} r^{k,l} - (kh)^2 U^{k,l}_x w^{k,l}. \end{aligned}$$

We choose the constant C_{20} so that $4\nu w^{k,l} C^2_{20}\eta^3 < 2C_{20}\nu(w^{k,l})^2 + \frac{1}{2}C_{20}\eta^2$ for $0 \leqq \eta \leqq \bar{\eta}^{k,l}_{n_0}$. It is obvious that with this choice of C_{20} and for T sufficiently small, when $kh \leqq \sqrt{T}$ we obtain $\Lambda^{k,l}(\kappa_{20}) < 0$.

By the conditions of Lemma 4 it follows from (45) that when T is sufficiently small we have $z^{k,l}(0) \leqq 0$ for $kh \leqq \sqrt{T}$. In addition we choose the constant C_{20} sufficiently small that the difference $\kappa^{k,l}_{20} - z^{k,l} = s^{k,l}$ is nonnegative at $\bar{\eta}^{k,l}_{n_0}$, and $\kappa^{0,l}_{20} \geqq z^{0,l}$. The last inequality is possible by (13). For $s^{k,l}$ we have the inequality

$$\begin{aligned} &\nu\,(w^{k,l})^2 s^{k,l}_{\eta\eta} + A^{k,l} s^{k,l}_{\eta} + B^{k,l} s^{k,l} + A^{k,l}_{\eta} s^{k,l} + 2\nu w^{k,l} \varkappa^{k,l}_{20\eta} s^{k,l} \\ &+ 2\nu w^{k,l} z^{k,l} s^{k,l}_{\eta} - \frac{k}{2} S^{k,l} - (\eta + \lambda h)(kh)^2 U^{k,l} \frac{s^{k,l} - s^{k,l-1}}{h} \leqslant -\frac{k}{2} s^{k-1,l}. \end{aligned}$$

(50)

The coefficient of $s^{k,l}$ in (50) is

$$(51)\quad (kh)^2 \eta U^{k,l}_x - 4C_{20}\nu\eta w^{k,l} - \frac{k}{2} + \frac{1}{2}.$$

We take T sufficiently small that $T(\max|U_x|) < 4\nu C_{20} C_8 (1 - \bar{\eta}^{k,l}_{n_0}) \sigma_\mu(\bar{\eta}^{k,l}_{n_0})$. Then the expression (51) is negative. Hence we can apply the maximum principle; we find that $z^{k,l} \leqq -C_{20}\eta^2$ for $0 \leqq \eta \leqq \bar{\eta}^{k,l}_{n_0}$. This proves the estimate for $z^{k,l}$. In (34) we can take $C_{12} \leqq \min(C_{19}, C_{20})$.

It follows from the equations (7) that

$$w^{k,l} w^{k,l}_{\eta\eta} = \frac{kh}{2\nu} \frac{\rho^{k,l}}{w^{k,l}} + (\eta + \lambda h)(kh)^2 U^{k,l} \frac{r^{k,l}}{\nu w^{k,l}} - \frac{A^{k,l} z^{k,l}}{\nu w^{k,l}} - \frac{B^{k,l}}{\nu}.$$

Hence when we take (33) and (34) into account we obtain (35), and for a sufficiently small T it easily follows that $w^{k,l} w^{k,l}_{\eta\eta} < -1/4\nu$ for $kh \leqq \sqrt{T}$. This proves Lemma 4.

Theorem 1. *We assume that the conditions of Lemma 4 are fulfilled. Then in the domain* $\Omega\{0 \leqq \tau \leqq \sqrt{T}, 0 \leqq \xi \leqq X, 0 \leqq \eta \leqq 1\}$ *there exists a solution of the problem* (5), (6) *with the following properties:* $w(\tau, \xi, \eta)$ *is continuous in* Ω, $C_8(1-\eta)\sigma_\mu \leqq w \leqq C_7(1-\eta)\sigma_{\mu_1}$, w_η *is continuous with respect to* η *for* $\eta < 1$, $-C_{11}\sigma_{\mu_1} \leqq w_\eta \leqq C_{12}\Phi(\eta)$, *where* $\Phi(\eta)$ *is the function in Lemma* 4, $|w_\xi| \leqq C_9(1-\eta)\sigma_{\mu_1}$, $|w_\tau| \leqq C_{10}(1-\eta)\sigma_{\mu_1}$, $ww_{\eta\eta}$ *is bounded, and* $ww_{\eta\eta} < -1/4\nu$. *In* Ω *the function* w *satisfies* (5) *almost everywhere, and it satisfies the conditions* (6). *A solution of the problem* (5), (6) *with these properties is unique. Here* X *is a given number, and* T *is a sufficiently small number that depends on* ν, U *and* v_0.

Proof. The existence of a solution of the problem (5), (6) follows from the lemmas in the same way as in the article [**4**]; we prove its uniqueness. We consider two solutions w_1 and w_2 of the problem (4), (5), and put $w_1 - w_2 = \bar{w}$. The function $\bar{w}$ satisfies the equation

$$\nu w_1 \bar{w}_{\eta\eta} - \frac{\tau}{2}\frac{\bar{w}_\tau}{w_1} - \tau^2\eta U \frac{\bar{w}_\xi}{w_1} + A\frac{\bar{w}_\eta}{w_1} + \frac{B\bar{w}}{w_1} + \nu\frac{w_{2\eta\eta}}{w_1}(w_1 + w_2)\bar{w} = 0 \tag{52}$$

and the boundary conditions

$$\bar{w}\Big|_{\eta=1} = 0, \quad \left(\nu\bar{w}_\eta - \frac{C\bar{w}}{w_1 w_2}\right)\Big|_{\eta=0} = 0. \tag{53}$$

We multiply the terms in (52) by $\bar{w}e^{-\alpha\tau}$, $\alpha > 0$, and integrate over Ω; when we integrate by parts we obtain

$$\int_\Omega \Big[-\nu w_1(\bar{w}_{1\eta})^2 + \nu w_{1\eta\eta}\frac{\bar{w}^2}{2} + \frac{\bar{w}^2}{4w_1} - \frac{\tau\bar{w}^2 w_{1\tau}}{4w_1^2} - \alpha\tau\frac{\bar{w}^2}{4w_1} + \tau^2\eta\frac{U_\xi\bar{w}^2}{2w_1}$$

$$- \tau^2\eta U\frac{\bar{w}^2 w_{1\xi}}{2w_1^2} - \frac{A_\eta\bar{w}^2}{2w_1} + \frac{A\bar{w}^2 w_{1\eta}}{2w_1^2} + \left(\frac{B}{w_1} + \nu w_{2\eta\eta}\frac{(w_1+w_2)}{w_1}\right)\bar{w}^2\Big] e^{-\alpha\tau}\,d\tau d\xi\,d\eta$$

$$- \int_{\eta=0}\frac{A\bar{w}^2}{2w_1}e^{-\alpha\tau}\,d\tau d\xi - \int_{\xi=X}\frac{\tau^2\eta U\bar{w}^2}{2w_1}e^{-\alpha\tau}\,d\tau\,d\eta - \int_{\tau=\sqrt{T}}\frac{\tau\bar{w}^2}{4w_1}e^{-\alpha\tau}\,d\xi d\eta$$

$$+ \int_{\eta=0}\left(\nu w_{1\eta}\frac{\bar{w}^2}{2} - \frac{\bar{w}^2 C}{w_2}\right)e^{-\alpha\tau}\,d\tau\,d\xi = 0. \tag{54}$$

The integrals over the boundary $\xi = X$ and over the boundary $\tau = \sqrt{T}$ are nonpositive. We note that $A|_{\eta=0} = -C$; then the sum of the integrals over the boundary $\eta = 0$ gives

$$\int_{\eta=0} \left(\frac{\nu w_{1\eta}}{2} - \frac{C}{w_2} + \frac{C}{2w_1} \right) \bar{w}^2 e^{-\alpha\tau} \, d\tau d\xi = \int_{\eta=0} \left(\frac{\tau}{2} v_0 - \frac{C}{w_2} \right) \bar{w}^2 e^{-\alpha\tau} \, d\tau d\xi < 0,$$

because by the choice of T we have $\tau v_0/2 - C/w_2 < 0$ for $\tau \leq \sqrt{T}$. Hence from (54) we obtain

$$\text{(55)} \quad \begin{aligned} \int_\Omega \Bigg[\left(\frac{\nu w_{1\eta\eta}}{2} + \frac{1}{4w_1} \right) - \frac{\tau w_{1\tau}}{4w_1^2} - \frac{\alpha\tau}{4w_1} + \tau^2 \eta \frac{U_\xi}{2w_1} - \eta \frac{\tau^2 U w_{1\xi}}{2w_1^2} - \frac{A_\eta}{2w_1} \\ + \frac{A w_{1\eta}}{2w_1^2} + \left(\frac{B}{w_1} + \nu w_{2\eta\eta} \frac{(w_1 + w_2)}{w_1} \right) \Bigg] \bar{w}^2 e^{-\alpha\tau} \, d\tau \, d\xi \, d\eta \geq 0. \end{aligned}$$

It follows from the estimates (32) – (35) and equation (5) that for small values of τ we have

$$\frac{\nu w_{1\eta\eta}}{2} + \frac{1}{4w_1} = 0 \left(\frac{\tau}{w_1} \right) \text{ and } \frac{B}{w_1} + \nu w_{2\eta\eta} \frac{(w_1 + w_2)}{w_1} = - \frac{1}{2w_1} + 0 \left(\frac{\tau}{w_1} \right).$$

The order as $\tau \to 0$ of the remaining terms in the left-hand member of (55) is not lower than the first, and as $\eta \to 1$ their order is not higher than $1/w_1$. Hence it is clear that for α sufficiently large we obtain

$$\begin{aligned} \left(\frac{\nu w_{1\eta\eta}}{2} + \frac{1}{4w_1} \right) - \frac{\tau w_{1\tau}}{4w_1^2} - \frac{\alpha\tau}{4w_1} + \frac{\tau^2 \eta U_\xi}{2w_1} - \eta \frac{\tau^2 U w_{1\xi}}{2w_1^2} \\ - \frac{A_\eta}{2w_1} + \frac{A w_{1\eta}}{2w_1^2} + \left(\frac{B}{w_1} + \nu w_{2\eta\eta} \frac{(w_1 + w_2)}{w_1} \right) < 0. \end{aligned}$$

In view of this it follows from (55) that $\bar{w}^2 = 0$ almost everywhere. Because $\bar{w}$ is continuous we have $\bar{w} \equiv 0$, and the theorem is proved.

It follows from (3) and (4) that

$$\text{(56)} \quad \frac{u(t, x, y)}{U(t, x)} \qquad \frac{y}{\sqrt{t}} = \int_0 \frac{ds}{w(\sqrt{t}, x, s)}.$$

This equality allows us to deduce from Theorem 1 the following result.

Theorem 2. *Assume that the functions U_x, U_{xx}, U_t/U, $(U_t/U)_x$, $\sqrt{t} U_{xt}$, $\sqrt{t}(U_t/U)_t$ and v_0 are bounded, $U_x(0,0) > 0$, $U_x + U_t/U > 0$, and $-M_1 U < U_t/U$ for small values of x. Assume also that $v_0 \leq M_2 t^{1/2+\epsilon}$, $|v_{0x}| \leq M_3 t^{1/2}$ and $\sqrt{t}|x_{0t}| \leq M_4$, where ϵ is a positive constant. Then for T sufficiently small there exists in D_T a unique solution u, v of the problem* (1), (2) *with the following properties: u/U is continuous for $t > 0$, $\sqrt{t} u_y/U$ is bounded and continuous, $u \to U$ as $y \to \infty$, u and v satisfy the conditions* (2), *and u_y, u_x, u_{yy}, u_t and v_y*

are bounded and continuous for $t>0$. *The equations* (1) *are satisfied almost everywhere in* D_T. *In addition the following inequalities hold*:

$$C_8\left(1-\frac{u}{U}\right)\sigma_\mu\left(\frac{u}{U}\right)\leqslant\frac{\sqrt{t}\,u_y}{U}\leqslant C_7\left(1-\frac{u}{U}\right)\sigma_{\mu_1}\left(\frac{u}{U}\right),\tag{57}$$

$$-C_{11}\sigma_{\mu_1}\left(\frac{u}{U}\right)\leqslant\frac{\sqrt{t}\,u_{yy}}{u_y}\leqslant C_{12}\Phi\left(\frac{u}{U}\right),\tag{58}$$

where $\Phi(s)=-s^2$ *for* $0\leqq s<1-\delta_0$, *and* $\Phi(s)=-\sigma_\mu(s)$ *for* $1-\delta_0\leqq s<1$, $\delta_0>0$; *and*

$$\left|\frac{2tu_{yy}}{u_y}\frac{tuU_t-u_tU}{U^2}+\frac{u_y}{U}+2t\left(\frac{u_{yt}U-u_yU_t}{U^2}\right)\right|\leqslant C_{10}\left(1-\frac{u}{U}\right)\sigma_{\mu_1}\left(\frac{u}{U}\right).\tag{59}$$

We omit the proof of this theorem; it can be proved in exactly the same way as Theorem 2 of [**4**] and Theorem 2 of [**5**].

The following estimates follow from (57):

$$\begin{aligned}U(t,x)\exp\left(-\frac{C_7^2y^2}{4t}-\frac{C_7y\sqrt{-\log\mu_1}}{\sqrt{t}}\right)&\leqslant U(t,x)-u\\ &\leqslant U(t,x)\exp\left(-\frac{C_8^2y^2}{4t}-\frac{C_8y\sqrt{-\log\mu}}{\sqrt{t}}\right).\end{aligned}\tag{60}$$

These estimates characterize the speed with which $u(t,x,y)$ tends to $U(t,x)$ as $y\to\infty$, and the behavior of u for small t and fixed x and y; the latter is connected with the speed of formation of the boundary layer. It also follows from (60) that u satisfies the conditions (2). Estimates of the form (60) are also valid for the sum of the first terms of the expansion of u in powers of the time variable that were found by Blasius [**1**].

In Lemma 3 we constructed approximate solutions of the problem (4), (5), and in Theorem 1 we proved their convergence. It follows from (56) that the approximate solution $u((kh)^2,lh,y)$ of the problem (1), (2) can be constructed, with the help of the solution $w^{k,l}$, from the formula

$$\frac{y}{\sqrt{kh}}=\int_0^{\frac{u((kh)^2,\,lh,\,y)}{U((kh)^2,\,lh)}}\frac{ds}{w^{k,l}(s)}.$$

It follows from Theorem 2 that $u((kh)^2,lh,y)$, constructed in this way, is convergent to u as $h\to 0$.

The author thanks Professor O. A. Oleĭnik for her interest in this work.

Received 12/DEC/70

Bibliography

[1] H. Blasius, *Grenzschichten in Flüssigkeiten mit kleiner Reibung*, Z. Math. Phys. **56** (1908), 1-37.

[2] L. G. Loĭcjanskiĭ, *The laminar boundary layer*, Fizmatgiz, Moscow, 1962. (Russian)

[3] H. Schlichting, *Grenzschicht-Theorie*, G. Braun, Karlsruhe, 1951; English transl., McGraw-Hill, New York; Pergamon Press, London, 1955. MR **13**, 177; **17**, 912.

[4] O. A. Oleĭnik, *Mathematical problems of boundary layer theory*, Uspehi Mat. Nauk **23** (1968), no. 3 (141), 3-65 = Russian Math. Surveys **23** (1968), no. 3, 1-66. MR **37** #4428.

[5] ———, *Formation of the boundary layer in the case of a gradual acceleration*, Sibirsk. Mat. Ž. **9** (1968), 1199-1220 = Siberian Math. J. **9** (1968), 898-915. MR **38** #6242.

[6] ———, *Approximate solutions and asymptotic expansions for the problem of boundary layer development during acceleration*, Prikl. Mat. Meh. **33** (1969), 441-455 = J. Appl. Math. Mech. **33** (1969), 428-442. MR **40** #7650.

Translated by
L. W. LONGDON

Trudy Moskov. Mat. Obšč.
Tom 28 (1973)

Trans. Moscow Math. Soc.
Vol. 28 (1973)

GENERALIZED MEASURES OR DISTRIBUTIONS ON HILBERT SPACE

D. N. DUDIN

Contents

Introduction

Interest has recently grown in differential equations on spaces of an infinite number of dimensions. Since the theory of distributions or generalized functions is a suitable apparatus for partial differential equations on R^n, it is natural to attempt to construct analogs of it on infinite-dimensional linear spaces. An idea for such a construction was presented by Fomin in [**17**]. As indicated there, because of the absence of Lebesgue measure in a sufficiently large class of infinite-dimensional spaces, the notion of a generalized function disintegrates. Generalized functions are naturally understood as functionals on a certain space of basic measures. However, another possible approach is to consider functionals on a certain space of test functions. Such functionals are naturally called generalized measures or, briefly, distributions.

The theory of generalized measures is important because (i) differential equations for measures naturally arise in infinite-dimensional spaces (see [**10**] and Example 5.5 below), and (ii) in a theory of generalized functions on an infinite-dimensional space it is difficult (if not impossible) to construct an analog of a δ-function, whereas the construction of a δ-measure is obvious. Having a δ-functional, one can determine fundamental solutions of differential operators, which, as Proposition 4.8 shows, turn out to be useful for the solution of both differential equations for generalized measures and differential equations for functions.

In the present paper we construct a theory of distributions on Hilbert

AMS (MOS) subject classifications (1970). Primary 46F05; Secondary 28A40, 35R15.

space in which there exists a wide class of continuous differential operators, including both the Laplace operator in the sense of Lévy (see [**9**] and Example 5.1 below) and the differential operators of second order introduced by Daleckiĭ [**13**–**15**]. The space of arguments is taken as a Hilbert space merely for the sake of simplicity, the majority of results being readily extendable to more general cases.

All of the numbered statements below are enumerated in succession, independently of whether they are a theorem, lemma, proposition, etc. Because of the present confusion in terminology let us agree to not distinguish between the terms "compactum" and "compact set," understanding both of them in the sense of being able to choose a finite subcovering from any open covering. The application of a linear functional f to a vector x will be denoted by $\langle f, x\rangle$ in those cases when the notation $f(x)$ is unsuitable.

The author wishes to express his gratitude to Professor S. V. Fomin for his direction of the work.

§1. The space $C(S, E)$

Let S be a complete separable metric space, let E be a locally convex space and let $C(S,E)$ denote the linear space of all continuous functions on S with a bounded range in E. In order to introduce a topology in $C(S,E)$ we consider the family $\Psi(S)$ of all functions ψ satisfying the following two conditions:

a) $\psi : S \to (0, \infty]$.

b) The set $\{s : s \in S, \psi(s) \leqq C\}$ is compact for any $C \in R^1$.

It follows from condition b) that every function of $\Psi(S)$ is lower semicontinuous and hence takes its own least value, which implies that $\inf_{s\in S}\psi(s) > 0$ for any $\psi \in \Psi(S)$. Consequently the set $\{\varphi : \varphi \in C(S,E),\ q(\varphi(s)) \leqq \psi(s)$ for all $s \in S\}$ will be absorbing for any continuous seminorm q on E and $\psi \in \Psi(S)$. In addition, it is obviously absolutely convex. From the assertion

$$(\psi_1 \in \Psi(S) \text{ and } \psi_2 \in \Psi(S) \text{ and } \psi_3(s) = \min(\psi_1(s), \psi_2(s))) \Rightarrow \psi_3 \in \Psi(S)$$

it follows that the intersection of two such sets is also such a set, which implies that these sets form a local base at the origin for a locally convex topology that is separable if E is separable. The same topology can be given by the seminorms $p_{\psi q}(\varphi) = \sup_{s\in S} q(\varphi(s))/\psi(s)$, where $\psi \in \Psi(S)$ and q is a continuous seminorm on E. We put $\alpha/\infty = 0$ for any $\alpha \in R^1$.

PROPOSITION 1.1. *If E is complete, so is $C(S,E)$.*

PROOF. Since for any compactum $K \subset S$ the function

$$\psi(s) = \begin{cases} 1 & \text{for } s \in K, \\ \infty & \text{for } s \notin K \end{cases}$$

belongs to $\Psi(S)$, every Cauchy net converges uniformly on each compactum.

Consequently the limit function of a Cauchy net is continuous and the completion of $C(S,E)$ can be realized as a subspace of the space of all continuous mappings of S into E. It remains to prove that every function in the completion of $C(S,E)$ is bounded.

Let $\varphi_1 : S \to E$ be an unbounded function. Then there exists a continuous seminorm q on E such that the numerical function $q \circ \varphi_1$ is unbounded, i.e. there exists a sequence $\{s_n\} \subset S$, $s_i \neq s_j$ for $i \neq j$, such that $q(\varphi_1(s_n)) \geqq n$. Let

$$\psi(s) = \begin{cases} n/2 & \text{for } s \in \{s_n\}, \\ \infty & \text{for } s \in S \setminus \{s_n\}. \end{cases}$$

Then the neighborhood $\{\varphi : p_{\psi q}(\varphi_1 - \varphi) < 1\}$ does not contain any bounded functions, i.e. φ_1 does not belong to the completion of $C(S,E)$.

PROPOSITION 1.2. *Suppose a net φ_α ($\varphi_\alpha \in C(S,E)$), $\alpha \in A$, converges uniformly on each compactum and the set $\bigcup_{\alpha \in A} \varphi_\alpha(S)$ is bounded in E. Then φ_α converges in $C(S,E)$ to the same limit.*

PROOF. Let q be a continuous seminorm on E. Then there exists an $L \in R^1$ such that $q(\varphi_\alpha(s)) \leqq L$ for all $\alpha \in A$ and $s \in S$. It can be assumed without loss of generality that $\varphi_\alpha \to 0$ uniformly on each compactum. Let $\psi \in \Psi(S)$. Then there exists a $\beta \in A$ such that the inequality $q(\varphi_\alpha(s)) < \psi(s)$ holds for all $\alpha \geqq \beta$ on the compactum $\{s : s \in S, \psi(s) \leqq L\}$ and hence everywhere on S. Q.E.D.

Proposition 1.2 admits a sharpening for sequences.

PROPOSITION 1.3. *In order for a sequence $\{\varphi_n\}$, $\varphi_n \in C(S,E)$, to be convergent in $C(S,E)$ it is necessary and sufficient that it converge uniformly on each compactum and that the set $\bigcup_1^\infty \varphi_n(S)$ be bounded in E.*

PROOF. The sufficiency follows from Proposition 1.2. We prove necessity. Suppose $\varphi_n \to 0$ in $C(S,E)$. In the proof of Proposition 1.1 we saw that then $\varphi_n \to 0$ uniformly on each compactum. Suppose that the set $\bigcup_1^\infty \varphi_n(S)$ is unbounded in E. Then there exists a continuous seminorm q on E such that the set

$$q\left(\bigcup_{n=1}^{\infty} \varphi_n(S)\right) = \bigcup_{n=1}^{\infty} q(\varphi_n(S))$$

is unbounded, which implies the existence of a sequence $\{s_k\} \subset S$ and a subsequence φ_{n_k} such that $q(\varphi_{n_k}(s_k)) \geqq k$. We again let

$$\psi_s(s) = \begin{cases} k/2 & \text{for } s = s_k, \\ \infty & \text{for } s \notin S \setminus \{s_k\} \end{cases}$$

Then the subsequence φ_{n_k} does not intersect the neighborhood $\{\varphi : \varphi \in C(S,E),\ q(\varphi(s)) < \psi(s)\}$ of 0, i.e. φ_n does not converge to 0. Q.E.D.

PROPOSITION 1.4. *A set $\Phi \subset C(S,E)$ is bounded if and only if the set*

$$A = \bigcup_{\varphi \in \Phi} \varphi(S) \subset E$$

is bounded.

PROOF. Suppose A is bounded. Then for any continuous seminorm q on E and any $\psi \in \Psi(S)$ we have

$$\sup_{\varphi \in \Phi} p_{\psi q}(\varphi) \leqslant \sup_{e \in A} q(e) \cdot \sup_{s \in S} 1/\psi(s) < \infty,$$

i.e. Φ is bounded. Conversely, suppose A is unbounded. Then there exists a continuous seminorm q on E that is unbounded on A, which implies the existence of sequences $\{s_n\} \subset S$ and $\{\varphi_n\} \subset \Phi$ such that $q(\varphi_n(s_n)) > n$. Let

$$\psi(s) = \begin{cases} \left(\min_{s=s_n} n\right)^{1/2} & \text{for } s \in \{s_n\}, \\ \infty & \text{for } s \in S \setminus \{s_n\} \end{cases},$$

Then $p_{\psi q}(\varphi_n) \to \infty$, i.e. Φ is unbounded.

REMARK 1.5. If S is compact, the topology introduced by us in $C(S, E)$ clearly coincides with the topology of uniform convergence.

REMARK 1.6. If S is not compact, $C(S, E)$ will be neither a barreled space nor a bornological space, even if E is a Banach space. In fact, the set $\{\varphi : \varphi \in C(S, E), \sup_{s \in S} \|\varphi(s)\| \leqq 1\}$ is a barrel and absorbs all bounded sets but is not a neighborhood of 0.

REMARK 1.7. The completeness of S has not been used in this section. Indeed, all of the results remain true when S is an arbitrary Hausdorff space if one assumes that $C(S, E)$ is the space of all bounded functions on S with values in E and whose restrictions to each compactum are continuous.

§2. The space $C(S, E)'$

We now describe the space $C(S, E)'$. We first confine ourselves to the case when $E = B$ is a normed space. A space $M(S, B')$ will be introduced for this purpose. Let μ be an additive set function defined on a family $\mathfrak{B}$ of sets and taking values in B'. By its variation $\|\mu\|(A)$ on a set $A \in \mathfrak{B}$ is meant $\sup \sum \langle \mu(A_i), x_i \rangle$, where the supremum is taken over (i) all partitionings of A into a finite number of pairwise disjoint sets $A_i \in \mathfrak{B}$ and (ii) finite collections $x_i \in B$, $\|x_i\| \leqq 1$; $i = 1, \cdots, n$; $n = 1, 2, \cdots$. Clearly $\|\mu\|(A) \geqq \|\mu(A)\| \geqq 0$.

Everywhere below $\mathfrak{B}$ will denote the σ-algebra of Borel subsets of S.

PROPOSITION 2.1. *Suppose $\mu : \mathfrak{B} \to B'$ is an additive set function that is countably additive in the topology $\sigma(B', B)$ and such that $\|\mu\|(S) < \infty$. Then $\|\mu\|$ is a finite countably additive nonnegative measure and μ is countably additive in the topology $\beta(B', B)$.*

PROOF. From the definition of the variation of a set function it directly follows that

$$\|\mu\|(A_1 \cup A_2) \geqq \|\mu\|(A_1) + \|\mu\|(A_2)$$

for $A_2, A_2 \in \mathfrak{B}$ and $A_1 \cap A_2 = \emptyset$. On the other hand, suppose

$$A_1 \cup A_2 = \bigcup_{i=1}^{n} C_i.$$

Then

$$\sum_{i=1}^{n} \langle \mu(C_i), x_i \rangle = \sum_{i=1}^{n} \langle C_i \cap A_1), x_i \rangle$$

$$+ \sum_{i=1}^{n} \langle \mu(C_i \cap A_2), x_i \rangle \leqslant \|\mu\|(A_1) + \|\mu\|(A_2),$$

which implies that $\|\mu\|(A_1 \cup A_2) \leqq \|\mu\|(A_1) + \|\mu\|(A_2)$, and hence that $\|\mu\|$ is an additive measure.

From the additiveness and nonnegativeness of $\|\mu\|$ it follows that

$$\sum_{i=1}^{\infty} \|\mu\|(A_i) \leqslant \|\mu\| \left(\bigcup_{i=1}^{\infty} A_i \right)$$

for any sequence of pairwise disjoint sets $A_i \in \mathfrak{B}$. On the other hand, suppose

$$\bigcup_{i=1}^{\infty} A_i = \bigcup_{k=1}^{n} C_k$$

with $C_k \in \mathfrak{B}$ and $C_k \cap C_l = \emptyset$ when $k \neq l$, and suppose $\{x_k\}$ ($\|x_k\| \leqq 1$) is such that

$$\sum_{k=1}^{n} \langle \mu(C_k), x_k \rangle \geqslant \|\mu\| \left(\bigcup_{i=1}^{\infty} A_i \right) - \varepsilon,$$

for some $\epsilon > 0$. Then

$$\sum_{i=1}^{\infty} \|\mu\|(A_i) \geqslant \sum_{i=1}^{\infty} \sum_{k=1}^{n} \langle \mu(C_k \cap A_i), x_k \rangle = \sum_{k=1}^{n} \langle \mu(C_k), x_k \rangle$$

$$\geqslant \|\mu\| \left(\bigcup_{i=1}^{\infty} A_i \right) - \varepsilon.$$

By virtue of the arbitrariness of ϵ we get that

$$\sum_{i=1}^{\infty} \|\mu\|(A_i) \geqslant \|\mu\| \left(\bigcup_{i=1}^{\infty} A_i \right),$$

which implies that $\|\mu\|$ is countably additive.

From the inequality $\|\mu\|(A) \geqq \|\mu(A)\|$ we get that for any sequence of pairwise disjoint sets $A_i \in \mathfrak{B}$ the series

$$\sum_{i=1}^{\infty} \mu(A_i)$$

converges absolutely and weakly, and therefore also converges in the topology $\beta(B', B)$. Q.E.D.

Let $M(S, B')$ denote the space of all weakly countably additive set functions $\mu: \mathfrak{B} \to B'$ of bounded variation. They will also be countably additive in the strong topology by virtue of Proposition 2.1.

PROPOSITION 2.2. *The function* $\mu \to \|\mu\|(S)$ *is a norm on* $M(S, B')$ *that converts it into a Banach space.*

The proof of this proposition is a word for word repetition of the corresponding proof for the case $B = B' = R^1$ (see [3]).

PROPOSITION 2.3. *Any measure* $\mu \in M(S, B')$ *defines a continuous functional on* $C(S, B)$ *by means of the formula* $F_\mu(\varphi) = \int \varphi d\mu$ [1].

PROOF. For $n = 1, 2, \cdots$ we choose a compactum K_n so that $\|\mu\|(S \backslash K_n) \leq n^{-2}$. Let χ_n be the characteristic function of the set $S \backslash K_n$ and let $\psi(s) = 1 + \sum_1^\infty \chi_n(s)$. For any $\varphi \in C(S, B)$ the function $\varphi_n = \chi_n \varphi$ is a Baire function with compact range and hence μ-integrable. Moreover, $\varphi_n \to \varphi$ in measure, so that φ is measurable and (by virtue of its boundedness) integrable. We have

$$\left|\int \varphi d\mu\right| \leqslant \int \|\varphi\| \, d\|\mu\| = \int (\psi \|\varphi\| / \psi) \, d\|\mu\|$$

$$\leqslant p_{\psi\|\cdot\|}(\varphi) \int \psi d\|\mu\| = p_{\psi\|\cdot\|}(\varphi) \left[\|\mu\|(S) + \sum_{n=1}^{\infty} \|\mu\|(S \diagdown K_n) \right]$$

$$\leqslant p_{\psi\|\cdot\|}(\varphi) \left[\|\mu\|(S) + \sum_{n=1}^{\infty} n^{-2} \right].$$

Since $\psi \in \Psi(S)$, the last inequality implies the continuity of the functional $F_\mu(\varphi) = \int \varphi d\mu$.

PROPOSITION 2.4. *Suppose* $\mu: \mathfrak{B} \to B'$ *is such that for any* $b \in B$ *the formula* $\mu_b(A) = \langle b, \mu(A) \rangle$ *defines a charge of finite variation. Then* $\|\mu\|(S) = \sup \int \varphi d\mu$, *where the supremum is taken over all* $\varphi \in C(S, B)$ *with a finite-dimensional range and such that* $\|\varphi(s)\| \leq 1$ *for all* $s \in S$.

PROOF. Suppose $\epsilon > 0$ and $i = 1, \cdots, n$. Let a partitioning of S into mutually disjoint Borel sets A_i and a collection of elements $x_i \in B$, $\|x_i\| \leq 1$, be such that

$$\sum_{i=1}^{n} \langle x_i, \mu(A_i) \rangle > \|\mu\|(S) - \varepsilon/2;$$

let a collection of compacta K_i and a collection of open sets G_i be such that $K_i \subset A_i \subset G_i$, $K_i \cap G_j = \emptyset$ for $i \neq j$ and

$$|\langle x_i, \mu(G_i \backslash K_i) \rangle| < \epsilon/2n;$$

and let φ_i be a continuous nonnegative function that vanishes outside G_i and does not vanish on A_i (for example, $\varphi_i(s) = \rho(s, S \backslash G_i)$). The norm of the expression

$$x_i \varphi_i(s) \Big/ \sum_{k=1}^{n} \varphi_k(s)$$

is everywhere less than or equal to one; similarly, the norm

$$\left\| \left(x_i\varphi_i(s) \Big/ \sum_{k=1}^{n} \varphi_k(s) \right) - x_i\chi_{A_i}(s) \right\|$$

$$= \| x_i \| \left| \left(\varphi_i(s) \Big/ \sum_{k=1}^{n} \varphi_k(s) \right) - \chi_{A_i}(s) \right| \leqslant 1$$

and is different from zero only on $G_i \setminus K_i$. Consequently

$$\left| \int \left(x_i\varphi_i \Big/ \sum_{k=1}^{n} \varphi_k \right) d\mu - \langle x_i, \mu(A_i) \rangle \right|$$

$$= \left| \int \left[\left(x_i\varphi_i \Big/ \sum_{k=1}^{n} \varphi_k \right) - x_i\chi_{A_i} \right] d\mu \right| \leqslant \frac{\varepsilon}{2n},$$

which implies

$$\left| \int \left(\sum_{i=1}^{n} x_i\varphi_i \Big/ \sum_{k=1}^{n} \varphi_k \right) d\mu - \sum_{i=1}^{n} \langle x_i, \mu(A_i) \rangle \right| \leqslant n\varepsilon/2n = \varepsilon/2.$$

In addition,

$$\left\| \sum_{i=1}^{n} x_i\varphi_i(s) \Big/ \sum_{k=1}^{n} \varphi_k(s) \right\|$$

$$\leqslant \sum_{i=1}^{n} \left\| x_i\varphi_i(s) \Big/ \sum_{k=1}^{n} \varphi_k(s) \right\| \leqslant \sum_{i=1}^{n} \varphi_i(s) \Big/ \sum_{k=1}^{n} \varphi_k(s) = 1.$$

From the arbitrariness of ϵ we get that

$$\sup \int \varphi d\mu \geqslant \| \mu \| (S).$$

Combining this estimate with the inequality

$$\int \varphi d\mu \leqslant \int \| \varphi(s) \| \; \| \mu \| (ds),$$

we obtain the required result. Q.E.D.

Theorem 2.5. $C(S,B)' = M(S,B')$.

Proof. Propositions 2.3 and 2.4 imply that $M(S,B') \subset C(S,B)'$. Let us prove the converse inclusion. Suppose $F \in C(S,B)'$. Then there exists a $\psi \in \Psi(S)$ such that $|F(\varphi)| \leq p_{\psi\|\cdot\|}(\varphi)$. Let $K_\alpha = \{s : s \in S, \psi(s) \leq \alpha\}$, and let χ_α be its characteristic function. We extend F onto the space of all Baire functions $S \to B$ with preservation of the inequality $|F(\varphi)| \leq p_{\psi\|\cdot\|}(\varphi)$. Suppose $b \in B$ and $\varphi \in C(S, R^1)$. Then for fixed b and α the formula

$$F_b^\alpha(\varphi) = F(b\chi_\alpha\varphi)$$

defines a linear functional on $C(K_\alpha, R^1)$. From the inequalities

$$0 < \inf_{s\in S} \psi(s) \leqslant \psi(s) \leqslant \alpha,$$

which hold on K_α, it follows that F_b^α is continuous in the topology of uniform convergence. By the Riesz representation theorem there exists a unique charge μ_b^α that is concentrated on K_α and is such that

$$F_\beta^\alpha(\varphi) = \int \varphi d\mu_b^\alpha$$

for $\varphi \in C(K_\alpha, R^1)$. Clearly μ_b^α depends on b linearly. In addition,

$$|\mu_b^\alpha|(S) = \sup_{|\varphi|\leqslant 1} |F(b\chi_\alpha\varphi)|$$

$$\leqslant \|b\| \sup_{|\varphi|\leqslant 1} p_{\varphi\|\cdot\|} \ (b\chi_\alpha\varphi/\|b\|) = \|b\| \sup_{|\varphi|\leqslant 1}\sup_{s\in S} \chi_\alpha(s)\varphi(s)/\psi(s)$$

$$\leqslant \|b\| \sup_{s\in S} 1/\psi(s) = \|b\| / \inf_{s\in S} \psi(s),$$

i.e. the dependence of μ_b^α on b is continuous, and one can therefore write

$$\mu_b^\alpha(A) = \langle b, \mu^\alpha(A)\rangle, \quad \mu^\alpha(A) \in B'.$$

The countable additivity of μ_b^α for any b implies the countable additivity of μ^α in the weak topology. Further,

$$\|\mu^\alpha\|(S) = \sup_{\varphi} \int \varphi d\mu^\alpha = \sup_{\varphi} F(\varphi\chi_\alpha)$$

$$\leqslant \sup_{\varphi} p_{\varphi\|\cdot\|} \ (\varphi\chi_\alpha) \leqslant 1/\inf_{s\in S} \psi(s) < \infty.$$

Here the supremum is the same as in Proposition 2.4. Thus $\mu_b^\alpha \in M(S, B')$. The validity of the equality

$$\int \varphi d\mu^\alpha = F(\chi_\alpha\varphi)$$

for all $\varphi \in C(S, B)$ will follow if we prove that the set of functions with a finite-dimensional range is dense in $C(K_\alpha, B)$ in the topology of uniform convergence. Suppose $\varphi \in C(K_\alpha, B)$. Then $\varphi(K_\alpha)$ is a compactum. We take $\epsilon > 0$ and construct an ϵ-net $b_1, \cdots, b_n$ in $\varphi(K_\alpha)$. Let

$$\varphi_i(s) = \rho(s, \{s : s \in K_\alpha, \|\varphi(s) - b_i\| \geqq \epsilon\}) \quad \text{and} \quad \varphi_0(s) = \sum_{i=1}^n b_i\varphi_i(s) \Big/ \sum_{k=1}^n \varphi_k(s),$$

We will then have

$$\|\varphi(s) - \varphi_0(s)\| \leqslant \sum_{i=1}^n \varphi_i(s) \|\varphi(s) - b_i\| \Big/ \sum_{k=1}^n \varphi_k(s).$$

If $\varphi_i(s) \neq 0$, then $\|\varphi(s) - b_i\| < \epsilon$, and hence the last expression is less than

$$\varepsilon \sum_{i=1}^{n} \varphi_i(s) \Big/ \sum_{k=1}^{n} \varphi_k(s) = \varepsilon.$$

Thus the formula

$$\int \varphi d\mu^{\alpha} = F(\chi_{\alpha}\varphi)$$

has been proved for all $\varphi \in C(S, B)$. Further, for $\beta \geqq \alpha$

$$\|\mu^{\alpha} - \mu^{\beta}\|(S) = \sup_{\|\varphi\| \leqslant 1} \int \varphi d(\mu^{\alpha} - \mu^{\beta}) = \sup_{\|\varphi\| < 1} F(\varphi \cdot (\chi_{\alpha} - \chi_{\beta}))$$

$$\leqslant \sup_{\|\varphi\| \leqslant 1} \sup_{s \in S} \|\varphi(s)\| (\chi_{\alpha}(s) - \chi_{\beta}(s)) / \psi(s) \leqslant 1/\alpha,$$

which implies by virtue of Proposition 2.2 that μ^{α} has a limit for $\alpha \to \infty$ with respect to the norm $|||\nu||| = \|\nu\|(S)$ and that this limit μ belongs to $M(S, B')$. From the inequality

$$\left| \int \varphi d\nu \right| \leqslant \int \|\varphi\| d \|\nu\|$$

it follows that as $\alpha \to \infty$

$$\int \varphi d\mu^{\alpha} \to \int \varphi d\mu$$

for any $\varphi \in C(S, B)$. On the other hand, $p_{\psi\|\cdot\|}(\varphi - \chi_{\alpha}\varphi) \to 0$, and hence $F(\varphi\chi_{\alpha}) \to F(\varphi)$ as $\alpha \to \infty$ for any $\varphi \in C(S, B)$, which implies

$$F(\varphi) = \int \varphi d\mu$$

for $\varphi \in C(S, B)$. Q.E.D.

PROPOSITION 2.6. *The polar set of the neighborhood* $U = \{\varphi : \varphi \in C(S, B), \|\varphi(s)\| \leqq \psi(s) \ \forall s \in S\}$ *of* 0, *where* $\psi \in \Psi(S)$, *is*

$$\{\mu : \mu \in M(S, B'), \ \int \psi d \|\mu\| \leqslant 1\}.$$

PROOF. Suppose $\int \psi d\|\mu\| \leqq 1$. Then for any $\varphi \in U$

$$\left|\int \varphi d\mu\right| \leqslant \int \|\varphi\| d \|\mu\| \leqslant \int \psi d \|\mu\| \leqslant 1,$$

i.e. $\mu \in U^0$. Conversely, suppose $\int \psi d\|\mu\| > 1$. Then there exists a simple function η such that $\eta(s) \leqq \psi(s)$ and $\alpha = \int \eta d\|\mu\| > 1$. It can be assumed in this connection that η takes the value t_k ($t_k \geqq t_{k-1}$) on the set $C_k = \{s : s \in S, t_k < \psi(s) \leqq t_{k+1}\}$ for $k = 1, \cdots, n-1$ and the value t_n on the set $\{s : s \in S, t_n < \psi(s)\}$. We construct collections of sets $A_{ki} \in \mathfrak{B}$, $i = 1, \cdots, m_k$, such that

$$\bigcup_{i=1}^{m_k} A_{ki} = C_k, \qquad A_{ki} \cap A_{kj} = \varnothing \ \text{ for } i \neq j$$

and collections of elements $x_{ki} \in B$ such that

$$\sum_{i=1}^{m_k} \langle x_{ki}, \mu(A_{ki}) \rangle > \|\mu\|(C_k) - (\alpha - 1)/2nt_k .$$

Further, we construct open sets $G_{ki} \supset A_{ki}$ and compacta $K_{ki} \subset A_{ki}$ such that

$$G_{ki} \cap \left(\bigcup_{l=1}^{k-1} C_l \right) = \varnothing$$

$G_{ki} \cap K_{kj} = \emptyset$ for $i \neq j$ and

$$\|\mu\|(G_{ki} \setminus K_{ki}) < \frac{\alpha - 1}{2nm_k t_k},$$

and let $\varphi_{ki}(s) = \rho(s, S \setminus G_{ki})$. Then for $s \in C_p$

$$\left\| \frac{\sum_{k=1}^{n} \sum_{i=1}^{m_k} x_{ki}\varphi_{ki}(s)\, t_k}{\sum_{k=1}^{n} \sum_{i=1}^{m_k} \varphi_{ki}(s)} \right\| \leqslant \frac{t_p \sum_{k=1}^{p} \sum_{i=1}^{m_k} \|x_{ki}\|\, \varphi_{ki}(s)}{\sum_{k=1}^{p} \sum_{i=1}^{m_k} \varphi_{ki}(s)} \leqslant t_p,$$

i.e.

$$\left\| \frac{\sum_{k=1}^{n} \sum_{i=1}^{m_k} x_{ki}\varphi_{ki}(s)\, t_k}{\sum_{k=1}^{n} \sum_{i=1}^{m_k} \varphi_{ki}(s)} \right\| \leqslant \eta(s) \leqslant \psi(s) .$$

In addition,

$$\int \frac{x_{ki}\varphi_{ki}}{\sum_{k=1}^{n} \sum_{i=1}^{m_k} \varphi_{ki}}\, d\mu > \langle x_{ki}, \mu(A_{ki}) \rangle - \frac{\alpha - 1}{2nm_k t_k} .$$

Summing over i, we get

$$\int \frac{\sum_{i=1}^{m_k} x_{ki}\varphi_{ki}}{\sum_{k=1}^{n} \sum_{i=1}^{m_k} \varphi_{ki}}\, d\mu > \sum_{i=1}^{m_k} \langle x_{ki}, \mu(A_{ki}) \rangle - \frac{\alpha - 1}{2nt_k}$$

$$> \|\mu\|(C_k) - \frac{\alpha - 1}{2nt_k} - \frac{\alpha - 1}{2nt_k} = \|\mu\|(C_k) - \frac{\alpha - 1}{nt_k} .$$

Multiplying this equality by t_k and summing over k, we get

$$\int \frac{\sum_{k=1}^{n}\sum_{i=1}^{m_k} x_{ki}\varphi_{ki}t_k}{\sum_{k=1}^{n}\sum_{i=1}^{m_k}\varphi_{ki}}\, d\mu > \sum_{k=1}^{n} t_k \|\mu\|(C_k) - \alpha + 1 = 1,$$

i.e. μ does not belong to the polar set of U. Q.E.D.

PROPOSITION 2.7. *Suppose $P \subset M(S, B')$ and either B is complete or P is absolutely convex. Then in order for P to be compact in the topology*

$$\sigma(M(S, B'), C(S, B))$$

it is necessary and sufficient that the following three conditions be satisfied:

a) *P is closed in the topology $\sigma(M(S, B'), C(S, B))$.*

b) *There exists a $C \in R^1$ such that $\|\mu\|(S) \leqq C$ for all $\mu \in P$.*

c) *For any $\epsilon > 0$ there exists a compactum $K \subset S$ such that $\|\mu\|(S \backslash K) \leqq \epsilon$ for all $\mu \in P$.*

PROOF. NECESSITY. The necessity of a) is obvious. We equip $C(S, B)$ with the norm $\|\varphi\| = \sup_{s \in S} \|\varphi(s)\|$. Then $C(S, B)$ becomes a normed space (Banach space if B is complete) and $M(S, B')$ becomes a subspace of its dual. From Proposition 2.4 it follows that the norm in $M(S, B')$ coincides with $\|\mu\|(S)$. But a weakly compact set in the dual of a Banach space is strongly bounded and an absolutely convex weakly compact set in the dual of a locally convex space is strongly bounded. Thus b) is proved.

Condition c) implies the condition

c$'$) *For any $\epsilon > 0$ and $\sigma > 0$ there exists a compactum K such that*

$$\|\mu\|(S \backslash K^{(\sigma)}) \leqq \epsilon$$

for all $\mu \in P$, where $K^{(\sigma)}$ denotes the σ-neighborhood of K.

Conversely, suppose condition c$'$) is satisfied and K_n is a compactum such that

$$\|\mu\|(S \backslash K_n^{(n-1)}) \leqq \epsilon/2^n$$

for all $\mu \in P$. Then

$$\bigcap_{n=1}^{\infty} \overline{K_n^{(n-1)}}$$

will be a compactum satisfying the requirement in c); i.e. c) and c$'$) are equivalent.

Suppose c) is not satisfied. Then an $\epsilon > 0$ and a $\sigma > 0$ exist such that for any compactum K there exists a $\mu \in P$ such that $\|\mu\|(S \backslash K^{(\sigma)}) > \epsilon$. Let $K_0 = \emptyset$, $\mu_0 = 0 \in M(S, B')$ and $\varphi_0 = 0 \in C(S, B)$. Suppose constructed for $i = 1, \cdots, n$ compacta K_i and elements $\mu_i \in P$, $\varphi_i \in C(S, B)$ with $\|\varphi_i\| \leqq 1$

such that $\rho(K_i, K_j) \geqq \sigma$ for $i \neq j$,

$$\|\mu_i\|(S \setminus \bigcup_{j=1}^{i-1} K_j^{(\sigma)}) > \varepsilon, \qquad \|\mu_i\|(S \setminus (\bigcup_{j=1}^{i-1} K_j^{(\sigma)} \cup K_i)) \leqslant \varepsilon/4;$$

$\|\mu_i\|(K_i) > \epsilon$, $\varphi_i = 0$ outside $K_i^{(\sigma/3)}$,

$$\int \varphi_i d\mu_i > \varepsilon. \text{ and } \int_{\bigcup_{l=1}^{j-2} K_l^{(\sigma)}} \|\varphi_j(s)\| \, \|\mu_i\|(ds) \leqslant \varepsilon\, 2^{i-j-2} \text{ for } j \geqslant i \geqslant 1.$$

Then there exists an element $\mu_{n+1} \in P$ such that

$$\|\mu_{n+1}\| \, (S \setminus \bigcup_{i=1}^{n} K_i^{(\sigma)}) > \varepsilon .$$

We construct with respect to μ_{n+1} a compactum

$$K_{n+1} \subset S \setminus \bigcup_{i=1}^{n} K_i^{(\sigma)}$$

such that $\|\mu_{n+1}\|(K_{n+1}) > \epsilon$ and

$$\|\mu_{n+1}\|(S \setminus (\bigcup_{i-1}^{n} K_i^{(\sigma)} \cup K_{n+1})) < \frac{\varepsilon}{4} .$$

We construct with respect to μ_{n+1} and K_{n+1} an element $\varphi_{n+1} \in C(S, B)$ that vanishes outside $K_{n+1}^{(\sigma/3)}$ and is such that $|||\varphi_{n+1}||| \leqq 1$,

$$\int \varphi_{n+1} d\mu_{n+1} > \varepsilon \quad \text{and} \quad \int_{\bigcup_{l=1}^{n-1} K_l^{(\sigma)}} \|\varphi_{n+1}\| d\|\mu_k\| \leqslant \varepsilon 2^{k-n-3} .$$

Thus we obtain a sequence of measures μ_n and functions φ_n, $n = 1, 2, \cdots$, satisfying the following condition: for any sequence of integers $\alpha_k = 0, +1$ or -1

$$\sum_{k=1}^{\infty} \alpha_k \varphi_k \in C(S, B) \quad \text{and} \quad \Big|\Big|\Big| \sum_{k=1}^{\infty} \alpha_k \varphi_k \Big|\Big|\Big| \leqslant 1,$$

$$\int \varphi_k d\mu_k > \varepsilon \quad \text{and} \quad \Big| \int \sum_{k=n+1}^{\infty} \alpha_k \varphi_k d\mu_n \Big| \leqslant \varepsilon/2 .$$

Suppose that the sequence $\{\mu_n\}$ has a limit point $\mu \in M(S, B')$. We choose N so large that

$$\int \sum_{k=N+1}^{\infty} \|\varphi_k(s)\| \, \|\mu\|(ds) \leqslant \varepsilon / 4.$$

We set $\alpha_{N+1} = 1$ and, having defined $\alpha_{N+1}, \cdots, \alpha_{N+n-1}$, we choose α_{N+n} so that $|\alpha_{N+n}| = 1$ and

$$\alpha_{N+n} \int \varphi_{N+n} d\mu_{N+n}$$

has the same sign as

$$\int \sum_{k=N+1}^{N+n-1} \alpha_k \varphi_k d\mu_{N+n}.$$

Then

$$\left| \int \sum_{k=N+1}^{\infty} \alpha_k \varphi_k d\mu \right| \leqslant \int \sum_{k=N+1}^{\infty} \| \varphi_k(s) \| \, \| \mu \| (ds) \leqslant \varepsilon / 4,$$

whereas

$$\left| \int \sum_{k=N+1}^{\infty} \alpha_k \varphi_k d\mu_{N+p} \right| \geqslant \left| \int \sum_{k=N+1}^{N+p-1} \alpha_k \varphi_k d\mu_{N+p} \right|$$

$$+ \left| \int \alpha_{N+p} \varphi_{N+p} d\mu_{N+p} \right| - \left| \int \sum_{k=N+p+1}^{\infty} \alpha_k \varphi_k d\mu_{N+p} \right|$$

$$\geqslant 0 + \varepsilon - \varepsilon / 2 = \varepsilon / 2.$$

Thus the neighborhood

$$\left\{ \nu : \nu \in M(S, B'), \left| \int \sum_{k=N+1}^{\infty} \alpha_k \varphi_k d(\mu - \nu) \right| \leqslant \varepsilon / 4 \right\}$$

of μ does not contain any terms of the sequence $\{\mu_n\}$ with index greater than N; i.e. the sequence $\{\mu_n\}$ has no limit points and P is not even countably compact.

SUFFICIENCY. If we prove that conditions b) and c) imply $P \subset U^0$, where U is a neighborhood of 0 in $C(S, B)$, the assertion will follow from the Tihonov-Alaoglu theorem.

Let K_n be a compactum such that $\|\mu\|(S \setminus K_n) \leqq 2^{-n}$ for all $\mu \in P$, let χ_n be the characteristic function of its complement and let $\psi = 1/2C + 2^{-1}\sum_1^\infty \chi_n$. Clearly $\psi \in \Psi(S)$ and

$$\int \psi d \|\mu\| = 2^{-1} \int \sum_{n=1}^{\infty} \chi_n \, d \|\mu\| + \|\mu\|(S)/2C \leqslant 2^{-1} \sum_{n=1}^{\infty} 2^{-n} + 2^{-1} = 1,$$

for any $\mu \in P$; i.e. $P \subset \{\varphi : \varphi \in C(S, B), \|\varphi(s)\| \leqq \psi(s)\}^0$. Q.E.D.

COROLLARY 2.8. *The topology of $C(S,B)$ is the Mackey topology*

$$\tau(C(S,B), M(S,B')).$$

Let us return to the space $C(S,E)$, where E is a locally convex space. It is known that E can be represented as a projective limit of spaces $E_q = E/q^{-1}(0)$ with norm $\tilde{q}(l + q^{-1}(0)) = q(l)$, where q is a continuous seminorm on E (see [4]). It follows that E' can be represented as an inductive limit of the family of spaces E_q' in the category of linear spaces. But the spaces E_q' have the structure of Banach (and hence of pseudotopological linear) spaces. One can therefore put $E' = \lim \operatorname{ind} E_q'$, where the inductive limit is taken in the category of pseudotopological linear spaces (see [5]). In other words, a filter converges in E' if and only if it is the image of a convergent filter under an inclusion $i_q: E'_q \to E'$.

Let $\mu: \mathfrak{B} \to E'$. We will say that μ is of bounded variation if it is representable as a composition of the form $\mu = i_q \circ \nu$, where $\nu: \mathfrak{B} \to E'_q$ is of bounded variation. We will write $\|\mu\|_q(A)$ to denote the variation of ν on a set A.

PROPOSITION 2.9. *Suppose $\mu: \mathfrak{B} \to E'$ is countably additive in the topology $\sigma(E', E)$ and of bounded variation. Then μ is countably additive in the pseudotopology introduced above.*

This proposition follows directly from Proposition 2.1. We note that this is a quite strong result since, as is easily seen, the constructed pseudotopology is not weaker than the (pseudo-) topology $\beta(E', E)$.

THEOREM 2.10. *The dual of $C(S, E)$ is the space $M(S, E')$ of all set functions $\mathfrak{B} \to E'$ of bounded variation that are countably additive in the topology $\sigma(E', E)$.*

PROOF. Suppose $F \in C(S, E)'$. Then there exist a $\psi \in \Psi(S)$ and a continuous seminorm q on E such that $|F(\varphi)| \leq p_{\psi q}(\varphi)$. The natural projection $E \to E_q$ defines a mapping $\kappa: C(S, E) \to C(S, E_q)$ whose kernel coincides with $C(S, q^{-1}(0))$. Clearly, F can be represented in the form $F(\varphi) = \widetilde{F}(\kappa\varphi)$, where $\widetilde{F} \in C(S, E_q)'$. According to Theorem 2.5 there exists a $\nu \in M(S, E'_q)$ such that $\widetilde{F}(\varphi) = \int \varphi d\mu$. Setting, by definition, $\int \kappa\varphi d\nu = \int \varphi d(i_q \circ \nu)$, we get $F(\varphi) = \int \varphi d\mu$, where $\mu = i_q \circ \nu \in M(S, E')$. Thus every element of $C(S, E)'$ can be identified with some $\mu \in M(S, E')$.

Conversely, if $\mu = i_q \circ \nu$ and $\nu \in M(S, E'_q)$ then ν defines a continuous linear functional $\widetilde{F}$ on $C(S, E_q)$, and, by setting $F(\varphi) = \widetilde{F}(\kappa\varphi)$, we obtain a functional $F \in C(S, E)'$.

Here a nonzero measure defines a nonzero functional.

REMARK 2.11. All of the results of this section except Proposition 2.7 and Corollary 2.8 remain valid in the case when S is an arbitrary normal topological space if $C(S, E)$ is defined as in Remark 1.7 and $M(S, E')$ is assumed to consist of all set functions $\mu: \mathfrak{B} \to E'$ satisfying the following conditions:

a) μ is countably additive in the topology $\sigma(E', E)$.

b) There exists a continuous seminorm q on E such that $\|\mu\|_q(S) < \infty$.

c) For any $\epsilon > 0$ there exists a compactum $K \subset S$ such that $\|\mu\|_q(S \backslash K) < \epsilon$.

d) $\|\mu\|_q$ is regular.

The proofs of Propositions 2.3 and 2.4 and Theorem 2.5 require slight modifications in this connection.

In addition to the case when S is complete separable metric space there are other cases for which (a) $\wedge$ b)) $\Rightarrow$ (c) $\wedge$ d)). Suppose there exists a metric on S that converts it into a complete separable metric space and defines a topology stronger than the original topology but inducing the same σ-algebra of Borel sets (for example, S is a separable Hilbert space equipped with the weak topology). Then (a) $\wedge$ b)) $\Rightarrow$ (c) $\wedge$ d)). In fact, for any $\epsilon > 0$ and $A \in \mathfrak{B}$ there then exists a set $K \subset A$ that is compact in the topology given by the metric and such that $\|\mu\|_q(A \setminus K) < \epsilon$. But K will also be compact in the weaker original topology.

A more general case is obtained by taking $S = \bigcup_1^\infty S_n$, where each S_n is a member of $\mathfrak{B}$, is a subset of S_{n+1} and satisfies the conditions of the preceding case. For any $\epsilon > 0$ and $A \in \mathfrak{B}$ there then exist a positive integer n and a compactum $K \subset A \cap S_n$ such that $\|\mu\|_q(A \setminus S_n) < \epsilon/2$ and $\|\mu\|_q((A \cap S_n)\setminus K) < \epsilon/2$, and hence $\|\mu\|_q(A \setminus K) < \epsilon$. This second case is important since it includes rigged Hilbert spaces [**20**].

§3. The spaces $\mathscr{D}(H)$ and $\mathscr{D}'(H)$

In this section we will consider in place of an arbitrary S a separable Hilbert space H. Let H^k denote the complete projective tensor product of k copies of H. It is known [**4**] that the dual of H^k is the space $\mathfrak{A}_k$ of all k-linear forms on $H \times \cdots \times H$ (k factors) that are continuous in the totality of variables with norm $\|a\| = \sup_{\|x_i\| \leq 1} a(x_1, \cdots, x_k)$. We equip $\mathfrak{A}_k$ with a locally convex topology which we require to be not weaker than the topology $\sigma(\mathfrak{A}_k, H^k)$ and not stronger than the topology $\beta(\mathfrak{A}_k, H^k)$ induced by the above norm. We note that the class of bounded sets is the same in all of these topologies. Everywhere below, unless otherwise stipulated, by a derivative we will mean a derivative with respect to a system of finite sets (see [**2**]).

We consider the k-multiple differentiation operator as an operator D^k: $C(H, R^1) \to C(H, \mathfrak{A}_k)$. We denote the intersection of the domains of definition of the operators D^l, $l = 0, 1, \cdots, k$, by $C^k(H, R^1)$ and equip it with the weakest topology in which these operators are continuous. Here and below we let $\mathfrak{A}_0 = H^0 = R^1$, $C^0(H, R^1) = C(H, R^1)$ and D^0 be the identity operator.

Let $C^\infty(H, R^1) = \lim\operatorname{proj} C^k(H, R^1)$.

PROPOSITION 3.1. *$C^\infty(H, R^1)$ is dense in $C(H, R^1)$.*

PROOF. Let $\{P_n\}$ be a sequence of orthogonal projections in H that strongly converges to the identity operator, with $P_{n+1} \geqq P_n$ and the range of P_n being n-dimensional. Then for any function $\varphi \in C(H, R^1)$ the sequence $\{\varphi_n(x)\} = \{\varphi(P_n, x)\}$ is bounded and converges uniformly on each compactum, and hence (by virtue of Proposition 1.2 or 1.3) converges in $C(H, R^1)$. We approximate φ_n uniformly on the ball $\mathscr{S}_n = \{x : x \in H, \|x\| \leqq n\}$ to within n^{-1} by an infinitely differentiable function $\tilde{\varphi}_n$. This can always be done by setting $\tilde{\varphi}_n(x) = \int_{R^n = P_n H} \varphi_n(y)\alpha(P_n x - y)\,dy$, where α is a suitable infinitely differentiable nonnegative function with compact support on R^n and such that $\int \alpha(x)\,dx = 1$. In this connection $\sup_{x \in H} |\tilde{\varphi}_n(x)| \leqq \sup_{x \in H} |\varphi_n(x)| \leqq \sup_{x \in H} |\varphi(x)|$. It is obvious that $\tilde{\varphi}_n$ also converges to φ uniformly on each compactum and hence in $C(H, R^1)$.

The following proposition readily follows from Proposition 1.1.

PROPOSITION 3.2. *If all of the* $\mathfrak{A}_k$ *are complete, so is* $C^\infty(H, R^1)$.

PROPOSITION 3.3. *The class of functions in* $C^\infty(H, R^1)$ *does not depend on the topologies in the* $\mathfrak{A}_k$; *it consists of the infinitely Fréchet differentiable functions all of whose derivatives are bounded. In other words, a derivative* $D^{k+1}\varphi$ *can be defined as the Fréchet derivative of a mapping* $D^k\varphi$ *acting from* H *into the space* $\mathfrak{A}_k$ *equipped with the norm* $\|a\| = \sup_{\|x_i\| \leq 1} \alpha(x_1, \cdots, x_k)$.

PROOF. The boundedness of $D^{k+1}\varphi$ in the topology $\sigma(\mathfrak{A}_{k+1} H^{k+1})$ implies its boundedness in the topology $\beta(\mathfrak{A}_{k+1} H^{k+1})$. Hence $D^k\varphi$, as a mapping of H into the space $\mathfrak{A}_k$ equipped with a norm, satisfies the Lipschitz condition and therefore has a Fréchet derivative (see [**16**]). Q.E.D.

THEOREM 3.4. *Every continuous linear functional on* $C^\infty(H, R^1)$ *can be represented in the form* $F(\varphi) = \sum_0^n \int \varphi^{(k)} d\mu_k$, *where* $\mu_k \in M(H, \mathfrak{A}_k')$.

PROOF. We construct an imbedding of $C^\infty(H, R^1)$ into $\prod_{k=0}^\infty C(H, \mathfrak{A}_k)$ according to the formula $\varphi \mapsto (\varphi^{(0)}, \varphi^{(1)}, \cdots)$. It is obvious that this imbedding is a linear homeomorphism of $C^\infty(H, R^1)$ onto a subspace of $\prod_{k=0}^\infty C(H, \mathfrak{A}_k)$. Hence a continuous linear functional on $C^\infty(H, R^1)$ can be extended to a continuous linear functional on $\prod_{k=0}^\infty C(H, \mathfrak{A}_k)$. Such an extension has the form $F(\varphi) = \sum_0^n F_k(\varphi_k)$, where $\varphi = (\varphi_0, \varphi_1, \cdots)$ and $F_k \in C(H, \mathfrak{A}_k)'$. Its restriction to $C^\infty(H, R^1)$ has the form required in the theorem. Q.E.D.

Let $\mathscr{S}_r$ denote the ball $\{x : x \in H, \|x\| \leq r\}$ and let $C_{0r}^\infty(H, R^1)$ denote the subspace of $C^\infty(H, R^1)$ consisting of those functions that vanish outside $\mathscr{S}_r$. Clearly $C_{0r}^\infty(H, R^1)$ is closed in $C^\infty(H, R^1)$ and therefore complete if all of the $\mathfrak{A}_k$ are complete. Let $\mathscr{D}(H) = \lim\operatorname{ind} C_{0n}^\infty(H, R^1)$. Propositions 3.2 and 3.3 readily imply

PROPOSITION 3.5. *Independently of the topologies in the* $\mathfrak{A}_k$, *the space* $\mathscr{D}(H)$ *consists of all of the infinitely Fréchet differentiable functions with bounded support all of whose derivatives are bounded. It is complete if all of the* $\mathfrak{A}_k$ *are complete.*

THEOREM 3.6. *Every continuous linear functional on* $\mathscr{D}(H)$ *can be represented in the form* $F(\varphi) = \sum_0^\infty \int \varphi^{(k)} d\mu_k$, *where* μ_k *is a set function defined on the* δ*-ring of bounded Borel sets that takes its values in* $\mathfrak{A}_k'$, *is of bounded variation on any ball* $\mathscr{S}_r \subset H$ *and is countably additive in the topology* $\sigma(\mathfrak{A}_k', \mathfrak{A}_k)$. *Moreover, any ball is a* μ_k*-null set for all* k *except a finite number.*

PROOF. A linear functional F is continuous on $\mathscr{D}(H)$ if and only if its restrictions F_n to the subspaces $C_{0n}^\infty(H, R^1)$ are continuous. Each functional F_n can be represented in the form $F_n(\varphi) = \sum_0^\infty \int \varphi^{(k)} d\mu_k^n$, where $\mu_k^n \in M(H, \mathfrak{A}_k')$ and for fixed n only a finite number of the μ_k^n are different from zero. Let $\alpha_n \in C_{0,n-1}^\infty(H, R^1)$ and $\alpha_n = 1$ on $\mathscr{S}_{n-2}$. We have

$$F(\varphi) = F(\alpha_n \varphi) + F((1 - \alpha_n)\varphi) = F_{n-1}(\alpha_n \varphi) + F((1 - \alpha_n)\varphi);$$

$$F_{n-1}(\alpha_n\varphi) = \sum_{k=0}^{\infty} \int_{\mathcal{S}_{n-1}} (\alpha_n\varphi)^{(k)}$$

$$d\mu_k^{n-1} = \sum_{k=0}^{\infty} \int_{\mathcal{S}_{n-2}} \varphi^{(k)}\, d\mu_k^{n-1} + \sum_{k=0}^{\infty} \int_{\mathcal{S}_n \setminus \mathcal{S}_{n-2}} (\alpha_n\varphi)^{(k)}\, d\mu_k^{n-1}.$$

The second sum is a continuous linear functional on $C_{0,n-1}^{\infty}(H, R^1)$ and hence can be represented in the form

$$\sum_{k=0}^{\infty} \int_{\mathcal{S}_n \setminus \mathcal{S}_{n-2}} \varphi^{(k)}\, d\tilde{\mu}_k^{n-1},$$

where only a finite number of the $\tilde{\mu}_k^{n-1}$ are different from zero. On the other hand,

$$F_n((1-\alpha_n)\varphi) = \sum_{k=0}^{\infty} \int_{\mathcal{S}_n \setminus \mathcal{S}_{n-2}} ((1-\alpha_n)\varphi)^{(k)}\, d\mu_k^n = \sum_{k=0}^{\infty} \int_{\mathcal{S}_n \setminus \mathcal{S}_{n-2}} \varphi^{(k)}\, d\tilde{\tilde{\mu}}_k^n.$$

Combining the obtained expressions, we get

$$F_n(\varphi) = \sum_{k=0}^{\infty} \int_{\mathcal{S}_{n-2}} \varphi^{(k)}\, d\mu_k^{n-1} + \sum_{k=0}^{\infty} \int_{\mathcal{S}_n \setminus \mathcal{S}_{n-2}} \varphi^{(k)}\, d(\tilde{\mu}_k^{n-1} + \tilde{\tilde{\mu}}_k^n).$$

It can be seen from this equality that, by redefining the μ_k^n, we can assume that μ_k^n and μ_k^{n-1} coincide on all of the Borel subsets of $\mathcal{S}_{n-2}$. If the double sequence $\{\mu_k^n\}$ satisfies this condition, we can put $\mu_k(A) = \lim_{n\to\infty} \mu_k^n(A) = \mu_k^{n+2}(A)$ for any Borel set $A \subset \mathcal{S}_n$. If $\varphi \in C_{0n}^{\infty}(H, R^1)$, then

$$F(\varphi) = F_{n+2}(\varphi) = \sum_{k=0}^{\infty} \int_{\mathcal{S}_n} \varphi^{(k)}\, d\mu_k^{n+2} = \sum_{k=0}^{\infty} \int \varphi^{(k)}\, d\mu_k.$$

REMARK 3.7. From Remark 1.5 it follows that when $H = R^n$ the space $\mathcal{D}(H)$ constructed by us coincides with the space $\mathcal{D}$ of L. Schwartz [7].

REMARK 3.8. All the results of this section can be extended to the case when H is a separable Fréchet space whose metric is infinitely differentiable for noncoincident arguments.

REMARK 3.9. The following modification of our construction can turn out to be useful: replace the Hilbert space H in the definitions of H^k and $\mathfrak{A}_k$ by a Hilbert space $H_1 \subset H$ and regard the differentiation operators as defined in the sense of H_1 differentiation (see [2]).

§4. Differential equations in the space $\mathcal{D}'(H)$

Let $a_k \in \mathfrak{A}_k'$, $k = 0, 1, \cdots, m$. We will call a symbol of the form $L(D) = \sum_0^m a_k D^k$ a differential operator with constant coefficients, and we set $(L(D)\varphi)(x) = \sum_0^m \langle a_k, \varphi^{(k)}(x)\rangle$. Clearly $L(D)$ maps $\mathcal{D}(H)$ continuously into itself. Its adjoint operator $L'(D)$ acts in the space $\mathcal{D}'(H)$.

It is natural to consider an equation of the form $L'(D)\mathcal{F} = \mathcal{G}$, where $\mathcal{F}$

and $\mathscr{G}$ are functionals belonging to $\mathscr{D}'(H)$ and the functional $\mathscr{G}$ is assumed to be known. As in the finite-dimensional case, a particularly important role is played by the equation $L'(D)\mathscr{E} = \delta$, where $\delta \in \mathscr{D}'(H)$ is defined by the equality $\delta(\varphi) = \varphi(0)$. The solutions of this equation will be called the fundamental solutions of the operator $L'(D)$.

We introduce the operation of convolution. Let $\mathscr{F} \in \mathscr{D}'(H)$, $\varphi \in \mathscr{D}(H)$ and $\varphi_x(y) = \varphi(x+y)$. Then the function $\mathscr{F} * \varphi = \mathscr{F}(\varphi_x)$ will be called the convolution of the functional $\mathscr{F}$ and the test function φ.

PROPOSITION 4.1. *For any $\mathscr{F} \in \mathscr{D}'(H)$ and $\varphi \in \mathscr{D}(H)$ the convolution $\mathscr{F} * \varphi$ is infinitely differentiable and*

$$\langle h_1 \otimes \ldots \otimes h_k, \quad D^k(\mathcal{F} * \varphi)(x)\rangle = (\mathcal{F} * \langle h_1 \otimes \ldots \otimes h_k D^k_\varphi\rangle)(x).$$

PROOF. It suffices to prove the assertion for $k = 1$. The rest of the proof will then follow by induction on k,

$$\langle h, D(\mathcal{F} * \varphi)(x)\rangle = \lim_{t\to 0}[(\mathcal{F} * \varphi)(x + th)$$

$$-(\mathcal{F} * \varphi)(x)]/t = \lim_{t\to 0}\mathcal{F}((\varphi_{x+th} - \varphi_x)/t)$$

$$(\varphi_{x+th}(y) - \varphi_x(y))/t = \langle h, D\varphi(x + y + \theta h\rangle, \text{ where } |\theta| \leqq |t|.$$

Since $D\varphi$ is continuous, $(\varphi_{x+th} - \varphi_x)/t \to \langle h, D\varphi_x\rangle$ uniformly on each compactum as $t \to 0$. For $|t| \leqq 1$ these functions have a common bounded support and are uniformly bounded. It can readily be seen that analogous assertions hold for all of the derivatives of these functions. We therefore have convergence in $\mathscr{D}(H)$, which implies

$$\langle h, D(\mathcal{F} * \varphi)(x)\rangle = \mathcal{F}(\langle h, D\varphi_x\rangle) = (\mathcal{F} * \langle h, D\varphi\rangle)(x). \qquad \text{Q.E.D.}$$

COROLLARY 4.2. *Suppose given a differential operator of the form $L(D) = \sum_0^m a_k D^k$, where $a_k \in H^k$. Then*

$$L(D)(\mathcal{F} * \varphi) = \mathcal{F} * (L(D)\varphi).$$

PROOF. Clearly it suffices to prove the assertion for $a_k D^k$. We make the substitution $a_k = \sum_1^\infty \lambda_l x_1^l \otimes \cdots \otimes x_k^l$, where $\lambda_l \geqq 0$, $\sum_1^\infty \lambda_l < \infty$ and $\|x_i^l\| = 1$. Then

$$\left\langle \sum_{l=1}^n \lambda_l x_1^l \otimes \ldots \otimes x_k^l, \quad D^k(\mathcal{F} * \varphi)\right\rangle = \mathcal{F} * \left\langle \sum_{l=1}^n \lambda_l x_1^l \otimes \ldots \otimes x_k^l, \quad D^k\varphi\right\rangle.$$

By virtue of the boundedness of all of the derivatives of φ,

$$\left\langle \sum_{l=1}^n \lambda_l x_1^l \otimes \ldots \otimes x_k^l, \quad D^k\varphi\right\rangle \underset{n\to\infty}{\to} \langle a_k, \quad D^k\varphi\rangle$$

uniformly together with all of its derivatives. In addition, the terms of the sequence have a common bounded support. We therefore have convergence in the topology of $\mathscr{D}(H)$, and hence

$$\mathcal{F} * \left\langle \sum_{l=1}^{n} \lambda_l x_1^l \otimes \ldots \otimes x_k^l, \quad D^k\varphi \right\rangle \underset{n\to\infty}{\to} \mathcal{F} * \langle a_k, D^k\varphi\rangle$$

at each point. On the other hand, at each point

$$\left\langle \sum_{l=1}^{n} \lambda_l x_1^l \otimes \ldots \otimes x_k^l, \quad D^k(\mathcal{F} * \varphi) \right\rangle \to \langle a_k, \quad D^k(\mathcal{F} * \varphi)\rangle. \quad \text{Q.E.D.}$$

We will assume everywhere below that the topology in $\mathfrak{A}_k$ is the topology of uniform convergence on a system $\mathfrak{M}_k$ of bounded subsets of H^k satisfying

CONDITION 4.3. *If $m_k \in \mathfrak{M}_k$ and $m_l \in \mathfrak{M}_l$, then the closed absolutely convex hull of the set $m_k \otimes m_l \subset H^k \otimes H^l = H^{k+l}$ belongs to $\mathfrak{M}_{k+l}$.*

For example, $\mathfrak{M}_k$ is the system of closed balls with center at 0 or the system of absolutely convex compacta.

PROPOSITION 4.4. *Let $\mathscr{F} \in C^\infty(H, R^1)$. Then the operator $\varphi \mapsto \mathscr{F} * \varphi$ continuously maps $\mathscr{D}(H)$ into $C^\infty(H, R^1)$.*

PROOF. According to Theorem 3.4 the functional $\mathscr{F}$ admits the representation $\mathscr{F}(\varphi) = \sum_0^n \int \varphi^{(k)} d\mu_k$, where $\mu_k \in M(H, \mathfrak{A}_k')$. It suffices to prove the continuity on each $C_{0r}^\infty(H, R^1)$ of each of the operators $\varphi \mapsto \mathscr{F}_k * \varphi$, where $\mathscr{F}_k \varphi = \int \varphi^{(k)} d\mu_k$. Let $\psi \in \Psi(H)$. We choose an $m_k \in \mathfrak{M}_k$ such that μ_k is of bounded variation relative to the seminorm $q_k(x) = \sup_{y \in m_k} |\langle x, y\rangle|$, defined on $\mathfrak{A}_k$. We next choose a $\psi_1 \in \Psi(H)$ such that $\int \psi_1 d\|\mu_k\|_{q_k} = 1$, and put $\psi_2(x) = \inf_{y\in H} \psi(y)\psi_1(x-y)$. The fact that $\psi_2 \in \Psi(H)$ will be proved in Lemma 4.6. We have

$$\begin{aligned}\langle h_1 \otimes \ldots \otimes h_l, \quad D^l(\mathcal{F}_k * \varphi)(x)\rangle &= (\mathcal{F}_k * \langle h_1 \otimes \ldots \otimes h_l, D^l\varphi\rangle)(x) \\ &= \int \langle h_1 \otimes \ldots \otimes h_l, \quad D^{l+k}\varphi(x+y)\rangle\, \mu_k(dy).\end{aligned}$$

We now take an arbitrary set $m_l \in \mathfrak{M}_l$ and let

$$q_l(x) = \sup_{y \in m_l} |\langle x, y\rangle| \quad \text{and} \quad q_{l+k}(x) = \sup_{y\in m_l \otimes m_k} |\langle x, y\rangle|$$

(q_l is defined on $\mathfrak{A}_l$ while q_{l+k} is defined on $\mathfrak{A}_{l+k}$). Suppose

$$q_k(\langle h^l, D^{l+k}\varphi(x)\rangle) \leqq \psi_2(x)$$

for all $x \in H$ and fixed $h^l \in H^l$. Then

$$\begin{aligned}|\langle h^l, D^l(\mathcal{F}_k * \varphi)(x)\rangle| &= \left|\int \langle h^l, D^{l+k}\varphi(x+y)\rangle \mu_k(dy)\right. \\ &\leqslant \int q_k\left(\left\langle h^l, \quad D^{l+k}\varphi(x+y)\right\rangle\right) \|\mu_k\| q_k(dy)\end{aligned}$$

$$\leqslant \int \psi_2(x+y)\, \|\mu_k\|_{q_k}(dy) \leqslant \int \psi(x)\, \psi_1(y)\, \|\mu_k\|_{q_k}(dy) = \psi(x).$$

And if

$$\sup_{h^l \in H^l} q_k\left(\left\langle h^l, D^{k+l}\varphi(x)\right\rangle\right) = q_{l+k}(D^{l+k}\varphi(x)) \leqslant \psi_2(x),$$

it follows by taking the supremum over all $h^l \in m_l$ that $q_l(D^l(\mathscr{F}_k * \varphi)(x)) \leqq \psi(x)$, which by virtue of the arbitrariness of ψ, l, q_l and k implies the continuity of the mapping $\varphi \mapsto \mathscr{F}_k * \varphi$ of $\mathscr{D}(H)$ into $C^\infty(H, R^1)$. Q.E.D.

We note that this mapping remains continuous if $\mathscr{D}(H)$ is equipped with the topology induced in it by the topology of $C^\infty(H, R^1)$. On the other hand, $\mathscr{D}(H)$ is dense in $C^\infty(H, R^1)$. For a proof we take an arbitrary $\varphi \in C^\infty(H, R^1)$ and $\alpha \in \mathscr{D}(H)$ such that $\alpha(x) = 1$ for $\|x\| \leqq 1$ and consider the sequence $\{\varphi_n\} = \{\alpha_n \varphi\}$, where $\alpha_n(x) = \alpha(x/n)$. This sequence together with all of its derivatives is bounded and converges to φ, uniformly on each bounded set and hence on each compactum. Consequently $\varphi_n \to \varphi$ in $C^\infty(H, R^1)$. This implies

PROPOSITION 4.5. *Under the conditions of Proposition 4.4 the mapping $\varphi \mapsto \mathscr{F} * \varphi$ can be uniquely extended to a continuous mapping of $C^\infty(H, R^1)$ into $C^\infty(H, R^1)$.*

LEMMA 4.6. *Let $\psi_1, \psi_2 \in \Psi(H)$ and $\psi_3(x) = \inf_{y \in H} \psi_1(y)\psi_2(x - y)$. Then $\psi_3 \in \Psi(H)$.*

PROOF. Clearly, $\psi_3(x) \geqq \inf_{y \in H} \psi_1(y) \inf_{z \in H} \psi_2(z)$. Therefore condition a) of §1 is satisfied. We introduce the operator $P(x, y) = x + y$ acting from $H \times H$ into H and the function $\psi_4(x, y) = \psi_1(x)\psi_2(y)$. If it is proved that $\psi_4 \in \Psi(H \times H)$, the set

$$\{x : x \in H, \psi_3(x) \leqq C\} = P\{(x, y) : (x, y) \in H \times H, \psi_4(x, y) \leqq C\}$$

will be compact by virtue of the continuity of P. It therefore suffices to prove the compactness of the set $B_C = \{(x, y) : (x, y) \in H \times H, \psi_4(x, y) \leqq C\}$. Clearly,

$$B_C \subset \{x : x \in H,\ \psi_1(x) \leqslant C/\inf_{y \in H} \psi_2(y)\} \times \{y : y \in H,\ \psi_2(y) \leqslant C/\inf_{x \in H} \psi_1(x)\}.$$

Therefore B_C is relatively compact. On the other hand, the set $H \times H \backslash B_C = \{(x, y) : (x, y) \in H \times H, \psi_4(x, y) > C\} = \bigcup_{\alpha \in (0, \infty)} \{x : x \in H, \psi_1(x) > \alpha\} \times \{y : y \in H, \psi_2(y) > C/\alpha\}$ is open, and hence B_C is closed and compact. Q.E.D.

We will say that a functional $\mathscr{F} \in \mathscr{D}'(H)$ (i) vanishes on an open set $U \subset H$ if $\mathscr{F}(\varphi) = 0$ for every $\varphi \in \mathscr{D}(H)$ with support contained in U, and (ii) has bounded support if it vanishes on the complement of some ball. If a functional $\mathscr{F}$ vanishes on the complement of a ball $\mathscr{S}_{r-\epsilon} = \{x : x \in H, \|x\| \leqq r - \epsilon\}$, it naturally extends to the space of all functions coinciding on $\mathscr{S}_r$ with the functions of $C^\infty(H, R^1)$. In particular, every functional with bounded support belongs to $C^\infty(H, R^1)'$.

We can now sharpen Proposition 4.1.

PROPOSITION 4.7. *The convolution of any $\mathscr{F} \in \mathscr{D}'(H)$ and $f \in \mathscr{D}(H)$ is infinitely Fréchet differentiable in the sense indicated in Proposition 3.3. Moreover, every derivative of it is bounded on every bounded set.*

PROOF. Suppose f vanishes outside a ball $\mathscr{S}_r$. We choose an arbitrary positive R, construct an $\alpha \in \mathscr{D}(H)$ that is equal to one on $\mathscr{S}_{R+r}$ and construct a functional F_α by means of the formula $F_\alpha(\varphi) = F(\alpha\varphi)$. Clearly F_α has bounded support and therefore maps $\mathscr{D}(H)$ into $C^\infty(H, R^1)$. Moreover, $(\mathscr{F} * f)(x) = (\mathscr{F}_\alpha * f)(x)$ for $x \in \mathscr{S}_R$. Q.E.D.

By the convolution of two functions $\mathscr{F}$ and $\mathscr{G} \in \mathscr{D}'(H)$ we will mean the functional acting according to the formula $(\mathscr{F} * \mathscr{G})(\varphi) = \mathscr{F}(\mathscr{G} * \varphi)$. This convolution is defined if the space $\mathscr{G} * \mathscr{D}(H)$ has a topology for which the inclusion mapping of $\mathscr{D}(H)$ or some subspace of it into $\mathscr{G} * \mathscr{D}(H)$ is continuous and such that its range is dense in $\mathscr{G} * \mathscr{D}(H)$ and the restriction of $\mathscr{F}$ to its domain is continuous in the topology induced in it by the topology in $\mathscr{G} * \mathscr{D}(H)$. In this case $\mathscr{F}$ can be uniquely extended to $\mathscr{G} * \mathscr{D}(H)$ with preservation of continuity. The composition of $\mathscr{F}$ and the operator $\varphi \mapsto \mathscr{G} * \varphi$ then turns out to be a continuous linear functional on $\mathscr{D}(H)$. In particular, the above convolution is defined if one of the functionals has bounded support or if both of them belong to $C^\infty(H, R^1)$.

PROPOSITION 4.8. *Suppose given a differential operator* $L(D) = \sum_0^m a_k D^k$ *with* $a_k \in H^k$ *whose adjoint has a fundamental solution* $\mathscr{E}$. *Then the following assertions are true*:

a) *For any* $\alpha \in \mathscr{D}(H)$ *the equation* $L(D)\varphi = \alpha$ *has a solution in ordinary functions, which is* $\mathscr{E} * \alpha$.

b) *The equation* $L'(\mathscr{D})\mathscr{F} = \mathscr{G}$ *has the solution* $\mathscr{E} * \mathscr{G} \in \mathscr{D}'(H)$ *for any* $\mathscr{G} \in \mathscr{D}'(H)$ *with bounded support.*

PROOF. a) $(L(D)(\mathscr{E} * \alpha))(x) = (\mathscr{E} * (L(D)\alpha))(x) = ((L'(D)\mathscr{E}) * \alpha)(x) = (\delta * \alpha)(x) = \alpha_x(0) = \alpha(x)$.

b)

$$\begin{aligned}(L'(D)(\mathscr{E} * \mathscr{G}))(\varphi) &= (\mathscr{E} * \mathscr{G})(L(D)\varphi) = \mathscr{E}(\mathscr{G} * (L(D)\varphi)) \\ &= \mathscr{E}(L(D)(\mathscr{G} * \varphi)) = (L'(D)\mathscr{E})(\mathscr{G} * \varphi) \\ &= \delta(\mathscr{G} * \varphi) = (\mathscr{G} * \varphi)(0) = \mathscr{G}(\varphi). \quad \text{Q.E.D.}\end{aligned}$$

§5. Examples

EXAMPLE 5.1. Let $\mathfrak{A}_2$ be equipped with the topology $\beta(\mathfrak{A}_2, H^2)$ (then Condition 4.3 implies that the topology in $\mathfrak{A}_k$ for any even k is $\beta(\mathfrak{A}_k, H^k)$) and let the symbol LIM denote a Banach limit (see [18]).

A Banach limit can be chosen so that $\mathrm{LIM}\,\xi_k = \lim n^{-1}\sum_1^n \xi_k$ for any bounded numerical sequence ξ_k for which the right side of this equality exists. If $\{e_i\}$ is an orthonormal system in H, then $\mathrm{LIM}\,a(e_k, e_k)$, $a \in \mathfrak{A}_2$, defines a continuous linear functional on $\mathfrak{A}_2$ and therefore $\mathrm{LIM}\langle e_k \otimes e_k, \varphi''(x)\rangle$ defines a differential operator that is an extension of the Laplace operator in the sense of Lévy (see [**9**], Russian p. 265). It is not known whether any of the assertions of §4 hold for this operator, but it would seem unlikely, since similarity operators, being second-order operators, have properties of first-order operators. (See in this connection [**10**], [**11**].) Taking this into account, the author finds it advisable to consider only operators with coefficients in the spaces H^k.

EXAMPLE 5.2. We will say that an operator $L(D)=\sum_0^m a_k D^k$ is finite-dimensional if its coefficients a_k belong to the algebraic tensor products $H\otimes\cdots\otimes H$ of the corresponding number of copies of H, i.e. a_k admits a representation of the form $a_k=\sum_{i=1}^{m_k} x_1^{ki}\otimes\cdots\otimes x_k^{ki}$. Let R^l be the linear span of the vectors x_n^{ki}, $k=1,\cdots,m$; $i=1,\cdots,m_k$; $n=1,\cdots,k$. The identity imbedding $\mathrm{id}: R^l\to H$ induces a restriction mapping $\tau:\mathscr{D}(H)\to\mathscr{D}(R^l)$ and its adjoint $\tau':\mathscr{D}'(R^l)\to\mathscr{D}'(H)$. Clearly the formula $\widetilde{L}(D)\tau=\tau L(D)$ defines a differential operator $\widetilde{L}(D):\mathscr{D}(R^l)\to\mathscr{D}(R^l)$ with constant coefficients. Let $\widetilde{\mathscr{G}}$ be a fundamental solution of the operator $\widetilde{L}'(D)$. It is easily seen that $\tau'\widetilde{\mathscr{G}}$ will be a fundamental solution for the operator $L'(D)$. In particular, every first-order operator has a fundamental solution.

EXAMPLE 5.3. The complete projective tensor product H^2 is isomorphic to the space of nuclear operators in H [**12**], while the space $\mathfrak{A}_2$ is isomorphic to the space of all bounded operators in H. If a nuclear operator $B\leftrightarrow b\in H^2$ and a bounded operator $A\leftrightarrow a\in\mathfrak{A}_2$ then $\langle b,a\rangle=\mathrm{tr}(BA)$. Therefore the second order differential operators with coefficients in H^2 coincide with the differential operators introduced by Daleckiĭ [**13**–**15**].

Let A be a positive definite selfadjoint nuclear operator in H, consider the differential operator $L(D)\varphi=2^{-1}\mathrm{tr}(A\varphi'')$ and let μ_{tA} be the Gaussian measure with Fourier transform $\exp\{-t(Ax,x)/2\}$. Then the formula $\mathscr{G}(B)=-\int_0^\infty \mu_{tA}(B)\,dt$ defines a countably additive measure on the σ-ring of bounded Borel subsets of H and therefore defines a functional $\mathscr{G}\in\mathscr{D}'(H)$ that is a fundamental solution for the operator $L(D)$. These assertions are a reformulation of some results of Gross [**16**].

REMARK 5.4. A direct reformulation of the results in [**16**] is possible only in the case when zero is not an eigenvalue of the operator A. In the general case one should consider the closed subspace $H_1\subset H$ generated by the eigenvectors of A with nonzero eigenvalues, the restriction mapping $\tau:\mathscr{D}(H)\to\mathscr{D}(H_1)$ and the operator $L_1(D):\mathscr{D}(H_1)\to\mathscr{D}(H_1)$ defined by the equality $L_1(D)\tau=\tau L(D)$ and take a fundamental solution $\mathscr{G}_1\in\mathscr{D}'(H_1)$ of the operator $L_1(D)$. Then $\tau'\mathscr{G}_1=\widetilde{\mathscr{G}}$ will be a fundamental solution of the operator $L(D)$. Since the measures μ_{tA} and $\widetilde{\mathscr{G}}$ are concentrated on H_1, the construction of Example 5.3 gives the same result.

EXAMPLE 5.5. Let μ_{tA} be the measure of Example 5.3, where A is a positive definite selfadjoint nuclear operator in H. In [**19**] it is deduced from heuristic considerations that μ_{tA} satisfies the Cauchy problem

$$\mu_{0A}=\delta;\quad \frac{d}{dt}\mu_{tA}=2^{-1}\,\mathrm{tr}\,(A\mu''_{tA})\ (t>0).$$

It can be shown that neither the left nor the right side of this equation exists in the sense of convergence on every Borel set.

In our theory of distributions it is possible to attach a strict meaning to this assertion. It is natural to identify the operator $\mathrm{tr}(A\mu'')$ with the adjoint of the operator $L(D): \varphi \mapsto \mathrm{tr}(A\varphi'')$.

Suppose given a one-parameter family $\{\mathscr{F}_t\}$ of distributions $\mathscr{F}_t \in \mathscr{D}'(H)$, $t \in R^1$. We will say that $\mathscr{F}_t$ is differentiable with respect to t if it is differentiable as a mapping of R^1 into the space $\mathscr{D}'(H)$ equipped with the topology $\sigma(\mathscr{D}'(H), \mathscr{D}(H))$. In other words, the derivative of $\mathscr{F}_t$ with respect to t is defined by the equality

$$\left(\frac{d}{dt}\mathcal{F}_t\right)(\varphi) = \frac{d}{dt}(\mathcal{F}_t(\varphi)).$$

Proposition 5.6. *Let $\mathscr{G} \in C^\infty(H, R^1)$. Then the convolution $\mathscr{F}_t = \mathscr{G} * \mu_{tA}$ is defined and satisfies the Cauchy problem*

$$\frac{d}{dt}\mathcal{F}_t = \frac{1}{2}L'(D)\mathcal{F}_t \quad (t > 0), \quad \lim_{t\to 0}\mathcal{F}_t = \mathcal{G}.$$

The limit exists in the topology $\sigma(\mathscr{D}'(H), \mathscr{D}(H))$.

Proof. We have

$$\begin{aligned}(L'(D)F_t)(\varphi) &= \mathcal{F}_t(L(D)\varphi) = (G * \mu_{tA})(L(D)\varphi) = G(L(D)(\mu_{tA} * \varphi)) \\ &= G\left(2\frac{d}{dt}(\mu_{tA} * \varphi)\right) = 2\frac{d}{dt}(G(\mu_{tA} * \varphi)) = 2\frac{d}{dt}((G * \mu_{tA})(\varphi)) \\ &= 2\left(\frac{d}{dt}\mathcal{F}_t\right)(\varphi).\end{aligned}$$

A proof of the equality $2^{-1}L(D)(\mu_{tA} * \varphi) = (d/dt)(\mu_{tA} * \varphi)$ can be found in [15]. Thus $d\mathscr{F}_t/dt = 2^{-1}L'(D)\mathscr{F}_t$, with both sides of the equality existing. Since $\mathscr{F}_t(\varphi) = \mathscr{G}(\mu_{tA} * \varphi)$, it suffices to prove that $\mu_{tA} * \varphi \to \varphi$ in $C^\infty(H, R^1)$. But

$$(\mu_{tA} * \varphi)(x) = \int \varphi(x + y)\,\mu_{tA}(dy) = \int \varphi(x + \sqrt{t}y)\,\mu_{1A}(dy),$$

$$D^k(\mu_{tA} * \varphi)(x) = \int D^k\varphi(x + \sqrt{t}y)\,\mu_{1A}(dy)$$

(the Bochner integral). By virtue of the continuity and boundedness of φ and all of its derivatives, $\varphi(x + \sqrt{t}y) \to \varphi(x)$ uniformly on each compactum with respect to x and on each ball with respect to y together with all of its derivatives, whence $\mu_{tA} * \varphi \to \varphi$ in $C^\infty(H, R^1)$. Q.E.D.

Example 5.7. Let $\{e_i\}$ be an orthonormal system in H and consider the differential operators $L_i(D) = e_i D$ and the system (*) $L_i'(D)F = 0$ for $i = 1, 2, \cdots$.

Proposition 5.8. *If the system $\{e_i\}$ is infinite, the system of equations* (*) *has no nonzero solutions in $D'(H)$.*

Proof. Let $\mathscr{F}$ be a solution of (*), let $\varphi \in \mathscr{D}(H)$, let r be the radius of a ball with center at the origin outside of which $\varphi = 0$ and let $\varphi_i(x) = \varphi(x + r\sqrt{2}e_i)$. Then the partial sums of the series $\sum_1^\infty \varphi_i$ have a common bounded support and together with all of their derivatives are bounded and converge uniformly on each compactum, so that the series converges in $\mathscr{D}(H)$.

From the equation $L_i'(D)F = 0$ it follows that $\mathscr{F}(\varphi) = \mathscr{F}(\varphi_i)$ and hence

$$\mathcal{F}\left(\sum_{i=1}^{\infty} \varphi_i\right) = \sum_{i=1}^{\infty} \mathcal{F}(\varphi_i) = \sum_{i=1}^{\infty} \mathcal{F}(\varphi).$$

The last series converges only when $\mathscr{F}(\varphi) = 0$. Q.E.D.

REMARK 5.9. The fact proved in Proposition 5.8 continues to hold in other theories of distributions. Suppose we have a theory of distributions, i.e. of continuous linear functionals on a space $\mathscr{K}$ consisting of functions defined on H, with $\mathscr{D}(H)$ dense in $\mathscr{K}$, and if a sequence $\varphi_n \in \mathscr{D}(H)$ has bounded support and converges uniformly together with all its derivatives, then φ_n converges in $\mathscr{K}$. Applying the arguments of Proposition 5.8 to the series $\sum_1^\infty \varphi_n/n$, we get that any solution of (*) vanishes on $\mathscr{D}(H)$ and hence, since $\mathscr{D}(H)$ is dense in $\mathscr{K}$, vanishes identically.

Received 20/SEPT/71

Bibliography

[1] R. G. Bartle, *A general bilinear vector integral,* Studia Math. **15** (1956), 337-352. MR **18**, 289.

[2] V. I. Averbuh and O. G. Smoljanov, *Differentiation theory in linear topological spaces,* Uspehi Mat. Nauk **22** (1967), no. 6 (138), 201-260 = Russian Math. Surveys **22** (1967), no. 6, 201-258. MR **36** #6933.

[3] P. R. Halmos, *Measure theory,* Van Nostrand, Princeton, N. J., 1950. MR **11**, 504.

[4] H. H. Schaefer, *Topological vector spaces,* Macmillan, New York, 1966. MR **33** #1689.

[5] H. Jarchow, *Marinescu-Räume,* Comment. Math. Helv. **44** (1969), 138-163. MR **40** #3260.

[6] A. N. Kolmogorov and S. V. Fomin, *Elements of the theory of functions and of functional analysis.* II: *Measure, Lebesgue integral, Hilbert space,* 2nd rev. ed., "Nauka", Moscow, 1966; English transl. of 1st ed., Graylock Press, Albany, N. Y.; Academic Press, New York, 1961. MR **22** #9566a,b; **38** #2559.

[7] L. Schwartz, *Théorie des distributions.* Vols. I, II, Actualités Sci. Indust., nos. 1091, 1122, Hermann, Paris, 1950, 1951. MR **12**, 31; 833.

[8] K. Yosida, *Functional analysis,* Die Grundlehren der math. Wissenschaften, Band 123, Academic Press, New York; Springer-Verlag, Berlin, 1965. MR **31** #5054.

[9] P. Lévy, *Problèmes concrets d'analyse fonctionnelle,* 2ième éd., Gauthier-Villars, Paris, 1951; Russian transl., "Nauka", Moscow, 1967. MR **12**, 834; **36** #6892.

[10] G. E. Šilov, *On certain questions of analysis in Hilbert space.* I, Funkcional. Anal. i Priložen. **1** (1967), no. 2, 81-90 = Functional Anal. Appl. **1** (1967), 158-165. MR **35** #4770.

[11] E. M. Polišcuk, *Linear equations in functional Laplacians,* Uspehi Mat. Nauk **19** (1964), no. 2 (116), 163-170. (Russian) MR **31** #2641.

[12] R. Schatten and J. von Neumann, *The cross-space of linear transformations.* I, II, III, Ann. of Math. (2) **47** (1946), 73-84, 608-630; **49** (1948), 557-582. MR **7**, 455; **8**, 31; **10**, 256.

[13] Ju. L. Daleckiĭ, *Differential equations with functional derivatives and stochastic equations for generalized random processes,* Dokl. Akad. Nauk SSSR **166** (1966), 1035-1038 = Soviet Math. Dokl. **7** (1966), 220-223. MR **35** #5788.

[14] ———, *Elliptic operators in functional derivatives and diffusion equations connected with them,* Dokl. Akad. Nauk SSSR **171** (1966), 21-24 = Soviet Math. Dokl. **7** (1966), 1399-1402. MR **35** #3749.

[15] ———, *Infinite-dimensional elliptic operators and the corresponding parabolic equations*, Uspehi Mat. Nauk **22** (1967), no. 4 (136), 3-54 = Russian Math. Surveys **22** (1967), no. 4, 1-53. MR **36** #6821.

[16] L. Gross, *Potential theory on Hilbert space*, J. Functional Analysis **1** (1967), 123-181. MR **37** #3331.

[17] S. V. Fomin, *Generalized functions of infinitely many variables and their Fourier transforms*, Uspehi Mat. Nauk **23** (1968), no. 2 (140), 215-216. (Russian)

[18] ———, *Differentiable measures in linear spaces*, Uspehi Mat. Nauk **23** (1968), no. 1 (139), 221-222. (Russian) MR **36** #4221.

[19] ———, *A Fourier transform method for functional differential equations*, Dokl. Akad. Nauk SSSR **181** (1968), 812-814 = Soviet Math. Dokl. **9** (1968), 951-953. MR **38** #3637.

[20] I. M. Gel′fand and N. Ja. Vilenkin, *Generalized functions*. Vol. 4: *Some applications of harmonic analysis*, Fizmatgiz, Moscow, 1961; English transl., Academic Press, New York, 1964. MR **26** #4173; **30** #4152.

Translated by
S. SMITH

Trudy Moskov. Mat. Obšč.
Tom 28 (1973)

Trans. Moscow Math. Soc.
Vol. 28 (1973)

EXAMPLE OF A MARTIN COMPACTUM WITH A NONNEGLIGIBLE IRREGULAR BOUNDARY POINT

M. G. ŠUR

Contents

Introduction

By a known result of Kellogg and Evans [2] the set of irregular (in the classical sense) boundary points of a bounded region in n-dimensional Euclidean space R^n has outer Newtonian (for $n \geqq 3$) or logarithmic (for $n = 2$) capacity zero. After the appearance of Naim's well-known work [7] there arose the supposition (see, for example, [1]) that if for the boundary of a region one adopts, not the Euclidean, but rather the Martin boundary then the cited result can be replaced by the following:

CONJECTURE $\mathscr{R}$. The set of irregular points of the Martin boundary has zero harmonic measure.

Doubt as to the validity of this conjecture was expressed by the author in [8]. In this work there is constructed an example of a three-dimensional Green space for which Conjecture $\mathscr{R}$ is false. In this example one of the irregular points of the Martin boundary is an atom of harmonic measure.

In the present work[1)] although the questions touched upon are similar in character to certain classical problems of potential theory, it turned out to be convenient to use probabilistic ideas. In the first section of the work there is constructed a certain three-dimensional Green space E admitting a sufficiently rich family of modifications. On this space a Wiener process is introduced. In §§2 and 3 certain ancillary results relative to a Wiener process on E are given. In §4 it is shown that for one of the modifications of the space E Conjecture $\mathscr{R}$ is false.

AMS (*MOS*) *subject classifications* (1970). Primary 60J50; Secondary 60J45, 60J65, 31B15.

[1)] A short version of this work was published in Teor. Verojatnost. i Primenen. **17** (1972), 366-370 [**9**].

§1. Construction of the space E

Let us recall the definition of a three-dimensional Green space. Let each point x of a connected metric space E with countable base possess a neighborhood isometric to an open ball in the space R^3. Then in E one naturally defines the Laplace operator, and then also harmonic and superharmonic functions. If E contains at least one nonnegative nonconstant superharmonic function, it is called a three-dimensional *Green space*. This definition is ordinarily somewhat extended by introduction of "points at infinity"; however, we shall not do this here (see [7]).

The principal element of the construction used in constructing the desired space are sets, called in this work *superstructures*. In order to introduce them consider the following objects:

a) arithmetic Euclidean space R^4 with points $x=(x^1,x^2,x^3,x^4)$, where the x^i are real numbers $(1 \leq i \leq 4)$;

b) (three-dimensional) hyperplanes $\mathfrak{E}$, $\mathfrak{E}^*$ and $\mathfrak{E}_1^*$ in R^4 defined, respectively, by the equations $x^4=0$, $x^4=d$ and $x^4=d'$, where $d'>d>0$;

c) a pair of equal disjoint closed three-dimensional cubes K and K' in $\mathfrak{E}$, each face of which is parallel to one of the coordinate hyperplanes in R^4 and which are symmetric to each other with respect to some hyperplane $\mathfrak{M}$ with equation of the form $x^3=c$ (in what follows such a pair of cubes will be called *admissible*); the lengths of the edges of these cubes we set equal to d;

d) three-dimensional cubes K^* and $K^{*\prime}$ lying in $\mathfrak{E}^*$ and having as (orthogonal) projections onto $\mathfrak{E}$, respectively, the cubes K and K';

e) hyperplanes $\mathfrak{M}^*$ and $\mathfrak{M}^{*\prime}$ parallel to $\mathfrak{M}$, symmetric to each other with respect to $\mathfrak{M}$ and such that the cubes K and K' lie on distinct sides of each of them;

f) the portion $\widetilde{\mathfrak{M}}^*$ of the hyperplane $\mathfrak{M}^*$ included between $\mathfrak{E}^*$ and $\mathfrak{E}_1^*$ (i.e. $\widetilde{\mathfrak{M}}^*=\{x\in\mathfrak{M}^*: d\leq x^4\leq d'\}$) and the analogous portion $\widetilde{\mathfrak{M}}^{*\prime}$ of $\mathfrak{M}^{*\prime}$, and also the portions $\widetilde{\mathfrak{E}}^*$ and $\widetilde{\mathfrak{E}}_1^*$ of the hyperplanes $\mathfrak{E}^*$ and $\mathfrak{E}_1^*$ included, respectively, between $\mathfrak{M}^*$ and $\mathfrak{M}^{*\prime}$;

g) any face J of K and the face J' of K' symmetric to it with respect to $\mathfrak{M}$, and also the faces J^* and $J^{*\prime}$ of K^* and $K^{*\prime}$, respectively, whose projections onto $\mathfrak{E}$ are J and J';

h) three-dimensional cubes Q and Q' in R^4, the first of which has as a pair of parallel faces the squares J and J^*, and the second, J' and $J^{*\prime}$.

Let us form the set $\mathfrak{N}^0$ by joining the sets $\mathfrak{E}^*\backslash\widetilde{\mathfrak{E}}^*$, $\widetilde{\mathfrak{M}}^*$, $\widetilde{\mathfrak{M}}^{*\prime}$ and $\widetilde{\mathfrak{E}}_1^*$ and then removing $K^*\cup K^{*\prime}$. If now from $\mathfrak{N}^0\cup Q\cup Q'$ we remove the faces of the cubes Q and Q' orthogonal to $\mathfrak{E}$, then we obtain the set $\mathfrak{N}$ which we shall call the *superstructure* in R^4 corresponding to the admissible pair of cubes K and K'.

For the sake of greater definiteness in what follows *we agree to consider only superstructures for which the distance between* $\mathfrak{M}^*$ *and* $\mathfrak{M}^{*\prime}$ *is equal to* 1.

Then for fixed centers of symmetry of K and K' the superstructure is uniquely determined by giving the two parameters d and d'.

We arrive at the definition of the Green space E. Consider the Hilbert space l^2 of sequences $x = (x^1, x^2, \cdots)$ of reals such that $\sum_{i \geqq 1} (x^i)^2 < \infty$; and, as usual, the square of the norm $\|x\|^2$ of $x \in l^2$ is equated to the sum of this series. Let H_n $(n \geqq 4)$ be the linear subspace in l^2 of those $x \in l^2$ for which $x^i = 0$ whenever i is different from 1, 2, 3 and n. The set H formed of the $x \in l^2$ with $x^i = 0$ for $i > 3$ is a subspace of any H_n $(n \geqq 4)$.

For the balance of this paper we fix a sequence of pairwise distinct points o_n $(n \geqq 1)$ of some line L lying in H, and require that $o_n \to 0$ as $n \to \infty$, where 0 is the zero in l^2. We enclose each o_n in a (also fixed from here on) three-dimensional open ball $U_n \subset H$ with center at o_n, where $U_n \cap U_m = \emptyset$ for $n \neq m$. Then we choose once and for all pairwise disjoint three-dimensional open balls U'_n with centers o'_n $(n \geqq 1)$ lying on the same line L; here we make sure that the following conditions hold:

a) the radius of U_n is equal to that of U'_n;

b) $\|o_n - o'_n\| > 1$;

c) $\lim_{n \to \infty} \|o'_n\| = \infty$.

The spheres bounding the balls U_n and U'_n, respectively, we denote by S_n and S'_n $(n \geqq 1)$. In U_n we select the three-dimensional cube $K(n)$ with center of symmetry at o_n, and in U'_n the cube $K'(n)$ with center at o'_n, requiring that the pair $K(n)$, $K'(n)$ be admissible (here H is identified with the hyperplane $\mathfrak{E}$) and that the distance between them exceed 1. The lengths of the edges of the cubes will be modified later on.

Having identified R^4 and H_4 in the obvious way, we construct the superstructure $\mathfrak{R}(1)$ in H_4 corresponding to the pair $K(1)$, $K'(1)$. Then identifying each point $(x^1, x^2, x^3, x^4) \in R^4$ with the point $(x^1, x^2, x^3, 0, x^4, 0, \cdots) \in H_5$, we construct the superstructure $\mathfrak{R}(2)$ in H_5 corresponding to the pair $K(2)$ and $K'(2)$. Further, identifying $(x^1, x^2, x^3, x^4) \in R^4$ with $(x^1, x^2, x^3, 0, 0, x^4, 0, \cdots)$ $\in H_6$, we construct the superstructure $\mathfrak{R}(3)$ in H_6 corresponding to $K(3)$ and $K'(3)$, and so on. For $n \neq m$ the superstructures $\mathfrak{R}(n)$ and $\mathfrak{R}(m)$ are disjoint $(n, m \geqq 1)$. We agree to denote by the symbols $Q(n)$, $Q'(n) \mathfrak{R}^0(n)$, $K^*(1)$, $J'(n)$, etc. the sets corresponding to the previously introduced sets Q, Q', $\mathfrak{R}^0$, K^*, J', etc. in the superstructure $\mathfrak{R}(n)$. By a_n we understand the center of symmetry of the square $J'(n)$ $(n \geqq 1)$. We define the parameters $d(n)$ and $d'(n)$ in analogy with d and d'.

Let us introduce a purely technical restriction: *we shall assume the parameter* $d(n)$ *less than* 1 *and not greater than a third of the radius of* U_n $(n \geqq 1)$.

Now we introduce the sets

$$E_0 = H; \; E_1 = \{E_0 \setminus [K(1) \cup K'(1)]\} \cup \mathfrak{R}(1);$$

$$E_{n+1} = \{E_n \setminus [K(n+1) \cup K'(n+1)]\} \cup \mathfrak{R}(n) \quad (n \geqslant 0);$$

and, finally, we define the desired space E as

$$E = \lim_{n\to\infty} E_n \setminus \{0\} = \{[E_0 \setminus \{0\}] \setminus \bigcup_{n\geq 1} [K(n) \cup K'(n)]\} \cup [\bigcup_{n\geq 1} \mathfrak{R}(n)].$$

On the set E we define a metric ρ differing from the metric induced by the standard metric of l^2. Namely, for any pair of points $x, y \in E$ we consider polygonal lines with a finite number of links joining these points and lying in E. The distance $\rho(x,y)$ between x and y we set equal to the greatest lower bound of the lengths (in the sense of the metric in l^2) of all possible polygonal lines of the indicated type. In what follows, in the absence of explicit comment, the space E will be assumed to be metrized by means of the metric ρ. Similarly one metrizes also the spaces E_n.

REMARK 1. In exactly the same way one can metrize any set obtained from R^3 by removing two cubes from it. Here each set $\mathfrak{R}^0(n)$ turns out to be isometric to a set of the indicated form.

From geometric considerations it is obvious that each point of E and E_n possesses a neighborhood isometric to an open ball in R^3. This permits one to naturally introduce in E and E_n the Laplace operator and to define harmonic and superharmonic functions.

By a *Wiener process* on E we shall understand a homogeneous continuous strong Markov diffusion process $X = (x_t, \zeta, \mathcal{M}_t, \mathbf{P}_x)$ on E with generating differential operator equal to half the Laplace operator (for terminology see [**5**]). Similarly we also define the Wiener process $X^{(n)} = (x_t^{(n)}, \zeta^{(n)}, \mathcal{M}_t^{(n)}, \mathbf{P}_x^{(n)})$ on E_n. The construction of Wiener processes in spaces of the form we are interested in was achieved by Doob [**4**].

Let us set down some notation connected with the process X. By the symbol $\pi(x, \Gamma)$ we denote the probability $\mathbf{P}_x\{\tau_\Gamma < \zeta\}$, where $\tau_\Gamma \equiv \tau(\Gamma)$ is the first instant (after $+0$) of the trajectory x_t reaching the Borel set $\Gamma \subset E$, and by $\pi_U(x, \Gamma)$ we denote the probability $\mathbf{P}_x\{x(\tau_U) \in \Gamma\}$, where Γ and U are Borel sets in E. We define the operator π_U on the family of nonnegative Borel functions f on E by the formula

$$\pi_U f(x) = \mathbf{E}_x f(x(\tau_U)) \qquad (x \in E).$$

We also set

$$Gf(x) = \mathbf{E}_x \int_0^\zeta f(x_t)\, dt,$$

if this mathematical expectation makes sense. For the processes $X^{(n)}$ variables of the type $\tau^{(n)}(\Gamma)$, $\pi^{(n)}(x, \Gamma)$, $\pi_U^{(n)} f$ and $G^{(n)}f$ are introduced in an analogous way.

In concluding this section let us prove the following result:

LEMMA 1. *The spaces E and E_n $(n \geq 1)$ are three-dimensional Green spaces.*

PROOF. For explicit verification we need only the existence in E and E_n $(n \geq 1)$ of nonnegative nonconstant superharmonic functions. Consider, for example, the space E and set the value of $g(x)$ at $x \in E$ equal to the $\mathbf{P}_x$-

probability that beginning with some moment $\tau < \zeta$ the Wiener trajectory will be found in $\mathfrak{R}^0(1)$ up to ζ. It is not hard to see that $g(x)$ is harmonic on E (see [5], §§12.18 and 13.14).

Using the isometricity of $\mathfrak{R}^0(1)$ to R^3 with two cubes removed (see Remark 1), we conclude from the properties of a classical Wiener process in R^3 that $g(x) \to 1$ as $\|x\| \to \infty$, $x \in \mathfrak{R}^0(1)$. If $x \in E$ converges (in the sense of the metric in l^2) to any point of the face $J(1)$ of the cube $K(1)$, then $g(x) \to 0$. These two facts prove that $g(x)$ is nonconstant. Thus E is a Green space.

From the lemma it follows that Gf is bounded whenever f is a bounded Borel function with compact support (see [4]).

§2. Ancillary results

The space E is a union of sets isometric to certain sets in R^3. Hence we preface the study of a Wiener process in E by consideration of certain properties of the Wiener process $\widetilde{X} = (\tilde{x}_t, \widetilde{\mathcal{M}}_t, \widetilde{\mathbf{P}}_x)$ in R^3. Here we naturally identify the points of R^3 with the points in H, which, in particular, permits us to look upon the previously introduced cubes $K(n)$ and $K'(n)$ and balls U_n and U_n' as sets in R^3.

Quantities of the type $\tilde{\pi}(x, \Gamma)$, $\tilde{\pi}_U(x, \Gamma)$ and $\widetilde{G}f$ are introduced for $\widetilde{X}$ exactly as the corresponding ones were for X. For a Borel set $\Gamma \subset R^3$ we have the estimate

$$\tilde{\pi}(x, \Gamma) \leqslant \frac{c(\Gamma)}{r(x, \Gamma)} \qquad (x \in R^3), \tag{1}$$

where r is the Euclidean metric on R^3 and $c(\Gamma)$ is the Newtonian capacity of Γ (see [3], Lemma 5.4 and the Remarks in §7). From this there follows

LEMMA 2. *Let*

$$\Gamma_0 = \bigcup_{n \geqslant 1} [U_n \cup U_n'] \quad \textit{and} \quad \Gamma_1 = \bigcup_{n \geqslant 1} [K(n) \cup K'(n)].$$

For any fixed point x_0 of the interior of $R^3 \backslash \Gamma_0$ one can find numbers $\gamma < 1$ and $\delta_n > 0$ $(n \geqq 1)$ such that

$$\tilde{\pi}(x_0, \Gamma_1) < \gamma \tag{2}$$

whenever

$$0 < d_n < \delta_n \qquad (n \geqq 1). \tag{3}$$

PROOF. Setting $r_0 = r(x_0, \Gamma_0) > 0$ we choose numbers $\delta_n > 0$ so small that for $d_n = \delta_n$ the capacity $c(K(n))$ does not exceed $r_0/2^{n+1}$ $(n \geqq 1)$. From this condition and from (1) it follows that

$$\tilde{\pi}(x_0, \Gamma_1) \leqslant \sum_{n \geqslant 1} [\tilde{\pi}(x_0, K(n)) + \tilde{\pi}(x_0, K'(n))] \leqslant 2r_0^{-1} \sum_{n \geqslant 1} c(K(n)) < 1.$$

We take for γ the value $\tilde{\pi}(x_0, \Gamma_1)$, which is admissible in this case. Then when condition (3) holds the corresponding value of $\tilde{\pi}(x_0, \Gamma_1)$ is found to be less than γ, i.e. (2) will hold.

Lemma 3. *Let f and g be bounded Borel functions on the two-dimensional sphere $S \subset R^3$, with $g > 0$. Then*

$$\lim_{\|x\|\to\infty} \frac{\tilde{\pi}_S f(x)}{\tilde{\pi}_S g(x)} = \frac{\int_S f(y)\,\sigma_S(dy)}{\int_S g(y)\,\sigma_S(dy)}, \tag{4}$$

where $\|\cdot\|$ is the Euclidean norm in R^3 and σ_S is Lebesgue measure on S.

Proof. For example, let the sphere S have radius r_0 and center 0, and let f be continuous on S. Then the function $\tilde{\pi}_S f(x)$, where $\|x\| > r_0$, is a solution of the exterior Dirichlet problem:

$$\Delta u = 0, \ u|_S = f \quad \text{and} \quad \lim_{\|x\|\to\infty} u(x) = 0.$$

Hence this function can be expressed by means of the Poisson integral for the exterior of the sphere:

$$\tilde{\pi}_S f(x) = \alpha \int_S f(y)\, I(x, y)\, \sigma_S(dy) \tag{5}$$

for $\|x\| > r_0$ (see [2], Chapter 1), where α is constant and

$$I(x, y) = \frac{\|x\|^2 - r_0^2}{\|x - y\|^3}.$$

The validity of (5) for continuous functions implies its validity for any bounded Borel functions f and S.

Note that $\|x\| \cdot I(x,y) \to 1$ as $\|x\| \to \infty$ uniformly for $y \in S$. Applying (5), for functions f and g satisfying the conditions of the lemma we obtain

$$\frac{\tilde{\pi}_S f(x)}{\tilde{\pi}_S g(x)} = \frac{\int_S f(y)\|x\|\, I(x, y)\,\sigma_S(dy)}{\int_S g(y)\|x\|\, I(x, y)\,\sigma_S(dy)} \to \frac{\int_S f(y)\,\sigma_S(dy)}{\int_S g(y)\,\sigma_S(dy)},$$

if $\|x\| \to \infty$. Thus the lemma is proved.

Lemma 4. *In the space R^3 let there be given a square D_1 with length of side h and a square D_2 with length of side kh, where $k > 0$ is fixed and the sides of D_1 and D_2 are parallel to each other, and the segment joining the centers of these squares has length h and is perpendicular to the planes containing D_1 and D_2. There exists a $\tilde{\gamma} > 0$ not depending on h and such that for all $x \in D_2$*

$$\tilde{\gamma} < \tilde{\pi}(x, D_1) < 1 - \tilde{\gamma}. \tag{6}$$

Proof. Let us subject R^3 to a similarity transformation so that D_1 becomes a square with unit side. Here the g.l.b. and l.u.b. of the values $\tilde{\pi}(x, D_1)$ for $x \in D_2$ do not change if by D_1 and D_2 we understand the images of the original squares D_1 and D_2. Moreover, these bounds will be positive and less than one, which proves the lemma.

Corollary 1. *Let the event Λ_n be that $\zeta = \infty$ and, in addition, that the* *trajectory* x_t *after some finite moment of time does not exit from* $\mathfrak{R}^0(n)$. *Let* $\Lambda_n' = \{\tau(S_n') < \tau(\mathfrak{R}^0(n))\}$. *Then for some constant* $\gamma_1 > 0$ *and all* $n \geq 1$ *the relations*

$$\mathbf{P}_{a_n}\{\Lambda_n\} > \gamma_1 \quad \text{and} \quad \mathbf{P}_{a_n}\{\Lambda_n'\} > \gamma_1 \tag{7}$$

hold independently of the choice of the system of parameters $d(n)$ *and* $d'(n)$ $(n \geq 1)$. (*Recall that* a_n *is the center of symmetry of the face* $J'(n)$ $(n \geq 1)$.)

PROOF. Let us imagine three equal three-dimensional rectangular parallelepipeds P_1, P_2, P_3 contained in E: the first lies in $Q'(n)$, and one of its faces is a square with side $d(n)/2$ and with center at a_n, and the face parallel to it lies in $\mathfrak{R}^0(x)$; the others P_2 and P_3 lie, respectively, in E_0 and $\mathfrak{R}^0(n)$ and have one common face with P_1. Let $F = P_1 \cup P_2 \cup P_3$ and let A_2 and A_3 be the faces of P_2 and P_3, respectively, which are parallel to the faces common to P_1. From similarity considerations (cf. Lemma 4) it is clear that for the Wiener particle emanating from a_n the probability of its being on A_3 at the moment $\tau(E\backslash F)$ is equal to some constant not depending on n. Moreover, by Lemma 4 a particle attaining A_3 with probability greater than some positive constant does not hit the faces $J^*(n)$ and $J^{*\prime}(n)$ of the cubes $K^*(n)$ and $K^{*\prime}(n)$, although it lives infinitely long. There follows the existence of a $\gamma_1 > 0$ satisfying the first of inequalities (7) and not depending on the choice of $d(n)$ and $d'(n)$ $(n \geq 1)$.

The second of inequalities (7) is treated in the same way, but with A_3 replaced by A_2.

Let us fix once and for all a nonnegative function v which belongs to the set $\mathscr{C}_0(E_0)$ of continuous functions on E_0 with compact support and which vanishes on the set Γ_0 $(v \not\equiv 0)$ introduced in Lemma 2. For the function Gv, which in what follows will be used in defining Martin operators, we have

COROLLARY 2. *Let* μ_n *and* μ_n' *be the restrictions of the measure* $\pi_{S_n \cup S_n'}(a_n, \cdot)$ *to the sets* S_n *and* S_n' $(n \geq 1)$, *respectively. One can select values of the parameters* $d'(n)$ *of the superstructures* $\mathfrak{R}(n)$ *such that for any choice of the parameters* $d(n)$ $(n \geq 1)$ *the relation*

$$\lim_{n\to\infty} \left[\int Gv d\mu_n \Big/ \int Gv d\mu_n'\right] = 0 \tag{8}$$

will hold on E.

PROOF. The function Gv is harmonic on $(E\backslash K) \cap E_0$, where K is the support of v. Hence the so-called Harnack inequality is applicable to it, and so, in particular, $Gv(x_1)/Gv(x_2) < \lambda_n$ for some $\lambda_n > 0$ and $x_1, x_2 \in S_n \cup S_n'$. Hence

$$\frac{\int Gv d\mu_n}{\int Gv d\mu_n'} \leqslant \frac{\max\limits_{S_n} Gv \cdot \mu_n(S_n)}{\min\limits_{S_n'} Gv \cdot \mu_n'(S_n')} < \lambda_n \frac{\mu_n(S_n)}{\mu_n'(S_n')}. \tag{9}$$

The denominator of the last fraction clearly is not less than $\mathbf{P}_{a_n}\{\Lambda_n'\}$, which by the preceding corollary exceeds some constant $\gamma_1 > 0$ not depending on the choice of the parameters of the superstructures $\mathfrak{R}(n)$.

As to the numerator, it can be made arbitrarily small by increasing the sole parameter $d'(n)$. In fact, a particle emanating from a_n and attaining S_n earlier than S_n' has from the outset (almost surely) been on $J^{*\prime}(n)$, and only next on $J^*(n)$ and afterwards on S_n. Hence

$$\mu_n(S_n) \leqslant \mathbf{P}_{a_n}^{(n)}\{\tau^{(n)}(S_n) < \tau^{(n)}[J^{*\prime}(n)] < \tau^{(n)}[J^*(n)] < \zeta^{(n)}\}$$
$$= \mathbf{E}_{a_n}^{(n)}\{\chi \cdot \mathbf{P}_{x^{(n)}(\tau)}^{(n)}\{\tau^{(n)}[J^*(n)] < \zeta^{(n)}\}\},$$

where $\tau = \tau^{(n)}[J^{*\prime}(n)]$ and χ is the indicator of the event $\{\tau^{(n)}(S_n) < \tau\}$. Now we must take into account that $x^{(n)}(\tau) \in J^{*\prime}(n)$ ($\mathbf{P}_{a_n}$-almost surely) and that for fixed n the probability $\mathbf{P}_x^{(n)}\{\tau^{(n)}[J^*(n)] < \zeta^{(n)}\} = \mathbf{P}_x\{\tau[J^*(n)] < \zeta\}$ by an estimate similar to (1) (see Remark 1) can be made arbitrarily small simultaneously for all $x \in J^{*\prime}(n)$ if the parameter $d'(n)$ is taken sufficiently large. From this we see that $\mu_n(S_n)$ also becomes small. Thus our assertion concerning the numerator in (9) is proved.

From the estimate (9) and the two preceding paragraphs it is apparent that for suitable choice of the parameters $d'(n)$ the relation (3) holds.

§3. The principal lemmas

There is a rather great deal of arbitrariness possible in choosing the parameters $d(n)$ and $d'(n)$. Throughout in what follows (even without specific recall) these parameters are subject to the following condition:[2)]

(A) *The parameters* $d(n)$ $(n \geqq 1)$ *are always chosen so that the assertion of Lemma* 2 *holds for some fixed* $x_0 \in E \setminus \Gamma_0$. *The parameters* $d'(n)$ *are fixed once and for all so that* (8) *holds for any choice of the parameters* $d(n)$ $(n \geqq 1)$.

Thus, henceforth only the parameters $d(n)$ can vary.

Let us attempt to estimate certain functions of the form $G^{(n)}f$.

LEMMA 5. *For* $1 \leqq n \leqq m$ *let the parameters* $d(n)$ *be fixed, and let there be given a function* $f \in \mathscr{C}_0(E_m)$ *with support contained in the set* $E_m \setminus \bigcup_{n>m}[U_n \cup U_n']$, *and* $f \geqq 0$ *but* $f \not\equiv 0$. *Here* $\mathscr{C}_0(E_m)$ *is the set of continuous functions on* E_m *with compact support;* $\mathscr{C}_0(E)$ *is defined similarly. For each compact set* K *lying in the indicated set one can find a constant* $c_0 > 0$ *such that everywhere on* K *for any* $n \geqq m$ *and any admissible set of values of the parameters* $d(k)$ $(k \geqq m+1)$ *there holds the estimate*

$$G^{(n)}f \geqq c_0. \tag{10}$$

Moreover, there exists a constant c_1 *such that everywhere on* E_n

2) The role of the second half of condition (A) will become fully apparent only in the proof of Theorem 2.

(11) $$G^{(n)}f \leqq c_1 \quad (n \geqq m)$$

for any admissible set of values $d(k)$ $(k \geqq m+1)$.

PROOF. Let $\hat{G}$ be the analog of the operator G for the part $\hat{X}$ of X on $E \cap E_m$. It is clear that $\hat{X}$ is also a part of any process $X^{(n)}$ on $E \cap E_m$ if $n \geqq m$. Hence $G^{(n)}f \geqq \hat{G}f$ on $E \cap E_m$ for $n \geqq m$, which for $c_0 = \inf_K \hat{G}f$ yields inequality (10).

The proof of the existence of the constant c_1 is harder. We restrict ourselves to the case when some compact set $F \subset E \cap E_m$ has simply connected interior containing the support K^* of the function f (in the contrary case one can represent f as the sum of a finite number of continuous functions with this property). Let $\tau_1 = \tau^{(n)}(K^*)$, and let τ_2 be the first exit moment after τ_1 of the trajectory $x_t^{(n)}$ from F, let τ_3 be the first moment after τ_2 of its occurence in K^*, let τ_4 be its first exit moment after τ_3 from F, and so on $(n \geqq m)$. Then for $n \geqq m$ and $x \in E_n$ we have

(12) $$G^{(n)}f(x) = \mathbf{E}_x^{(n)} \int_0^{\zeta(n)} f(x_t^{(n)})\, dt = \sum_{k \geqslant 1} \mathbf{E}_x^{(n)} \int_{\tau_{2k-1}}^{\tau_{2k}} f(x_t^{(n)})\, dt.$$

Let $\alpha_k(x)$ denote the kth term of the last series. Since the parts of the processes $X^{(n)}$ and $X^{(m)}$ coincide on F, it follows that

$$\alpha_1(x) \leqq G^{(m)}f(x) \leqq C,$$

where $C = \sup_{E_m} G^{(m)}f$. Moreover, by the strong Markov property we have

$$\alpha_k(x) = \mathbf{E}_x^{(n)} \alpha_1(x^{(n)}(\tau_{2k-2})) \quad (k > 1).$$

From this and (12), having set $\tau_0 \equiv 0$, we obtain

(13) $$G^{(n)}f(x) \leqslant C \sum_{k \geqslant 0} \mathbf{P}_x^{(n)}\{\tau_{2k} < \zeta^{(n)}\} \quad (x \in E_n).$$

Let us estimate the terms of the series on the right side of (13). For this we first establish the existence of a $\beta > 0$ not depending on the choice of the parameters $d(k)$, where $k \geqq m+1$, and such that

(14) $$\mathbf{P}_x^{(n)}\{\tau_1 = \zeta^{(n)}\} \geqq \beta,$$

where $n \geqq m$ and $x \in F_1$, where F_1 is the boundary of F. To this end, consider an arbitrary closed ball $U \subset E_0 \backslash (\Gamma_0 \cup F)$ and set

$$\eta = x^{(n)}(\tau^{(n)}(U)), \quad \Lambda = \{\tau^{(n)}(U) < \tau^{(n)}(K^*)\},$$

and denote the indicator of the event Λ by χ (Γ_0 and Γ_1 are defined in the text of Lemma 2). Then from the strong Markov property for $x \in F_1$ we find that

$$\mathbf{P}_x^{(n)}\{\tau_1 = \zeta^{(n)}\} \geqslant \mathbf{P}_x^{(n)}\{\Lambda;\ \tau_1 = \zeta^{(n)}\} = \mathbf{E}_x^{(n)}[\chi \cdot \mathbf{P}_\eta^{(n)}\{\tau_1 = \zeta^{(n)}\}]$$

$$\geqslant \mathbf{E}_x^{(n)}[\chi \cdot \mathbf{P}_\eta^{(n)}\{\tau^{(n)}(K^* \cup \Gamma_1) = \zeta^{(n)}\}] = \mathbf{E}_x^{(n)}[\chi \cdot \mathbf{P}_\eta^{(0)}\{\tau^{(0)}(K^* \cup \Gamma_1) = \zeta^{(0)}\}$$

(here the equivalence of the parts of the processes $X^{(n)}$ and $X^{(0)}$ on $E_0\backslash\Gamma_1$ is taken into consideration). The probability under the last mathematical expectation sign can be written as $1-\pi^{(0)}(\eta, K^*\cup\Gamma_1)$. Moreover, remember that the process $\hat{X}$ introduced at the beginning of the proof is a part of $X^{(n)}$. Hence from the last chain of relations it follows that for $x\in F_1$

$$\mathbf{P}_x^{(n)}\{\tau_1=\zeta^{(n)}\}\geqslant\beta_0\mathbf{E}_x^{(n)}\chi\geqslant\beta_0\hat{\mathbf{E}}_x\hat{\chi}=\beta_0\hat{\mathbf{P}}_x\{\hat{\tau}(U)<\hat{\tau}(K^*)\}\geqslant\beta_0\beta_1,$$

where the terms under the sign ^ are defined by means of $\hat{X}$ in exactly the same way as the analogous ones for X, and where

$$\beta_0=\inf_{x\in U}[1-\pi^{(0)}(x, K^*\cup\Gamma_1)] \text{ and } \beta_1=\inf_{x\in F_1}\hat{P}_x\{\hat{\tau}(U)<\hat{\tau}(K^*)\}.$$

The probability under the last "inf" is a function of x which is harmonic and positive on the set $[E\cap E_m]\backslash[U\cup K^*]$. Thus $\beta_1>0$. Moreover, by condition (A), $\pi^{(0)}(x,\Gamma_1)<1$ for some $x=x_0\in E_0\backslash\Gamma_0$. By the maximum principle this inequality must hold everywhere on $E_0\backslash\Gamma_0$. From the properties of the classical Wiener walk it follows that also $\pi^{(0)}(x,K^*\cup\Gamma_1)<1$, where the function in this inequality is harmonic on $[E\cap E_m]\backslash[K^*\cup\Gamma_1]$. Hence $\beta_0>0$, too. Now setting $\beta=\beta_0\beta_1$ we arrive at estimate (14).

Further, for $k\geqq 1$ and $x\in E_n$ we find that

$$\mathbf{P}_x^{(n)}\{\tau_{2k}<\zeta^{(n)}\}\leqslant\mathbf{P}_x^{(n)}\{\tau_{2k-1}<\zeta^{(n)}\}=\mathbf{E}_x^{(n)}\{\chi_{k-1}\cdot\mathbf{P}_\xi^{(n)}\{\tau_1<\zeta^{(n)}\}\},$$

where $\xi=x^{(n)}(\tau_{2k-2})$, with $\xi\in F_1$ (almost surely), and χ_{k-1} is the indicator of the event $\{\tau_{2k-2}<\zeta\}$. By (14) the probability in the last expectation does not exceed $1-\beta$. Thus

$$\mathbf{P}_x^{(n)}\{\tau_{2k}<\zeta^{(n)}\}\leqslant(1-\beta)\,\mathbf{P}_x^{(n)}\{\tau_{2k-2}<\zeta^{(n)}\}\leqslant\ldots\leqslant(1-\beta)^n.$$

Combining this fact with (13) and taking $c_1=C\sum_{n\geq 0}(1-\beta)^n$, we obtain estimate (11). The proof is complete.

Lemma 6. *Let there be given an* $\epsilon>0$ *and a function* $\varphi\in\mathscr{L}_0(E_0)$ *vanishing on the set* $\Gamma_0=\bigcup_{n\geq 1}[U_n\cup U_n']$. *One can select values of the parameters* $d(n)$ $(n\geqq 1)$ *so that in addition to condition* (A) *the following conditions also hold*:

(B$_1$) *For any function* $f\in\mathscr{L}_0(E)$, *everywhere on* E

$$Gf(x)=\lim_{n\to\infty}G^{(n)}f(x), \tag{15}$$

where the convergence is uniform in $x\in E_0\backslash\Gamma_0$.

(B$_2$) *For* $f\in\mathscr{L}_0(E)$,

$$\lim_{\|x\|\to\infty,\ x\in E_0\backslash\Gamma_0}Gf(x)=0.$$

(B$_3$) *Everywhere on* $E_0\backslash\Gamma_0$,

$$|G^{(0)}\varphi-G\varphi|<\epsilon \quad\textit{and}\quad |G^{(0)}v-Gv|<\epsilon. \tag{16}$$

(B$_4$) *For every compact set* $K\subset E_0\backslash\Gamma_0$,

$$\inf_V Gv > 0, \tag{17}$$

where $V = K \cup [\bigcup_{n \geq 1} S_n]$.

(B_5) *Everywhere on* $E_0 \backslash \Gamma_0$,

$$\pi^{(n)}(x, K(n+1) \cup K'(n+1)) \leqslant \varepsilon \cdot 2^{-n-1} Gv(x) \quad (n \geqslant 0). \tag{18}$$

PROOF. By Lemma 5, on the union V_n of the support of v and the set $\bigcup_{1 \leq m \leq n} [S_m \cup S_m']$ the function $G^{(k)}v$ takes on values exceeding some constant $\beta_n' > 0$, which is the same for all values $k \geq n$ and any admissible modifications of E. Hence one can choose a numerical sequence $\{\beta_n > 0;\ n \geq 1\}$ for which

$$\beta_n \leqq 2^{-n} \min[\beta, \epsilon\beta_n'], \tag{19}$$

where $\beta > 0$ is some constant whose choice will be specified later $(n \geq 1)$.

Let us proceed to construct the sequence $\{d(n);\ n \geq 1\}$. We choose $d(1)$ $(0 < d(1) < \delta_1)$ so that $\pi^{(0)}(x, K(1) \cup K'(1)) < \beta_1$ for all $x \in S_1 \cup S_1'$ (the existence of such is guaranteed by relation (1)). The probability

$$\pi^{(1)}(x, K(2) \cup K'(2))$$

converges to 0 as $d(2) \to 0$ for any $x \in E_1 \backslash [U(2) \cup U'(2)]$, for otherwise the Wiener trajectory on E_1, with positive probability, would contact the center of at least one of the balls U_2 or U_2', which is impossible. Moreover, this probability converges to 0 as $d(2) \to 0$ uniformly on each compact set lying in $E_1 \backslash [U(2) \cup U'(2)]$, since as a function of x it is harmonic on this set. Hence one can find $d(2)$ $(0 < d(2) < \delta_2)$ for which $\pi^{(1)}(x, K(2) \cup K'(2)) < \beta_2$ for $x \in S_2 \cup S_2'$.

Extending this argument, for each $n \geq 1$ we can find $d(n)$ $(0 < d(n) < \delta_n)$ with the property that

$$\pi^{(n-1)}(x, K(n) \cup K'(n)) < \beta_n \qquad (x \in S_n \cup S_n'). \tag{20}$$

This holds for all $x \in \widetilde{E}_{n-1}$, where $\widetilde{E}_{n-1} = E_{n-1} \backslash [U_n \cup U_n']$, since the maximum principle is applicable to its left side (which we can consider as the potential of some measure concentrated on $K(n) \cup K'(n)$). Thus for $x \in \widetilde{E}_{n-1}$ we have

$$\begin{aligned} \pi^{(n-1)}\Big(x, \bigcup_{m \geqslant n} [K(m) \cup K'(m)]\Big) &\leqslant \sum_{m \geqslant n} \pi^{(n-1)}(x, K(m) \cup K'(m)) \\ &= \sum_{m \geqslant n} \pi^{(m-1)}(x, K(m) \cup K'(m)) \leqslant \sum_{m \geqslant n} \beta_m, \end{aligned} \tag{21}$$

where we have used the equivalence of the parts of the processes $X^{(n)}$ and $X^{(m)}$ on the set $\widetilde{E}_{n-1}$.

Now let $f \in \mathscr{L}_0(E)$ and so also $f \in \mathscr{L}_0(E_n)$ for some $n = n_0$. By Lemma 5 the values of $|G^{(n)}f|$ for $n \geq n_0$ do not exceed some constant $C > 0$. Increasing C, if necessary, we may assume that also $|Gf| < C$. Moreover, everywhere on E_{n-1} we have the equality

$$G^{(n-1)}f(x) = \mathbf{E}_x \int_0^\tau f(x_t)\,dt + \mathbf{E}_x^{(n-1)} G^{(n-1)} f(x_\tau) \quad (n \geqslant n_0),$$

where $\tau = \tau(\bigcup_{m \geq n}[K(m) \cup K'(m)])$ and where we have used the equivalence of the parts of X and $X^{(n-1)}$ on $\widetilde{E}_{n-1}$. We can write the analogous equality also for Gf and then subtract it from the former. Then with (21) we obtain for $n \geqq n_0 + 1$ that on the set $\widetilde{E}_{n-1} \cap E$

$$|G^{(n-1)}f(x) - Gf(x)| = |\mathbf{E}_x^{(n-1)}[G^{(n-1)}f(x_\tau) - Gf(x_\tau)]|$$

(22)
$$\leqslant 2C \cdot \mathbf{P}_x^{(n-1)}\{\tau < \zeta^{(n-1)}\} = 2C \cdot \pi^{(n-1)}(x, \bigcup_{m \geqslant n}[K(m) \cup K'(m)]) \leqslant 2C \sum_{m \geqslant n} \beta_m.$$

Now replacing Gf by $G^{(n)}f$ we arrive at the inequality

(23)
$$|G^{(n-1)}f(x) - G^{(n)}f(x)| \leqslant 2C\beta_n \quad (n \geqslant n_0 + 1).$$

Now we note that by (19) the estimate (22) implies assertion (B_1). From (B_1) and (23) we get

(24)
$$|G^{(n)}f(x) - Gf(x)| \leqslant \sum_{m \geqslant n+1} |G^{(m)}f(x) - G^{(m-1)}f(x)| \leqslant 2C \sum_{m \geqslant n+1} \beta_m \leqslant 2C\beta$$

on the set $E_0 \backslash \Gamma_0$ if the support of f is contained in $E_0 \backslash \Gamma_0$. To underline the dependence of C on f let us denote it by C_f and next subject the choice of $\beta > 0$ to the condition

$$\beta < \min[\varepsilon/2C_v;\ \varepsilon/2C_\varphi;\ \inf_V G^{(0)}v/4C_v].$$

Then from (24) it follows that on $E_0 \backslash \Gamma_0$

$$|G^{(0)}\varphi - G\varphi| \leqslant 2C_\varphi\beta < 2C_\varphi\varepsilon/2C_\varphi = \varepsilon,$$

i.e. the first of inequalities (16) holds. We see the validity of the second one similarly. Thus (B_3) holds.

Again applying (24) we obtain on $E_0 \backslash \Gamma_0$ the inequality

$$|G^{(0)}v - Gv| \leqslant 2C_v\beta < 2C_v \inf_V G^{(0)}v/4C_v = \frac{1}{2}\inf_V G^{(0)}v$$

and hence

$$\inf_V Gv \geqslant \frac{1}{2}\inf_V G^{(0)}v > 0,$$

so that we have verified (B_4) as well.

Now let us turn to (19) and (20) and note that

$$\beta_n' \leqslant \varliminf_{k\to\infty}[\inf_V G^{(k)}v] \leqslant \inf_{V_n} Gv$$

by (B_1). Therefore on $E_0 \backslash \Gamma_0$ we have

$$\pi^{(n-1)}(x, K(n) \cup K'(n)) < \beta_n \leqslant 2^{-n}\varepsilon\beta_n' \leqslant \varepsilon 2^{-n} \inf_{V_n} Gv.$$

These relations permit us to use (18) on the boundary of $E_0\backslash\Gamma_0$ and on the support of v. Hence, by the maximum principle, (18) is valid everywhere on $E_0\backslash\Gamma_0$. Thus (B_5) is established.

It remains to verify (B_2). Here by (B_1) it suffices to prove the equality

$$\lim_{\|x\|\to\infty,\ x\in E_0\backslash\Gamma_0} G^{(n)}f(x)=0 \quad (n>0), \tag{25}$$

if $f\in\mathscr{C}_0(E_n)$. We assume n fixed. By means of the strong Markov property we find that on $E_0\backslash\Gamma_0$

$$G^{(n)}f=\pi_F^{(n)}G^{(n)}f=\pi_F^{(0)}G^{(n)}f,$$

where F is the union of the support of f and the set $\bigcup_{1\leqq m\leqq n}[K(m)\cup K'(m)]$. Hence on $E_0\backslash\Gamma_0$ we have

$$|G^{(n)}f(x)|\leqslant C_f\pi^{(0)}(x, F),$$

and from the properties of classical Brownian motion it follows that the right side of this inequality converges to 0 as $\|x\|\to\infty$. Inequality (25) is proved, and thus also the entire lemma.

§4. The main result

Let us now consider Martin operators corresponding to the processes X and $X^{(n)}$. For X the Martin operator $\mathscr{K}$ is defined by the formula

$$\mathscr{K}f(x)=Gf(x)/Gv(x) \qquad (x\in E)$$

on the class of bounded Borel functions f on E with compact support, where v is the function introduced in §2. Replacing G here by $G^{(n)}$ and E by $E^{(n)}$ we obtain the definition of the operators $\mathscr{K}^{(n)}$.

THEOREM 1. *Let there be given a number* $\epsilon_0>0$ *and a function* $\varphi\in\mathscr{C}_0(E_0)$ *vanishing on the set* Γ_0. *There exist values of the parameters* $d(n)$ $(n\geqq 1)$ *such that in addition to conditions* (A), (B_2) *and* (B_4) *the following conditions also hold*:

(C_1) *For every* $f\in\mathscr{C}_0(E)$

$$\mathscr{K}f(x)=\lim_{n\to\infty}\mathscr{K}^{(n)}f(x) \quad (x\in E), \tag{26}$$

where the convergence is uniform in $x\in E_0\backslash\Gamma_0$.

(C_2) *Everywhere on* $E_0\backslash\Gamma_0$

$$|\mathscr{K}^{(0)}\varphi-\mathscr{K}\varphi|<\epsilon_0. \tag{27}$$

PROOF. Let us take an arbitrary value of $\epsilon>0$ and applying Lemma 6 let us select values of $d(n)$ $(n\geqq 1)$ for which assertions (B_1) – (B_5) hold. Using the computation of (22) from the preceding proof and condition (B_5) we easily obtain that everywhere on $E_0\backslash\Gamma_0$

$$\begin{aligned}|G\varphi-G^{(0)}\varphi|&<2C_1\sum_{n\geqslant 0}\pi^{(n)}(x, K(n+1)\cup K'(n+1))\\ &\leqslant 2C_1\varepsilon Gv(x)\sum_{n\geqslant 0}2^{-n-1}=2C_1\varepsilon Gv(x),\end{aligned} \tag{28}$$

and similarly

$$|Gv - G^{(0)}v| < 2C_1\epsilon Gv,$$

where C_1 is some constant, the same for all possible modifications of E. Further, let us represent $\mathscr{K}\varphi - \mathscr{K}^{(0)}\varphi$ in the form

$$\frac{G\varphi}{Gv} - \frac{G^{(0)}\varphi}{G^{(0)}v} = \frac{G\varphi - G^{(0)}\varphi}{Gv} + \frac{G^{(0)}\varphi}{G^{(0)}v} \cdot \frac{G^{(0)}v - Gv}{Gv} \tag{29}$$

and set $\alpha = \sup_{E_0} |\mathscr{K}^{(0)}\varphi|$. From (29) and the two inequalities preceding it it follows that on $E_0 \backslash \Gamma_0$ we have

$$|\mathscr{K}\varphi - \mathscr{K}^{(0)}\varphi| \leqslant 2C_1\varepsilon Gv \left(\frac{1}{Gv} + \frac{\alpha}{Gv}\right) = 2C_1\varepsilon(1+\alpha). \tag{30}$$

Thus (C_2) holds if $\epsilon > 0$ is taken so small that the last expression is less than ϵ_0. Let us fix one of these values of $\epsilon > 0$ and the function $f \in \mathscr{L}_0(E)$. Recalling (B_1) one can write

$$\mathscr{K}f = \frac{Gf}{Gv} = \frac{\lim_{n\to\infty} G^{(n)} f}{\lim_{n\to\infty} G^{(n)} v} = \lim_{n\to\infty} \frac{G^{(n)} f}{G^{(n)} v} = \lim_{n\to\infty} \mathscr{K}^{(n)} f,$$

whence (26) follows. It remains only to show the uniform convergence in this formula in $x \in E_0 \backslash \Gamma_0$. To this end let us represent the difference $\mathscr{K}\varphi - \mathscr{K}^{(n)}\varphi$ on $E_0 \backslash \Gamma_0$ in a form similar to (29) and note the estimates

$$|Gf - G^{(n)}f| < 2C\varepsilon Gv 2^{-n} \quad \text{and} \quad |Gv - G^{(n)}v| < 2C_1\varepsilon Gv 2^{-n},$$

analogous to estimate (28), where $C > 0$ is a suitable constant. Then, as in (30), we shall have

$$|\mathscr{K}\varphi - \mathscr{K}^{(n)}\varphi| \leqslant 2\varepsilon Gv 2^{-n}\left(\frac{C}{Gv} + \frac{\alpha C_1}{Gv}\right) = 2(C + C_1\alpha)\varepsilon Gv 2^{-n}.$$

Thus we have proved the remaining part of (C_1), and hence the theorem in its entirety.

Let us consider the Martin compactification of E. The Martin compactum E^M is defined (cf. [**1**, **2**, **6**]) as the compactum $E^M \supset E$ with the following properties: a) E is dense in E^M; b) any of the functions $\mathscr{K}f$, where $f \in \mathscr{L}_0(E)$, admits a continuous extension to E^M (which will be denoted by the same symbol $\mathscr{K}f$; and c) if for some x and y in E^M the values $\mathscr{K}f(x)$ and $\mathscr{K}f(y)$ coincide for every choice of $f \in \mathscr{L}_0(E)$, then $x = y$. The set $\Delta(E) = E^M \backslash E$ is called the *Martin boundary*.

As to an unbounded set $F \subset E$ we shall say that it determines a unique point at infinity z of the boundary $\Delta(E)$ (notation: $z = \mathfrak{z}(F)$) if

$$\lim_{\|x\|\to\infty,\, x\in F} \mathscr{K}f(x) = \mathscr{K}f(z) \tag{31}$$

for all $f \in \mathscr{L}_0(E)$. In order that there be a point $z \in \Delta(E)$ with the cited property it suffices to require only the existence of the limit on the left side of (31) for all $f \in \mathscr{L}_0(E)$.

Analogous definitions are adopted for the case of the spaces E_n ($n \geqq 0$), and, in particular, the Martin boundaries $\Delta(E_n)$ are thus introduced.

The next two lemmas indicate some sets determining unique points at infinity on the corresponding Martin boundaries.

LEMMA 7. *For any $n \geqq 1$ each of the sets*

$$\mathfrak{R}(1), \cdots, \mathfrak{R}(n) \quad \textit{and} \quad E_0 \setminus \bigcup_{1 \leqq k \leqq n} [U_k \cup U_k']$$

determines (depending on the set) a unique point at infinity on the Martin boundary $\Delta(E_n)$. Moreover, each of the first n sets determines a unique point at infinity of the boundary $\Delta(E)$.

PROOF. Let us verify that $\mathfrak{R}(1)$ determines a unique point of $\Delta(E)$. Let $f \in \mathscr{L}_0(E)$, and recall that $\mathfrak{R}^{(0)}(1)$ is identified with some three-dimensional hyperplane with two cubes deleted (see Remark 1). Consider the two-dimensional sphere $S \subset E$ contained in the (three-dimensional) hyperplane in which lie the cubes $K^*(1)$ and $K^{*\prime}(1)$. Here we require that the interior of S contain both these cubes as well as the portion of the support of f which is in that hyperplane. Since $Gf(x) = \pi_S Gf(x)$ and $Gv(x) = \pi_S Gv(x)$, if the point $x \in \mathfrak{R}(1)$ lies outside the ball bounded by the sphere S, then for each such point

$$\mathscr{K}f(x) = \frac{Gf(x)}{Gv(x)} = \frac{\pi_S Gf(x)}{\pi_S Gv(x)}.$$

This fact together with Lemma 3 ensures the existence of the limit (31) for $F = \mathfrak{R}(1)$, so that the lemma holds in the case being considered. The other cases are treated similarly.

LEMMA 8. *If condition* (C_1) *holds, the set $E_0 \setminus \Gamma_0$ determines a unique point at infinity of the boundary $\Delta(E)$.*

PROOF. Let us observe two facts: first, by the preceding lemma the limit $\lim_{\|x\| \to \infty; x \in E_0 \setminus \Gamma_0} \mathscr{K}^{(n)} f(x)$ exists for all $f \in \mathscr{L}_0(E_n)$, where $n \geqq 0$, and each function $f \in \mathscr{L}_0(E)$ belongs to $\mathscr{L}_0(E_n)$ for sufficiently large n; and second, by condition (C_1) the sequence $\{\mathscr{K}^{(n)} f(x)\}$ converges to $\mathscr{K}f(x)$ as $n \to \infty$ uniformly for $x \in E_0 \setminus \Gamma_0$. These two facts imply the existence of the limit (31) for $F = E_0 \setminus \Gamma_0$ if $f \in \mathscr{L}_0(E)$. The lemma is proved.

Let us outline the subsequent exposition. Let $\mathfrak{z}_n = \mathfrak{z}(\mathfrak{R}(n))$ and $\mathfrak{z} = \mathfrak{z}(E_0)$ be points at infinity of the boundary $\Delta(E)$ defined by the corresponding sets ($n \geqq 1$). We shall prove (see Theorem 2) that under certain conditions $\mathfrak{z}$ is not a regular point of $\Delta(E)$. However, to do this we first need to answer the question: can $\mathfrak{z}$ not be the limit of the sequence $\{\mathfrak{z}_n; n \geqq 1\}$?

In order to answer this question (see Lemma 9) we fix the $d(n)$ ($n \geqq 1$) as follows. Note that the kernel defining the functional $\mathscr{K}^{(0)} f(0)$, in contrast with the kernel defining $\mathscr{K}^{(0)} f(\hat{\mathfrak{z}})$, is not identically constant on any region of E_0 where $f \in \mathscr{L}_0(E_0)$. Hence there exists a nonnegative function $\varphi \in \mathscr{L}_0(E_0)$ whose support lies in $E_0 \backslash \Gamma_0$ and for which

$$(32) \qquad \varepsilon_0 \equiv \frac{1}{5} \,|\, \mathcal{K}^{(0)}\varphi(\hat{\mathfrak{z}}) - \mathcal{K}^{(0)}\varphi(0) \,| > 0,$$

where $\hat{\mathfrak{z}}$ is the unique point of the Martin boundary $\Delta(E_0)$ (it corresponds to the point at infinity (in the usual sense) of the hyperplane E_0). Relying on Theorem 1 we choose for this pair φ and ϵ_0 parameters $d(n)$ $(n \geqq 1)$ which ensure conditions (A), (B_2), (B_4), (C_1) and (C_2), and keep them fixed in all that follows. Thus, as of now, *all parameters defining the space E are fixed.*

LEMMA 9. *The point $\mathfrak{z}$ is not a limit point (in the topology of the compactum $\Delta(E)$) for the sequence $\{\mathfrak{z}_n;\ n \geqq 1\}$.*

PROOF. Suppose for the moment that we have already proved the relations

$$(33) \qquad |\, \mathcal{K}\varphi(\mathfrak{z}) - \mathcal{K}^{(0)}\varphi(\hat{\mathfrak{z}}) \,| < 2\varepsilon_0 \quad \text{and} \quad |\, \mathcal{K}\varphi(\mathfrak{z}_n) - \mathcal{K}^{(0)}\varphi(0) \,| < 2\varepsilon_0,$$

where n is assumed sufficiently large. Then

$$|\, \mathcal{K}\varphi(\mathfrak{z}_n) - \mathcal{K}\varphi(\mathfrak{z}) \,| = |\, [\mathcal{K}\varphi(\mathfrak{z}_n) - \mathcal{K}^{(0)}\varphi(0)]$$
$$+ [\mathcal{K}^{(0)}\varphi(\hat{\mathfrak{z}}) - \mathcal{K}\varphi(\mathfrak{z})] + [\mathcal{K}^{(0)}\varphi(0) - \mathcal{K}^{(0)}\varphi(\hat{\mathfrak{z}})] \,| > \varepsilon_0,$$

since the moduli of the first two expressions in square brackets are less than $2\epsilon_0$ (see (33)), while the modulus of the third expression is equal to $5\epsilon_0$ (see (32)). From the last inequality $\mathcal{K}\varphi(\mathfrak{z}_n)$ does not converge to $\mathcal{K}\varphi(\mathfrak{z})$ as $n \to \infty$, and therefore $\mathfrak{z}$ is not a limit point of $\{\mathfrak{z}_n;\ n \geqq 1\}$.

Thus to prove the lemma it suffices to show that the inequalities (33) hold. For the first of them we need only take into account that if $x_n \in E_0 \backslash \Gamma_0$ $(n \geqq 1)$ and $\|x_n\| \to \infty$ as $n \to \infty$, then from (C_2) we have

$$|\, \mathcal{K}\varphi(\mathfrak{z}) - \mathcal{K}^{(0)}\varphi(\hat{\mathfrak{z}}) \,| = \lim_{n\to\infty} |\, \mathcal{K}\varphi(x_n) - \mathcal{K}^{(0)}\varphi(x_n) \,| \leqslant \varepsilon_0.$$

Let us look at the second inequality. In E_0 let us choose a disk K with center at the origin and radius small enough so that for $x \in K$

$$(34) \qquad |\, \mathcal{K}^{(0)}\varphi(0) - \mathcal{K}^{(0)}\varphi(x) \,| < \frac{\varepsilon_0}{2}.$$

Till the end of the proof let us agree to consider (without special reference) only the values of n for which $S_n \subset K$. This permits us to write

$$(35) \qquad \begin{aligned} |\, \mathcal{K}^{(0)}\varphi(0) - \mathcal{K}\varphi(x) \,| &\leqslant |\, \mathcal{K}^{(0)}\varphi(0) - \mathcal{K}^{(0)}\varphi(x) \,| \\ &+ |\, \mathcal{K}^{(0)}\varphi(x) - \mathcal{K}\varphi(x) \,| < \frac{3\varepsilon_0}{2}, \end{aligned}$$

if $x \in S_n$ (see (27)).

Note also that by (B_2) and (B_4) the quotient

$$(36) \qquad \sup_{S_n'} Gf \,/\, \inf_{S_n} Gv$$

tends to 0 as $n \to \infty$ for any $f \in \mathscr{L}_0(E)$. Hence on the values of n being considered we have the right to impose yet another restriction, requiring that for $f = \varphi$ the quotient (36) be less than $\epsilon_0/4$.

Let us fix any of the values of n satisfying the above restrictions and consider the sequence $\{y_k \in \mathfrak{R}^{(0)}(n); k \geqq 1\}$ such that $y_k \to \mathfrak{z}_n$ as $k \to \infty$. Let the symbols ν_k and ν_k' denote, respectively, the restrictions of the measures $\pi_{S_n \cup S_n'}(y_{(k)}, \cdot)$ to S_n and S_n'. Requiring additionally that the sequence y_k be entirely contained in the hyperplane $\mathfrak{M}(n)$ of the superstructure $\mathfrak{R}(n)$, we obtain from symmetry considerations that

$$\nu_k(S_n) = \nu_k'(S_n) \qquad (k \geqq 1). \tag{37}$$

Now we derive the estimate

$$|\mathscr{K}\varphi(y_k) - \mathscr{K}^{(0)}\varphi(0)| < \frac{15\varepsilon_0}{8} \tag{38}$$

for sufficiently large values of k. For this we shall start from the equalities

$$\mathscr{K}\varphi(y_k) = \frac{\pi_{S_n \cup S_n'} G\varphi(y_k)}{\pi_{S_n \cup S_n'} Gv(y_k)} = \frac{\int G\varphi d\nu_k}{\int Gvd[\nu_k + \nu_k']} + \frac{\int G\varphi d\nu_k'}{\int Gvd(\nu_k + \nu)}, \tag{39}$$

the first of which was used analogously in proof of Lemma 7.

Let $A(k)$ and $B(k)$ be the first and second quotients in the sum (39). Since for $f = \varphi$ the quantity (36) is less than $\epsilon_0/4$ and since (37) holds, we have

$$B(k) \leqslant \frac{\sup_{S_n'} G\varphi \cdot \nu_k'(S_n)}{\inf_{S_n} Gv \cdot \nu_k(S_n)} = \frac{\sup_{S_n'} G\varphi}{\inf_{S_n} Gv} < \frac{\varepsilon_0}{4}. \tag{40}$$

Now let us consider the quantities $A(k)$, representing them in the form

$$A(k) = A_1(k)/A_2(k), \tag{41}$$

where $A_1(k) = \int G\varphi d\nu_k / \int Gvd\nu_k$ and

$$A_2(k) = \frac{\int Gvd[\nu_k + \nu_k']}{\int Gvd\nu_k} = 1 + \frac{\int Gvd\nu_k'}{\int Gvd\nu_k}. \tag{42}$$

The second term in (42) is estimated above by the expression (36) for $f = v$, and hence it converges to 0 as $k \to \infty$. The estimate of $A_1(k)$ is based on the known fact that for any two integrable functions g and $h \geqq 0$, where $\inf g > 0$, which are defined on some measure space $(E, \mathscr{B}, \mu)$, there hold the inequalities[3)]

$$\inf \frac{h}{g} \leqslant \frac{\int h d\mu}{\int g d\mu} \leqslant \sup \frac{h}{g}. \tag{43}$$

[3)] For example, the first of inequalities (43) is easily obtained if in the numerator of the fraction to be estimated in (43) the function h is replaced by $g \cdot \inf[h/g]$.

These inequalities applied to our case ($\mu = \nu_k$) yield

$$\inf_{S_n} \mathcal{K}\varphi = \inf_{S_n} \frac{G\varphi}{Gv} \leqslant A_1(k) \leqslant \sup_{S_n} \frac{G\varphi}{Gv} = \sup_{S_n} \mathcal{K}\varphi, \tag{44}$$

whence from (35) it follows that

$$|A_1(k) - \mathcal{K}^{(0)}\varphi(0)| \leqslant \frac{3\varepsilon_0}{2}.$$

Combining the last inequality with the fact that $A_2(k) \to 1$ as $k \to \infty$ and recalling the representation (41), we obtain the estimate

$$|A(k) - \mathcal{K}^{(0)}\varphi(0)| < \frac{3\varepsilon_0}{2} + \frac{\varepsilon_0}{8} = \frac{13}{8}\varepsilon_0 \tag{45}$$

for sufficiently large values of k. This and (39) immediately imply the desired inequality (38) if k is sufficiently large.

Passing to the limit in (38) as $k \to \infty$ and using the convergence of $\mathscr{K}\varphi(y_k)$ to $\mathscr{K}\varphi(\mathfrak{z}_n)$, we obtain the second inequality in (33) for all n satisfying the conditions cited earlier. Thus the proof is complete.

Before formulating the main result we recall that a point $x \in \Delta(E)$ is called *regular* if for any of its open neighborhoods $U \subset E^M$

$$\lim_{y \to x} \mathbf{P}_y\{x_{\zeta-0} \in U\} = 1, \tag{46}$$

where the limit $x_{\zeta-0}$ is understood in the sense of the Martin topology (this limit exists almost surely). By an *irregular* point we mean a point $x \in \Delta(E)$ which is not regular. A point $x \in \Delta(E)$ is called *extremal* if the function $k_x(\cdot)$ corresponding to it in the Martin expansion is minimal (i.e. if for a harmonic function $u(\cdot) \geqq 0$ on E the inequality $u \leqq k_x$ is possible only when u differs from k_x only by a constant factor). We shall call a point $x \in \Delta(E)$ a point *having positive harmonic mass* if $\mathbf{P}_y\{x_{\zeta-0} = x\} \neq 0$ for $y \in E$.

THEOREM 2. *The point $\mathfrak{z}$ is an extremal irregular point of the Martin boundary $\Delta(E)$, and it has positive harmonic mass.*

The proof is divided into two parts.

1°. Condition (A) ensures the satisfaction of inequality (2) of Lemma 2. Thus $\mathbf{P}_{x_0}\{\Lambda\} > 0$, where $\Lambda = \{\zeta = \infty;\ \tau_{E_0/\Gamma_1} = \infty\}$, and x_0 is some point of E_0/Γ_0. Since E_0/Γ_0 is a quotient of three-dimensional Euclidean space E_0, from properties of a classical Wiener process it follows that

$$\mathbf{P}_{x_0}\{\Lambda; \lim_{t\to\infty} \|x_t\| = \infty\} = \mathbf{P}_{x_0}\{\Lambda\} = 0.$$

From this we make two deductions: first, the point $\mathfrak{z}$ must be extremal since $x_{\zeta-0}$ almost surely does not appear in the set of extremal points; second, the point $\mathfrak{z}$ must have positive harmonic mass.

2°. It remains for us to show that $\mathfrak{z}$ is irregular. To this end we consider the sequence $\{a_n; n \geqq 1\}$ whose properties were studied in the corollaries to

Lemma 4. According to the first of inequalities (7), with $\mathbf{P}_{a_n}$-probability exceeding some $\gamma_1 > 0$ the trajectory x_t does not stop and beginning with some moment of time always remains in the set $\mathfrak{R}^{(0)}(n)$, which can be identified with some set of Euclidean space (see Remark 1). Hence, in exactly the same way as in our derivation of (47), we conclude that

$$\mathbf{P}_{a_n}\{x_{\zeta-0} = \mathfrak{z}_n\} > \gamma_1 \tag{48}$$

for any $n \geqq 1$.

Suppose for the moment that we have established the relation

$$\lim_{n\to\infty} a_n = \mathfrak{z}. \tag{49}$$

Using the preceding lemma we choose a neighborhood U of $\mathfrak{z}$ not containing any of the $\mathfrak{z}_n$ $(n \geqq 1)$. By (48) we have

$$\overline{\lim_{n\to\infty}} \mathbf{P}_{a_n}\{x_{\zeta-0} \in U\} \leqslant \overline{\lim_{n\to\infty}} \mathbf{P}_{a_n}\{x_{\zeta-0} \neq \mathfrak{z}_n\} \leqslant 1 - \gamma_1.$$

These relations, together with (49), prove that for $x = \mathfrak{z}$ the relation (46) is impossible, i.e. that $\mathfrak{z}$ is irregular. Thus the theorem will be proved as soon as we have proved (49).

For the verification of (49) we use arguments quite similar to those we used in proving (38). Namely, consider a function $f \in \mathscr{L}_0(E)$ and choose n so large that the support of this function lies in $E \cap E_{n-1}$. Just as with (39) we find that

$$\mathscr{K}f(a_n) = \frac{\pi_{S_n \cup S'_n} Gf(a_n)}{\pi_{S \cup S'_n} Gv(a_n)} = \frac{\int Gf d\mu_n}{\int Gvd[\mu_n + \mu'_n]} + \frac{\int Gf d\mu'_n}{\int Gvd[\mu_n + \mu'_n]}, \tag{50}$$

where μ_n and μ'_n are the restrictions of the measure $\pi_{S_n \cup S'_n}(a_n, \cdot)$ to S_n and S'_n, respectively. Let $A(n)$ and $B(n)$ denote the first and second terms in (50). Then

$$A(n) = \frac{\int Gf d\mu_n}{\int Gv d\mu_n} \cdot \frac{1}{A_1(n)}, \tag{51}$$

where

$$A_1(n) = \frac{\int Gvd[\mu_n + \mu'_n]}{\int Gv d\mu_n} = 1 + \frac{\int Gv d\mu'_n}{\int Gv d\mu_n}.$$

By relation (8) in Corollary 2 the last quotient is infinitely large as $n \to \infty$. At the same time the quotient playing the role of the first factor in the product (51), by an estimate similar to (44), is found to lie between the quantities

$$\inf_{S_n} \mathscr{K}f = \inf_{S_n} \frac{Gf}{Gv} \text{ and } \sup_{S_n} \mathscr{K}f = \sup_{S_n} \frac{Gf}{Gv}.$$

Hence its modulus cannot exceed $\sup_E |\mathscr{K}f| < \infty$. Hence

(52) $$\lim_{n\to\infty} A(n) = 0.$$

Let us now look at $B(n)$:

(53) $$B(n) = \frac{\int G f d\mu_n'}{\int G v d\mu_n'} \cdot \frac{1}{B_1(n)},$$

where

$$B_1(n) = \frac{\int G v d(\mu_n + \mu_n')}{\int G v d\mu_n'} = 1 + \frac{\int G v d\mu_n}{\int G v d\mu_n'}.$$

Again recalling (8) we see that

(54) $$\lim_{n\to\infty} B_1(n) = 1,$$

and a new application of type (44) yields

(55) $$\inf_{S_n'} \mathcal{K} f \leqslant \frac{\int G f d\mu_n'}{\int G v d\mu_n'} \leqslant \sup_{S_n'} \mathcal{K} f.$$

Since $S_n' \subset E_0 \setminus \Gamma_0$ and $\inf_{S_n'} \|x\| \to \infty$ as $n \to \infty$, by Lemma 8 the first and last expressions in (55) tend to $\mathcal{K} f(\mathfrak{z})$. Hence the expression lying between them also tends to $\mathcal{K} f(\mathfrak{z})$ as $n \to \infty$. Combining this fact with (53) and (54), we obtain

(56) $$\lim_{n\to\infty} B(n) = \mathcal{K} f(\mathfrak{z}).$$

Relations (52) and (56) together with (50) lead to the equalities

$$\lim_{n\to\infty} \mathcal{K} f(a_n) = \lim_{n\to\infty} [A(n) + B(n)] = \mathcal{K} f(\mathfrak{z}).$$

Thus the first and last terms in this chain of equalities coincide for all $f \in \mathscr{L}_0(E)$, and this is equivalent to (49).

Theorem 2 is completely proved.

Theorem 2, clearly, is a contradiction to Conjecture $\mathscr{R}$.

In conclusion we note that it would be interesting to attempt to construct an example of a bounded region of Euclidean space for which Conjecture $\mathscr{R}$ would be false. It would also be interesting to construct an example of a Green space in which the set of irregular extremal points x of the Martin boundary with unbounded functions k_x would have positive harmonic measure (in our example, as one can check, a bounded function $k_{\mathfrak{z}}$ corresponds to the point $\mathfrak{z}$).

Received 1/DEC/71

Bibliography

[1] M. Brelot, *On Martin boundaries*, Lectures at Univ. Hiroshima, 1962.
[2] ———, *Éléments de la théorie classique du potentiel*, 2ième éd., Centre de Documentation, Paris, 1961. MR **31** #1387.

[3] J. L. Doob, *Semimartingales and subharmonic functions,* Trans. Amer. Math. Soc. **77** (1954), 86-121. MR **16**, 269.

[4] ———, *Brownian motion on a Green space,* Teor. Verojatnost. i Primenen. **2** (1957), 3-33 = Theor. Probability Appl. **2** (1957), 1-20. MR **21** # 5240.

[5] E. B. Dynkin, *Markov processes,* Fizmatgiz, Moscow, 1963; English transl., Vols. I, II, Die Grundlehren der math. Wissenschaften, Bände 121, 122, Academic Press, New York; Springer-Verlag, Berlin, 1965. MR **33** # 1886; # 1887.

[6] H. Kunita and T. Watanabe, *Markov processes and Martin boundaries.* I, Illinois J. Math. **9** (1965), 485-526. MR **31** # 5240.

[7] L. Naim, *Sur le rôle de la frontière de R. S. Martin dans la théorie du potentiel,* Ann. Inst. Fourier (Grenoble) **7** (1957), 183-281. MR **20** # 6608.

[8] M. G. Šur, *Boundary properties of excessive and co-excessive functions,* Proc. First Soviet-Japanese Sympos. on Probability Theory (Khaborovsk, 1969), part I, Novosibirsk, 1969, pp. 284-293. (Russian) RŽMat. **1970** # 4Б75.

[9] ———, *A Martin compactum with a non-negligible irregular boundary point,* Teor. Verojatnost. i Primenen. **17** (1972), 366-370 = Theor. Probability Appl. **17** (1972), 351-354. MR **46** # 377.

Translated by
J. MALEK

Trudy Moskov. Mat. Obsc.
Tom 28 (1973)

Trans. Moscow Math. Soc.
Vol. 28 (1973)

ON CRITICAL SETS OF MORSE-SMALE SYSTEMS

V. S. AFRAĬMOVIČ and L. P. ŠIL'NIKOV

Contents

§1. Introduction. Formulation of basic results

One of the fundamental problems of the qualitative theory of differential equations is the clarification of the basic elements which determine the structure of the decomposition of phase space by the trajectories. For two-dimensional systems on the plane this problem was completely solved in papers of Andronov, Pontrjagin, Leontovič and Maĭer.[1)] In the structurally stable case [1] such elements turn out to be trajectories which are Ljapunov unstable on at least one side, and in the general case [2], orbitally-unstable ones.

Below, the concept of critical trajectory is introduced, and we base our study on it.

Let (T_t, M) be a dynamical system given on a compactum M, i.e. a one-parameter group of homeomorphisms of the space M onto itself. Let $\{g_\alpha\}$ be the set of homeomorphisms of (T_t, M) onto itself with the topology

$$d(g_1, g_2) = \sup_{x\in M} d_0(g_1x, g_2x),$$

where $d_0(x,y)$ is a distance giving a topology equivalent to the original topology of the space M.

Definition 1.1 (cf. [3], [4]). We shall call a trajectory L *critical* if there exists an $\epsilon > 0$ such that for all $g \in \{g_\alpha\}$ which differ from the identity Id

---

AMS (MOS) subject classifications (1970). Primary 58F99, 34C35; Secondary 60J10.

[1)] Leontovič and Maĭer made their investigation under the assumption of a finite number of orbitally-unstable trajectories, which, for example, holds in the analytic case.

by a distance less than ϵ one has the condition

$$gL = L. \tag{1.1}$$

REMARK 1.1. *In the case when $t \in Z$, i.e. when the system (T_t, M) is generated by a homeomorphism $\mathscr{T}_1$, any homeomorphism g of the system onto itself has the following property: the diagram*

$$\begin{array}{ccc} M & \xrightarrow{\mathscr{T}_1} & M \\ g\downarrow & \mathscr{T}_1 & \downarrow g \\ M & \rightarrow & M \end{array} \tag{1.2}$$

is commutative.

It is clear that isolated equilibrium states and periodic motions are critical trajectories. It is easy to see that the trajectories of topological Markov chains and of Smale's basic sets are critical.

DEFINITION 1.2. We shall call the set of all critical trajectories of a system (T_t, M) its *critical set.*

We shall denote the critical set by $\mathfrak{M}$, and the restriction of the system (T_t, M) to $\mathfrak{M}$ by $(T_t, \mathfrak{M})$. It is not hard to see that $(T_t, \mathfrak{M})$ is topologically invariant, i.e. under a homeomorphism h which establishes the topological equivalence of two systems (T_t, M) and $(\widetilde{T}_t, \widetilde{M})$ the critical set of one system will be carried into the critical set of the second and conversely. In general, let $\{h_\beta\}$ be the set of all homeomorphisms of the system (T_t, M) onto $(\widetilde{T}_t, \widetilde{M})$. It is easy to see that $\{g_\alpha\} = \{h_{\beta_1}^{-1} \cdot h_{\beta_2}\}$.

The following conjecture turns out to be true: In the case when the system (T_t, M) is structurally stable, there exists an ϵ_0 such that for all homeomorphisms g_α which are ϵ_0-close to Id the condition $g_\alpha L = L$ will be satisfied for each critical trajectory L.

The problem of the present paper is the description of the critical sets of Morse-Smale systems. We recall [5–6] that a system (X, M), where X is a smooth vector-field on a C^∞-smooth closed $(m+n+1)$-dimensional manifold M, is called a *Morse-Smale system* if the set of its nonwandering points is the union of a finite number of structurally stable equilibrium states and periodic motions whose stable and unstable manifolds intersect transversely. As was recently established by Palis and Smale, such systems are structurally stable [7].

In Morse-Smale systems, besides equilibrium states and periodic motions which are nonwandering critical trajectories, there can be critical asymptotic trajectories. It is not hard to see that critical asymptotic trajectories can occur only in the case when one can find at least one pair of nonwandering motions such that the stable manifold W^+ of one has nonempty intersection with the unstable manifold W^- of the other and the dimension of this intersection is equal to 1. We shall ascribe to a critical asymptotic trajectory the type $(\dim W^+, \dim W^-)$. The set of all critical asymptotic trajectories of type $(m+1, \dim M - m)$ and critical nonwandering motions which are limits of them will be denoted by $\mathfrak{M}_{m+1}$. Obviously $\mathfrak{M} \setminus \bigcup_m \mathfrak{M}_{m+1}$, if it is not empty, consists of those nonwandering motions which are not limits of critical asymptotic trajectories. In the general case $\mathfrak{M}_{m+1}$ is not connected and consists of a finite number of components, which will be denoted by $\mathfrak{M}_{m+1}^{(1)}$, $\mathfrak{M}_{m+1}^{(2)}, \cdots$.

In this paper it is proved that the set of all trajectories of the system $(T_t, \mathfrak{M}_{m+1}^{(\kappa)})$ is in one-to-one correspondence with the set of all trajectories of some topological Markov chain generated by a graph G with a finite number of states. We shall denote a topological Markov chain by (G, Ω, σ), where Ω is the space of admissible symbolic sequences with respect to the graph G with distinguished zero element, and σ is the shift to the left by one.[2] We note especially that here the correspondence will preserve asymptotic properties of trajectories, so that, in particular, nonwandering trajectories of the system $(T_t, \mathfrak{M}_{m+1}^{(\kappa)})$ (equilibrium states and periodic motions) will correspond to fixed points of (G, Ω, σ). In this paper we give a constructive determination of the graph G of the topological Markov chain (G, Ω, σ) whose states are edges and whose transitions are vertices. However, the process of construction of (G, Ω, σ) is ambiguous in the sense that the graphs obtained will be different and corresponding Markov chains will, in general, not be conjugate.

Hence we shall introduce the concept of suspension over a discrete system, which in our case gives an adequate (and unique up to equivalence) description of $(T_t, \mathfrak{M}_{m+1}^{(\kappa)})$. The concept of suspension is not new.[3] However, our definition of suspension is somewhat different from the usual one in that we introduce equilibrium states in the suspension.

Let $f: \Omega \to R^1$ be a continuous, nonnegative function which can vanish only at the fixed points of a discrete dynamical system $(\mathscr{T}_t, \Omega)$, given by a homeomorphism $\mathscr{T}_1$ on a compactum Ω. We consider the set Σ_0 of points (x, s), $x \in \Omega$, $0 \leqq s \leqq f(x)$. In Σ_0 we identify the points $(x, f(x))$ and $(\mathscr{T}_1 x, 0)$. We denote the space obtained by Σ. We define on Σ a continuous flow in the following way:

1) if $f(x_0) = 0$, then $T_t(x_0, 0) = (x_0, 0)$ for all $t \in R^1$;

2) if $f(x) > 0$, then for $0 \leqq t \leqq 1$

$$T_t(x, s) = \begin{cases} (x, s + f(x) \cdot t), & \text{if } t < 1 - \dfrac{s}{f(x)}, \\ \left(\left(\mathscr{T}_1 x, \left(t - 1 + \dfrac{s}{f(x)}\right) \cdot f(\mathscr{T}_1 x)\right)\right), & \text{if } t \geqslant 1 - \dfrac{s}{f(x)}, \end{cases}$$

---

[2] In explicit form, the concept of topological Markov chain was introduced by Sinaĭ and Alekseev [**8, 9**].

[3] Apparently the first suspension over the topological Bernoulli scheme was constructed in an unpublished paper of Pontrjagin and Šnirel′man (cf. [**10**]). The suspension over a diffeomorphism was introduced by Anosov [**11**] and Smale [**6**]. In ergodic theory the suspension carries the appelation of *special flow*.

and for all other t the flow is uniquely given by the condition that $\{T_t\}$ is a one-parameter group of transformations.

DEFINITION 1.3. The flow (T_t, Σ) on the space Σ will be called the *suspension* of $(\mathscr{T}_1, \Omega)$.

In the case when $(\mathscr{T}_1, \Omega)$ is a topological Markov chain (G, Ω, σ), the flow over it will be denoted by (G, Σ, f, T_t); if $f \equiv 1$, by (G, Σ, T_t).

The basic results of the present paper are the following assertions: the subsystem $(T_t, \mathfrak{M}^{(\kappa)}_{m+1})$ of a Morse-Smale system is homeomorphic to a flow (G, Σ, f, T_t), where f vanishes only at those fixed points of (G, Ω, σ) which correspond to equilibrium states of the system $(T_t, \mathfrak{M}^{(\kappa)}_{m+1})$.

In §2 some properties of Morse-Smale systems are considered which are immediate consequences of results of Smale's paper [**5**]. Here we introduce the concepts of order of an asymptotic critical trajectory and of length of a connected component of the critical set $\mathfrak{M}^{(\kappa)}_{m+1}$. In §3 properties of solutions in the neighborhood of a fixed point of saddle type are described, at the basis of the study of which lies a boundary value problem. It is proved that the solution of the boundary value problem depends continuously on the change of boundary conditions and C^1-smoothly on perturbation of the vector field. Later, metric spaces and their maps are introduced. In §4 the preliminary classification of critical asymptotic trajectories from $\mathfrak{M}^{(\kappa)}_{m+1}$ is established. In §5 a sequence of properties of the suspension over a topological Markov chain is studied. The last section is devoted to the completion of the proof of the basic results.

The present paper is one of a cycle of papers which the authors will develop further. In particular, the immediate continuation of this paper will be a paper devoted to the study of critical sets of structurally stable systems with a countable set of periodic motions, close to Morse-Smale systems.

§2. Some properties of critical sets of Morse-Smale systems

We assume that the Morse-Smale system (X, M^{m+n+1}) has two saddle-like nonwandering motions L_1 and L_γ. Let the dimension of the stable manifold W_1^+ of the saddle or periodic motion L_1 be equal to $m+1$, and the dimension of the unstable manifold W_γ^- of the saddle or periodic motion L_γ be $n+1$. We assume also that $W_1^+ \cap W_\gamma^- \neq \emptyset$.

LEMMA 2.1. *In order that $W_1^+ \cap W_\gamma^-$ have an infinite set of trajectories, it is necessary and sufficient that in the closure of $W_1^+ \cap W_\gamma^-$ there be a periodic motion of saddle type (different from L_1 or L_γ, if L_1 or L_γ is a periodic motion), the dimensions of the stable and unstable manifolds of which are $m+1$ and $n+1$ respectively.*

PROOF. We shall prove the necessity first. In a sufficiently small neighborhood of L_1 in the usual way with the help of a Ljapunov function we construct a surface without contact S_1 for trajectories belonging to W_1^+, and in a sufficiently small neighborhood of L_γ a surface without contact

S_γ for trajectories belonging to W_γ^-. We assume that $W_1^+ \cap W_\gamma^-$ contains an infinite set of trajectories $\{\Gamma_\rho\}$. We denote by P_ρ^+ the point of intersection of the trajectory Γ_ρ with S_1, by P_ρ^- its intersection with S_γ, and by t_ρ the time of passage from P_ρ^+ to P_ρ^- along the trajectory Γ_ρ.

Let P^* be a limit point of the set $\{P_\rho^+\}$, and $\{P_{\rho_n}^+\} \subset \{P_\rho^+\}$ a sequence converging to P^* as $n \to \infty$. Two cases are possible:

a) $t_{\rho_n} \leqq \text{const}$;

b) $t_{\rho_n} \to \infty$ as $n \to \infty$.

In case a) it follows from the theorem on the continuous dependence on the initial conditions that through the point P^* passes a trajectory $\Gamma^* \subset W_1^+ \cap W_\gamma^-$. From the transversal character of the intersection of W_1^+ and W_γ^- at the point P^* it follows that P^* is an isolated point on S_1, which contradicts the convergence of $P_{\rho_n}^+$ to P^*.

In case b) through the point P^* there will also pass a trajectory Γ^*, asymptotic to L_1 as $t \to +\infty$. However, as $t \to -\infty$ it will tend to some element L, a saddle or periodic motion, whose unstable manifold we denote by W^-. The fact that $W_1^+ \cap W^- \neq \emptyset$ and properties of Morse-Smale systems imply the inequality

$$\dim W_1^+ + \dim W^- - (m + n + 1) \geqq 1,$$

i.e. $\dim W^- \geqq n+1$. On the other hand, since $\partial W_\gamma^- \cap W^- \neq \emptyset$, by Lemma 3.7 of [5] we have

$$n + 1 = \dim W_\gamma^- \geqq \dim W^-.$$

Hence $\dim W_\gamma^- = \dim W^-$, and by this same lemma L is a periodic motion. The sufficiency follows from Lemma 3.2 of [5] or Lemma 3.4 of §3 below.

By $\cdots L_i, \cdots, L_j, \cdots$ we shall denote the periodic motions belonging to $\overline{W_1^+ \cap W_\gamma^-}$, and by W_i^+ (W_i^-) the stable (unstable) manifold of the periodic motion L_i.

Lemma 2.2. *In order that the point* $P \in \overline{W_1^+ \cap W_\gamma^-} \setminus (W_1^+ \cap W_\gamma^-)$, *it is necessary and sufficient that* $P \in W_i^+ \cap W_j^-$ (*in particular* L_i *or* L_j *respectively can coincide with* L_1 *or* L_γ).

Proof. **Necessity.** Let the point $P \in \overline{W_1^+ \cap W_\gamma^-} \setminus (W_1^+ \cap W_\gamma^-)$. Just as in the proof of Lemma 2.1 we consider the surfaces S_1 and S_γ, where we choose the neighborhoods of L_1 and L_γ in which they are contained sufficiently small that P does not belong to these neighborhoods. We consider a sequence of points $P_n \to P$, $P_n \in \Gamma_{\rho_n} \subset W_1^+ \cap W_\gamma^-$. It will be assumed that P_n also does not belong to these neighborhoods. By $t^+(P_n)$ we denote the time of passage from P_n to the surface S_1, and by $t^-(P_n)$ the time of passage from S_γ to the point P_n along the trajectory Γ_{ρ_n}.

We consider the following logical possibilities:

(1) $\limsup_{n\to\infty} t^+(P_n) \leqslant \text{const} < \infty$, $\limsup_{n\to\infty} t^-(P_n) = \infty$,

(2) $\limsup_{n\to\infty} t^{+}(P_n) = \infty, \ \limsup_{n\to\infty} t^{-}(P_n) \leqslant \text{const} < \infty,$

(3) $\limsup_{n\to\infty} t^{+}(P_n) = \infty, \ \limsup_{n\to\infty} t^{-}(P_n) = \infty,$

(4) $\limsup_{n\to\infty} t^{+}(P_n) \leqslant \text{const} < \infty, \ \limsup_{n\to\infty} t^{-}(P) \leqslant \text{const} < \infty.$

Let, for example, (3) hold. Hence $P \in \partial W_\gamma^-$ (and also $P \in \partial W_1^+$). We denote the α- (ω-) limit set of the point P by β_j (β_j), the stable (unstable) manifold of the saddle or periodic motion β_j by W_j^+ (W_j^-), and analogously for β_i by W_i^+ (W_i^-). Since $W_i^+ \cap W_j^- \neq \emptyset$, one has

$$\dim W_i^+ + \dim W_\gamma^- - (m+n+1) \geqslant 1. \tag{2.1}$$

Moreover, since

$$P \in \partial W_\gamma^- \cap W_j^- \ (P \in \partial W_1^+ \cap W_i^+),$$

by Lemma 3.7 of [5] we get

$$n+1 = \dim W_\gamma^- \geqslant \dim W_j^-, \quad (m+1 = \dim W_1^+ \geqslant \dim W_i^+).$$

Taking inequality (2.1) into account, we get that $\dim W_j^- = n+1$ ($\dim W_i^- = m+1$), and by this same lemma β_i and β_j must be periodic motions, which we denote by L_i and L_j. One considers cases (1) and (2) analogously.

The possibility (4) does not occur, since it would imply that P belongs to the intersection $W_1^+ \cap W_\gamma^-$.

It is easy to see that all remaining nonwandering motions from the closure of $W_1^+ \cap W_\alpha^-$ are periodic of the same type.

The sufficiency follows also from Lemma 3.2 of [5] or Lemma 3.4 of §3. The lemma is proved.

From the statement made at the beginning of the section it follows that the critical set $\mathfrak{M}_{m+1}$ is nonempty. Let $\mathfrak{M}_{m+1}^{(\kappa)}$ be one of the connected components of $\mathfrak{M}_{m+1}$, containing an asymptotic trajectory. A nonwandering motion appearing in $\mathfrak{M}_{m+1}^{(\kappa)}$ will be denoted by L_i, its stable and unstable manifolds by W_i^+ and W_i^-, and its asymptotic trajectory by Γ_ρ. In $\mathfrak{M}_{m+1}^{(\kappa)}$, as was established, only periodic motions can be simultaneously α- and ω-limit motions. Equilibrium states, if there are any, will only be either α-limits for Γ_ρ (the dimension of their unstable manifolds will be $n+1$) or ω-limits for Γ_ρ, (the dimension of their stable manifolds will be $m+1$). Analogously to Smale [6] we introduce the concept of partial ordering in $\mathfrak{M}_{m+1}^{(\kappa)}$: we write $L_i \leqq L_j$ if $W_i^+ \cap W_j^- \neq \emptyset$; in particular, $L_i \leqq L_i$.

We also introduce the following relation: we write $L_i < L_j$ if $W_i^+ \cap W_j^- \neq \emptyset$ and contains only a finite number (different from zero) of asymptotic trajectories.

Analogously to Lemma 2.2, one proves

LEMMA 2.3. *In order that* $P \in \overline{W_i^+ \cap W_j^-} \setminus (W_i^+ \cap W_j^- \cup L_i \cup L_j)$, *it is necessary and sufficient that* $P \in W_k^+ \cap W_l^-$, *where* $|k-i| + |l-j| \neq 0$ *and* $L_i \leqq L_k \leqq L_l \leqq L_j$.

In turn, Lemma 2.3 and properties of Morse-Smale systems imply

LEMMA 2.4. *If* $W_i^+ \cap W_j^-$ *contains an infinite set of trajectories, then there exists a chain of periodic motions* $L_{i_1}, \cdots, L_{i_k}$ *such that*

$$L_i < L_{i_1} < \cdots < L_{i_k} < L_j. \tag{2.1}$$

By virtue of the fact that in Morse-Smale systems the same periodic motion in the chain (2.1) is not met twice, from all chains "connecting" L_i and L_j one can choose a chain containing the largest number of periodic motions—a chain of greatest length.[4]

DEFINITION 2.1. Let $\Gamma_p \subset W_i^+ \cap W_j^-$. We shall call Γ_p a *trajectory of kth order* if the length of the maximal chain connecting L_i and L_j is equal to k.

DEFINITION 2.2. The *length* of $\mathfrak{M}_{m+1}^{(\kappa)}$ will be the maximal possible order of an asymptotic trajectory in $\mathfrak{M}_{m+1}^{(\kappa)}$.

It is clear that the length of $\mathfrak{M}_{m+1}^{(\kappa)}$ is some finite number which we denote by γ_κ. Obviously, in order that $\mathfrak{M}_{m+1}^{(\kappa)}$ consist of a finite number of trajectories, it is necessary and sufficient that its length be equal to 0.

A proof by contradiction using Lemma 2.3 easily establishes

LEMMA 2.5. *In* $\mathfrak{M}_{m+1}^{(\kappa)}$ *there are only a finite number of trajectories of order* $j+1$ *not entirely contained in an arbitrary neighborhood of all trajectories of order* j, $j = 0, 1, \cdots$.

§3. Behavior of trajectories in the neighborhood of periodic motions

Let L_i be a periodic motion of saddle type, the dimension of whose stable manifold W_i^+ is equal to $m+1$, that of the unstable one W^- being equal to $n+1$. In some sufficiently small neighborhood V_i of L_i we construct an $(m+n)$-dimensional surface without contact S_i, transversal to L_i, on which will be defined a map T along trajectories passing through V_i. As is known, on S_i one can introduce euclidean coordinates in which the map $T: S_i \to S_i$ is written in the form

$$\bar{x} = Ax + f(x,y), \qquad \bar{y} = By + g(x,y), \tag{3.1}$$

where x is an m-vector, y is an n-vector, A is an $m \times m$ matrix, B is an $n \times n$ matrix, $\|A\| < 1$, $\|B^{-1}\| < 1$, and f and g are sufficiently smooth vector-valued functions. f has the form $f_1 \cdot x$, where f_1 vanishes for $x = y = 0$, and g has the form $g_1 \cdot y$, where g_1 vanishes for $x = y = 0$. In the coordinates (x,y) the traces of W_i^+ and W_i^- on the plate S_i will have the respective equations $y = 0$ and $x = 0$.

[4] A concept of length of chains "beh" in a wider sense was introduced by Palis ([12], Definition (1.7)).

We consider the system

$$x_j = A^j x_0 + \sum_{s=0}^{j} A^{j-s} f(x_s, y_s),$$

(3.2)

$$y_j = B^{j-k} y_k - \sum_{s=j}^{k} B^{j-k} g(x_s, y_s),$$

$$j = 0, 1, \dots, k,$$

where we have set $A^0 f = 0$ and $B^{k-k} g = 0$.

LEMMA 3.1. *There exists an $\bar{r} > 0$ such that for $\|x_0\| \leqq \bar{r}/2$ and $\|y_k\| \leqq \bar{r}/2$ the system* (3.2) *for each $k \geqq 1$ has a unique solution*

$$\{(x_j^0, y_j^0)\}_{j=0}^k,$$

where $\|x_j^0\| \leqq \bar{r}$ and $\|y_j^0\| \leqq \bar{r}$.

PROOF. We consider the space $R(r)$ of collections $z = \{(x_j, y_j)\}_0^k$, where $\|x_0\| \leqq r/2$ and $\|y_k\| \leqq r/2$; x_0 and y_k are fixed, and $\|x_j\| \leqq r$ and $\|y_j\| \leqq r$ for the remaining numbers j. We introduce in $R(r)$ the following metric: for any $z_1 = \{(x_j^1, y_j^1)\}_0^k$ and $z_2 = \{(x_j^2, y_j^2)\}_0^k$

$$d(z_1, z_2) = \max_{0 \leqslant j \leqslant k} \{\|x_j^1 - x_j^2\|, \quad \|y_j^1 - y_j^2\|\}.$$

Metrized in this way $R(r)$ will obviously be a complete space. We introduce an operator Φ, which to each element $z \in R(r)$, $z = \{(x_j, y_j)\}_0^k$, associates an element $\bar{z} = \{(\bar{x}_j, \bar{y}_j)\}_0^k$ by the following formula:

$$\bar{x}_j = A^j x_0 + \sum_{s=0}^{j} A^{j-s} f(x_s, y_s),$$

$$\bar{y}_j = B^{j-k} y_k - \sum_{s=j}^{k} B^{s-k} g(x_s, y_s).$$

We shall show that there exists an $r_0 > 0$ such that for $r \leqq r_0$ the operator Φ carries the space $R(r)$ into itself. In fact, $\bar{x}_0 = x_0$ and $\bar{y}_k = y_k$. Further,

$$\|\bar{x}_j\| \leqslant \|A\|^j \cdot \frac{r}{2} + \sum_{s=0}^{j} \|A\|^{j-s} \left(\left\| \frac{\partial \tilde{f}}{\partial x} \right\| \cdot r + \left\| \frac{\partial \tilde{f}}{\partial y} \right\| \cdot r \right),$$

$$\|\bar{y}_j\| \leqslant \|B^{-1}\|^{k-j} \cdot \frac{r}{2} + \sum_{s=j}^{k} \|B^{-1}\|^{k-s} \cdot \left(\left\| \frac{\partial \tilde{g}}{\partial x} \right\| \cdot r + \left\| \frac{\partial \tilde{g}}{\partial y} \right\| \cdot r \right),$$

where the derivatives of f and g are taken at some intermediate point. From the form of the map T we get the required assertion. We shall show now that there exists an $\bar{r} \leqq r_0$ such that for $r \leqq \bar{r}$ the map $\Phi: R(r) \to R(r)$ is a contraction. We set

$$c(r) = \max_{\|x\| \leqslant r,\ \|y\| \leqslant r} \left\{ \left\| \frac{\partial f}{\partial x} \right\|, \left\| \frac{\partial f}{\partial y} \right\|, \left\| \frac{\partial g}{\partial x} \right\|, \left\| \frac{\partial g}{\partial y} \right\| \right\}.$$

Then

$$\| \bar{x}_j^1 - \bar{x}_j^2 \| \leqslant \sum_{s=0}^{j} \| A \|^{j-s} \cdot \left(\left\| \frac{\partial \tilde{f}}{\partial x} \right\| \cdot \| x_s^1 - x_s^2 \| + \left\| \frac{\partial \tilde{f}}{\partial y} \right\| \cdot \| y_s^1 - y_s^2 \| \right)$$

$$\leqslant d(z_1, z_2) \cdot \frac{2c(r)}{1 - \| A \|},$$

$$\| \bar{y}_j^1 - \bar{y}_j^2 \| \leqslant \sum_{s=j} \| B^{-1} \|^{k-s} \left(\left\| \frac{\partial \tilde{g}}{\partial x} \right\| \cdot \| x_s^1 - x_s^2 \| + \left\| \frac{\partial \tilde{g}}{\partial y} \right\| \cdot \| y_s^1 - y_s^2 \| \right)$$

$$\leqslant d(z_1, z_2) \cdot \frac{2c(r)}{1 - \| B^{-1} \|},$$

where the derivatives of f and g are taken at some intermediate point. From the fact that $c(r) \to 0$ as $r \to 0$ follows the needed assertion. A fixed point of the operator Φ is a solution of the system (3.2), i.e. the lemma is proved.

Let $z_0 = \{ (x_j^0, y_j^0) \}_0^k$ be a fixed point of the operator Φ. It is easy to see that $x_0^0 \equiv x_0$, $y_k^0 \equiv y_k$ and

$$(x_0^0, y_0^0) \overset{T}{\to} (x_1^0, y_1^0) \overset{T}{\to} \cdots \overset{T}{\to} (x_k^0, y_k^0).$$

The following two lemmas were established in [**13**] and [**14**].

LEMMA 3.2. *The estimates*

$$\| x_k^0 \| + \left\| \frac{\partial x_k^0}{\partial x_0} \right\| + \left\| \frac{\partial x_k^0}{\partial y_k} \right\| < c \cdot \| A \|^k,$$

(3.4)

$$\| y_0^0 \| + \left\| \frac{\partial y_0^0}{\partial x_0} \right\| + \left\| \frac{\partial y_0^0}{\partial y_k} \right\| < c \| B^{-1} \|^k,$$

hold for some constant.

We fix points P_{ip} with coordinates $(x_{ip}, 0)$ and P_{iq} with coordinates $(0, y_{iq})$, where x_{ip} and y_{iq} satisfy the conditions of Lemmas 3.1 and 3.2. Let

$$U_{ip} = \{ (x_0, y_0) : \| x_0 - x_{ip} \| \leqslant \rho_{ip}, \quad \| y_0 \| \leqslant \rho_{ip} \},$$

$$U_{iq} = \{ (x_1, y_1) : \| x_1 \| \leqslant \rho_{iq}, \quad \| y_1 - y_{iq} \| \leqslant \rho_{iq} \}.$$

We choose ρ_{ip} sufficiently small that in the neighborhood considered we have $TU_{ip} \cap U_{ip} = \emptyset$. We impose analogous restrictions on ρ_{iq}. From Lemma 3.1 follows

LEMMA 3.3. *For any k larger than some $\bar{k}$ the map $T^k : U_{ip} \to U_{iq}$ is defined and can be written in the form*

(3.5) $$x_1 = x_k^0 (x_0, y_1),$$

(3.6) $$y_0 = y_0^0 (x_0, y_1).$$

We denote by $\sigma_{ik}^0(p,q)$ the domain of definition of $T^k : U_{ip} \to U_{iq}$. From the form of T^k and Lemma 3.2 it follows that $\sigma_{ik}^0(p,q)$ is stratified on the surface (3.6), smoothly tending to $y_0 = 0$ as $k \to \infty$. We note that

$$\sigma^0_{ik_1}(p,\ q) \cap \sigma^0_{ik_2}(p,\ q) = \varnothing, \qquad \sigma^1_{ik_1}(k,\ q) \cap \sigma^1_{ik_2}(p,\ q) = \varnothing,$$

where $\sigma^1_{ik}(p,q) = T^k \sigma^0_{ik}(p,q)$ and $k_1 \neq k_2$.

We assume that the stable manifold W_i^+ of the periodic motion L_i intersects the unstable manifold W_j^- of some other periodic motion L_j and the equation of the trace of W_j^- on S_i (which we denote by F), lying in U_{ip}, has the form

$$x_0 = \varphi(y_0), \tag{3.7}$$

where φ is a smooth function.

If ρ_{ip} is sufficiently small, then by virtue of the transversality of the intersection the equation of the trace will always have the form (3.7). We denote the equation of the intersection of the surface (3.7) with $\sigma^0_{ik}(p,q)$ by

$$x_0 = \varphi_k(y_0). \tag{3.8}$$

The image of (3.8) under the map T^k, which we denote by F_k, will obviously lie in $\sigma^1_{ik}(p,q)$, and, taking account of (3.5) and (3.6), the equation of this image will be

$$x_1 = x^0_k(\varphi_k(y_0), y_1), \tag{3.9}$$

$$y_0 = y^0_0(\varphi_k(y_0), y_1). \tag{3.10}$$

Solving (3.10) with respect to y_0 (which can be done for $\bar{k}$ sufficiently large, using the estimate (3.4)) and substituting in (3.9), we get the equation of this image in the explicit form $x_1 = \psi_k(y_1)$, while ψ_k by virtue of the estimate (3.4) tends to 0 together with its derivatives as $k \to \infty$.

Thus we have proved

LEMMA 3.4.[5] *For any sufficiently small neighborhood U_{iq} of the point $P_{iq} \in W_i^-$, lying on S_i, there exists a number $\bar{k}$ such that for all $k \geq \bar{k}$ the following assertions are true*:

1) *The intersection of the image of the surface* (3.7) *under the map T^k with U_{iq} is not empty.*

2) *The equation of this intersection can be written in the form $x_1 = \psi_k(y_1)$, where ψ_k tends to zero as $k \to \infty$ in the metric C^1.*

We note that the number $\bar{k}$ is such that

$$c \cdot \|B^{-1}\|^{\bar{k}} \cdot \|\partial\varphi/\partial y_0\| < 1.$$

We now consider an altered system (X', M). We assume $\|X - X'\|_{C^1} < \delta_0$, where δ_0 is a sufficiently small positive quantity. A periodic motion of the system (X', M), lying in the neighborhood V_i, we denote by L_i^1. It will be assumed that δ_0 is sufficiently small that S_i remains a cross-section for the trajectories of the system (X', M). Without loss of generality one can assume

[5] In implicit form this lemma was used in [**13**] and [**14**].

that in coordinates (x,y) the map $T':S_i\to S_i$ along trajectories of the altered system will have the form

$$\bar{x}=A'x+f'(x,y),\qquad \bar{y}=B'y+g'(x,y),\tag{3.11}$$

where f' and g' are smooth functions. f' has the form $f_1'\cdot x$, where f_1' vanishes for $x=y=0$, and g' has the form $g_1'\cdot y$, where g_1' vanishes for $x=y=0$ and

$$\|A-A'\|+\|B^{-1}-B'^{-1}\|+\|f-f'\|_{C^1}+\|g-g'\|_{C^1}<\delta(\delta_0),$$

where $\delta(\delta_0)\to 0$ as $\delta_0\to 0$. We pose a boundary problem for the map T':

$$\begin{aligned} x_j&=A'^j x_0+\sum_{s=0}^{j}A'^{j-s}f'(x_s,y_s),\\ y_j&=B'^{j-k}-\sum_{s=j}^{k}B'^{s-k}g'(x_s,y_s).\end{aligned}\tag{3.12}$$

For sufficiently small δ_0 this boundary problem has a unique solution z_0' for the same $\bar{r}$ as the boundary problem (3.2). In fact, for $r\leqq\bar{r}$ and sufficiently small $\delta_0>0$ the operator Φ', which is introduced analogously to Φ, will carry the space $R(r)$ into itself and will be a contraction. Let such a number δ_0 be fixed. One has

LEMMA 3.5. *For any $\epsilon>0$ one can find a $\bar{\delta}\leqq\delta_0$ such that $d(z_0',z_0)<\epsilon$ for $\|X-X'\|_{C^1}<\bar{\delta}$.*

PROOF. It is sufficient to show that for any $z\in R(r)$ we have $d(\Phi z,\Phi' z)<\delta_1(\bar{\delta})$, and $\delta_1(\bar{\delta})\to 0$ as $\bar{\delta}\to 0$. Let $z\in R(r)$ have the form $\{(x_j,y_j)\}_0^k$. Then Φz and $\Phi' z$ are written respectively by the formulas

$$\begin{cases}\bar{x}_j=A^j x_0+\sum_{s=0}^{j}A^{j-s}f(x_s,y_s),\\ \bar{y}_j=B^{j-k}y_k-\sum_{s=j}^{k}B^{s-k}g(x_s,y_s),\end{cases}$$

$$\begin{cases}\bar{x}_j'=A'^j x_0+\sum_{s=0}^{j}A'^{j-s}f'(x_s,y_s),\\ \bar{y}_j'=B'^{j-k}y_k-\sum_{s=j}^{k}B'^{s-k}g'(x_s,y_s).\end{cases}$$

We note the following facts:

1) From the form of the map T and the estimate (3.4) it follows that

$$\sum_{s=0}^{k}\|f(x_s,y_s)\|+\sum_{s=0}^{k}\|g(x_s,y_s)\|\leqslant c_1(r),\ \text{where } c_1(r)\to 0\quad\text{as}\quad r\to 0.$$

2) One has the estimates

$$\|A^s - A'^s\| \leqslant \|A - A'\| (\|A\|^{s-1} + \|A\|^{s-2}\|A'\| + \dots + \|A'\|^{s-1}),$$

$$\|B^{-s} - B'^{-s}\| \leqslant \|B^{-1} - B'^{-1}\| (\|B^{-1}\|^{s-1} + \|B^{-1}\|^{s-2} \cdot \|B'^{-1}\| + \dots + \|B'^{-1}\|^{s-1}),$$

which are easily proved by induction. Hence, setting

$$q = \max\{\|A\|, \|A'\|, \|B^{-1}\|, \|B'^{-1}\|\}$$

(obviously $q < 1$ for small $\bar{\delta}$), we get

$$\|A^s - A'^s\| \leqslant \|A - A'\| \cdot sq^{s-1} \leqslant D \cdot \|A - A'\|,$$

$$\|B^{-s} - B'^{-s}\| \leqslant \|B^{-1} - B'^{-s}\| sq^{s-1} \leqslant D \|B^{-1} - B'^{-1}\|,$$

where D is some constant.

We estimate $\|\bar{x}_j - \bar{x}'_j\|$. We have

$$\|\bar{x}_j - \bar{x}'_j\| \leqslant \sum_{s=0}^{j} \|A^{j-s} - A'^{j-s}\| \cdot \|f(x_s, y_s)\| + \sum_{s=0}^{j} \|A'\|^{j-s} \|f' - f\| \leqslant$$

$$\leqslant \|A - A'\| \cdot D \cdot c_1(r) + \frac{\delta(\bar{\delta})}{1 - \|A'\|} \leqslant \left(D \cdot c_1(r) + \frac{1}{1-q}\right) \delta(\bar{\delta}),$$

$$\delta(\bar{\delta}) \to 0 \quad \text{as} \quad \bar{\delta} \to 0.$$

Analogously

$$\|\bar{y}_j - \bar{y}'_j\| \leqslant \left(Dc_1(r) + \frac{1}{1-q}\right) \delta(\bar{\delta}).$$

Consequently $d(\Phi z, \Phi' z) \leqq 2(D \cdot c_1(r) + 1/(1-q))\delta(\bar{\delta})$, which proves the lemma.

Lemma 3.5 asserts that the solution of the boundary value problem depends continuously on the vector field. We shall show that it also depends continuously on the boundary conditions.

Let $z'_0 = \{(x'_j, y'_j)\}_0^k$ be a solution of the boundary value problem with conditions x'_0 and y'_k, i.e.

$$(3.13) \qquad \begin{aligned} x'_j &= A^j x'_0 + \sum_{s=0}^{j} A^{j-s} f(x'_s, y'_s), \\ y'_j &= B^{j-k} y'_k - \sum_{s=j}^{k} B^{s-k} g(x'_s, y'_s). \end{aligned}$$

One has

Lemma 3.6. *For any $\epsilon > 0$ one can find a $\delta_1 > 0$ such that $d(z_0, z'_0) < \epsilon$ for $\|x_0^0 - x'_0\| < \delta_1$ and $\|y_k^0 - y'_k\| < \delta_1$, where*

$$d(z_0, z'_0) = \max_{0 \leqslant j \leqslant k} \{\|x'_j - x_j^0\|, \ \|y'_j - y_j^0\|\}.$$

Proof. We introduce the following notation:

$$\Delta = \max_{0 \leqslant j \leqslant k} \{\| x_j^0 - x_j' \|, \ \| y_j^0 - y_j' \|\}.$$

We estimate

$$\| x_j^0 - x_j' \|, \ \| y_j^0 - y_j' \|.$$

$$\| x_j' - x_j^0 \| \leqslant \| A \|^j \cdot \| x_0' - x_0^0 \| + \sum_{s=0}^{j} \| A \|^{j-s} \left(\left\| \frac{\partial \tilde{f}}{\partial x} \right\| \cdot \| x_s' - x_s^0 \| \right.$$

$$\left. + \left\| \frac{\partial \tilde{f}}{\partial y} \right\| \cdot \| y_s' - y_s^0 \| \right) \leqslant \| A \|^j \cdot \| x_0' - x_0^0 \| + \Delta \frac{2c(r)}{1-q}.$$

Analogously

$$\| y_j' - y_j^0 \| \leqslant \| B^{-1} \|^{k-j} \cdot \| y_k' - y_k^0 \| + \Delta \cdot \frac{2c(r)}{1-q},$$

where $c(r)$ denotes, as in Lemma 3.1, the maximum of the norm of the derivatives of f and g, and $q = \max\{ \| A \|, \| B^{-1} \| \}$. Hence

$$\Delta \leqslant \frac{\| x_0' - x_0^0 \| + \| y_k' - y_k^0 \|}{2\left(1 - \frac{2c(r)}{1-q}\right)},$$

which proves the lemma.

We now consider the surface F', the equation of which is $x_0 = \varphi'(y_0)$, where φ' is a smooth function whose graph lies in U_{ip}. The intersection of the image of this surface under the map T'^k with U_{iq} we denote by F_k'.

Lemma 3.7. *For any $\epsilon > 0$ one can find a δ_2, $0 < \delta_2 < \bar{\delta}$, such that $d_{C^0}(F, F') < \delta_2$ for $\| X - X \|_{C^1} < \delta_2$ and for any $k \geqq \bar{k}$ one has the inequality $d_{C^0}(F_k, F_k') < \epsilon$, where $\bar{k}$ satisfies the hypotheses of Lemma* 3.4.

Proof. Let $x_1 = x_k^0(x_0, y_1)$ and $y_0 = y_0^0(x_0, y_1)$ be the kth and zeroth components of the solution of the boundary value problem (3.2), and let $x_1 = x_k'(x_0, y_1)$ and $y_0 = y_0'(x_0, y_1)$ be the kth and zeroth components of the solution of the boundary value problem (3.12) with boundary conditions x_0 and y_1. The equation of F_k in parametric form is (3.9), (3.10), while the equation of F_k' in parametric form is

$$x_1 = x_k'(\varphi_k'(y_0), \ y_1), \tag{3.14}$$

$$y_0 = y_0'(\varphi_k'(y_0), \ y_1), \tag{3.15}$$

where $x_0 = \varphi_k'(y_0)$ is the equation of the intersection of F' with $\sigma_{ik}^{0\prime}(p, q)$, i.e. with the domain of definition of the map $T'^k: U_{ip} \to U_{iq}$. As was already said above, (3.10) defines a function $y_0 = y_0(y_1)$ and (3.15) analogously defines a function $y_0 = y_0'(y_1)$, upon substitution of which in (3.14) we get the function $x_1 = \psi_k'(y_1)$, i.e.

$$y_0(y_1) \equiv y_0^0(\varphi(y_0(y_1)), \ y_1), \tag{3.16}$$

$$y_0'(y_1) \equiv y_0'(\varphi'(y_0'(y_1)), \ y_1), \tag{3.17}$$

(3.18) $$\psi_k(y_1) \equiv \overset{0}{x}_k(\varphi(y_0(y_1)), y_1),$$

(3.19) $$\psi'_k(y_1) \equiv x'_k(\varphi'(y'_0(y_1), y_1).$$

Subtracting (3.17) from (3.16) and estimating this difference in norm, we get

$$\begin{aligned}\|y_0(y_1) - y'_0(y_1)\| &\leqslant \|\overset{0}{y}_0(\varphi(y_0(y_1)), y_1) - \overset{0}{y}_0(\varphi'(y_0(y_1)), y_1)\| \\ &+ \|\overset{0}{y}_0(\varphi'(y_0(y_1)), y_1) - y'_0(\varphi'(y_0(y_1)), y_1)\| + \|y'_0(\varphi'(y_0(y_1)), y_1) \\ &- y'_0(\varphi'(y'_0(y_1)), y_1)\| \leqslant \left\|\frac{\partial \tilde{y}^0_0}{\partial x_0}\right\| \cdot \|\varphi(y_0(y_1)) - \varphi'(y_0(y_1))\| \\ &+ \|\overset{0}{y}_0(\varphi'(y_0(y_1)), y_1) - y_0(\varphi'(y_0(y_1)), y_1)\| \\ &+ \left\|\frac{\partial \tilde{y}'_0}{\partial x_0}\right\| \cdot \left\|\frac{\partial \tilde{\varphi}'}{\partial y_0}\right\| \cdot \|y_0(y_1) - y'_0(y_1)\|,\end{aligned}$$

where the derivatives are taken at some intermediate point. Let $q = \max\{\|A\|, \|A'\|, \|B^{-1}\|, \|B'^{-1}\|\}$, as in Lemma 3.5, and let $\|\partial\varphi'/\partial y_0\| \leqq L$ and $\|\partial\varphi/\partial y_0\| \leqq L$. Then, using the estimate (3.4) and the hypothesis of the lemma, we get

$$\|y_0(y_1) - y'_0(y_1)\| \leqslant \frac{c \cdot q^k \cdot \|\varphi - \varphi'\| + \|\overset{0}{y}_0(\varphi'(y_0(y_1)), y_1) - y'_0(\varphi'(y_0(y_1)), y_1)\|}{1 - c \cdot \|B'^{-1}\|^k \cdot L},$$

and since $\|\varphi - \varphi'\| < \delta_2$ by virtue of the assumptions, and

$$\|y_0(\varphi'(y_0(y_1)), y_1) - y'_0(\varphi'(y_0(y_1)), y_1)\| \leqq \delta_1(\delta_2),$$

where $\delta_1(\delta_2) \to 0$ as $\delta_2 \to 0$ by virtue of Lemma 3.5, one has

(3.20) $$\|y_0(y_1) - y'_0(y_1)\| \leqslant \frac{cq^k\delta_2 + \delta_1(\delta_2)}{1 - c \cdot \|B'^{-1}\|^k \cdot L}.$$

We note that by virtue of Lemma 3.4 one has the inequality

$$c \cdot \|B'^{-1}\|^{\bar{k}} \cdot L < 1.$$

Subtracting (3.18) from (3.19) and estimating the difference in norm, we get

$$\begin{aligned}\|\psi'_k(y_1) - \psi_k(y_1)\| &\leqslant \|x'_k(\varphi'(y'_0(y_1)), y_1) - x'_k(\varphi(y'_0(y_1)), y_1)\| \\ &+ \|x'_k(\varphi(y'_0(y_1)), y_1) - \overset{0}{x}_k(\varphi(y'_0(y_1)), y_1)\| + \|\overset{0}{x}_k(\varphi(y'_0(y_1)), y_1) \\ &- \overset{0}{x}_k(\varphi(y_0(y_1)), y_1)\| \leqslant \left\|\frac{\partial \tilde{x}'_k}{\partial x_0}\right\| \cdot \|\varphi'(y'_0(y_1)) - \varphi(y'_0(y_1))\| \\ &+ \|x'_k(\varphi(y'_0(y_1)), y_1) - \overset{0}{x}_k(\varphi(y'_0(y_1)), y_1)\| + \left\|\frac{\partial \tilde{x}^0_k}{\partial x_0}\right\| \cdot \left\|\frac{\partial \varphi}{\partial y_0}\right\| \cdot \|y'_0(y_1) - y_0(y_1)\|,\end{aligned}$$

where the derivatives are taken at some intermediate point.

By virtue of Lemma 3.5

$$\| x_k'(\varphi(y_0'(y_1)),\ y_1) - x_k^0(\varphi(y_0'(y_1)),\ y_1) \| \leqslant \delta_1(\delta_2).$$

Taking this into account, and also the estimates (3.4) and (3.20), we get

$$\text{(3.21)} \qquad \| \psi_k'(y_1) - \psi_k(y_1) \| \leqslant [cq^k \delta_2 + \delta_1(\delta_2)] \left[1 + \frac{cq^k L}{1 - c \| B'^{-1} \|^k L} \right],$$

where $\delta_1(\delta_2) \to 0$ as $\delta_2 \to 0$. The lemma is proved.

For any periodic motion $L_i \in \mathfrak{M}_{m+1}^{(\kappa)}$ the trajectories ω-asymptotic to L_i belonging to $\mathfrak{M}_{m+1}^{(\kappa)}$ will be denoted by $\{\Gamma_{ip}\}$, and those which are α-asymptotic, by $\{\Gamma_{iq}\}$. For each trajectory Γ_{ip} we fix some point P_{ip} of the intersection of Γ_{ip} with S_i with coordinates $(x_{ip}, 0)$, and for each trajectory Γ_{iq} we fix a point P_{iq} of the intersection of Γ_{iq} with S_i with coordinates $(0, y_{iq})$. Without loss of generality one can assume that the distance from the point $(0,0)$ to the set $\{P_{ip}\} \cup \{P_{iq}\}$ is bounded below by some constant depending only on r. For each pair x_{ip}, x_{iq} we pose the boundary value problem (3.2) and we choose an r-neighborhood V_i small enough that r satisfies the hypotheses of Lemma 3.1, i.e. so that the boundary value problem has a solution. Let U_{ip} be a neighborhood of the point P_{ip} and U_{iq} a neighborhood of the point P_{iq} on S_i, and let U_{ip} and U_{iq} satisfy the hypotheses of Lemma 3.3. From Lemma 3.3 it follows that there exists a $k_{pq}^i > 0$ such that for $k \geqq k_{pq}^i$ the map T^k: $U_{ip} \to U_{iq}$ is defined and described in the form (3.5), (3.6). We denote this map by $T_{ipq}^+(k)$. The unstable manifold containing the trajectory Γ_{ip} intersects S_i and also W_i^+ transversally for sufficiently small r, and consequently the equation of this intersection has the form $x_0 = \varphi_{ip}(y_0)$, where φ_{ip} is a smooth function and $\| \partial \varphi_{ip} / \partial y_0 \| < L_{ip}$. The stable manifold containing the trajectory Γ_{iq} intersects transversally with W_i^- and with S_i. The equation of this intersection will be $y_1 = \varphi_{iq}(x_1)$, where φ_{iq} is a smooth function and $\| \partial \varphi_{iq} / \partial x_1 \| < L_{iq}$.

Let $L_j \subset \mathfrak{M}_{m+1}^{(\kappa)}$ be a saddle, the dimension of whose stable manifold W_j^+ is equal to $m+1$. In a sufficiently small neighborhood of L_j we construct an m-dimensional smooth manifold without contact D_j, homeomorphic to S^m, and in a sufficiently small neighborhood of D_j an $(m+n)$-dimensional smooth surface S_j, the intersection of which with W_j^+ coincides with D_j, without contact for trajectories of the system (X, M), homeomorphic with $S^m \times B^n$ (B^n is an n-dimensional ball). Let $\{P_{jp}\}$ be the set of points of the intersection of $\mathfrak{M}_{m+1}^{(\kappa)}$ with S_j, and let $\{\Gamma_{ip}\}$ be the set of trajectories passing through P_{jp}. In some sufficiently small neighborhood of P_{jp} on S_j, which will be denoted by U_{jp} as before, we introduce euclidean coordinates (x_0, y_0), x_0 an m-dimensional vector, y_0 an n-dimensional vector, in which the equation of $D_j \cap U_{jp}$ will be $y_0 = 0$. Through the point P_{jp} passes a trajectory $\Gamma_{jp} \subset \mathfrak{M}_{m+1}^{(\kappa)}$. The unstable manifold containing Γ_{jp} meets transversally with D_j and with S_j, and hence the equation of this intersection in U_{jp} will have the form $x_0 = \varphi_{jp}(y_0)$, where φ_{jp} is a smooth function and $\|\partial\varphi_{jp}/\partial y_0\| < L_{jp}$. Analogously, if L_k is a saddle, the dimension of whose unstable manifold W_k^- is equal to $n+1$, then in a sufficiently small neighborhood of L_k we construct an n-dimensional smooth sphere D_k without contact for trajectories belonging to W_k^-, and in a sufficiently small neighborhood of D_k an $(m+n)$-dimensional surface S_k, homeomorphic with $S^n \times B^m$ (B^m is an m-dimensional ball), the intersection of which with W_k^- coincides with D_k. By $\{P_{kq}\}$ we denote the set of points of the intersection of $\mathfrak{M}_{m+1}^{(\kappa)}$ with S_k; by $\{\Gamma_{kq}\}$, the set of trajectories passing through the points $\{P_{kq}\}$.

In a sufficiently small neighborhood of each point P_{kq}, which will be denoted by U_{kq} as before, we introduce euclidean coordinates (x_1, y_1), x_1 an m-dimensional vector, y_1 an n-dimensional vector, in which the equation of $D_k \cap U_{kq}$ will be $x_1 = 0$. Through the point P_{kq} passes the trajectory $\Gamma_{kq} \subset \mathfrak{M}_{m+1}^{(\kappa)}$. The stable manifold containing Γ_{kq} meets transversally with D_k and with S_k, and hence the equation of this intersection in U_{kq} has the form $y_1 = \varphi_{kq}(x_1)$, where φ_{kq} is a smooth function and $\|\partial\varphi_{kq}/\partial x_1\| < L_{kq}$.

We introduce the following spaces:

1) The space H_{ip}^+ of surfaces F_{ip}^+ whose equations are written in the form $x_0 = \varphi(y_0)$, where φ is a continuous function, whose graph lies in U_{ip}, whose Lipschitz constant does not exceed $L_{ip}+1$, and $\|\varphi\| \leqq \rho_1 < \rho$.

2) The space $H_{iq}^+(\rho_1)$ of surfaces $F_{iq}^+(\rho_1)$ whose equations are written in the form $x_1 = \varphi(y_1)$, where φ is a continuous function whose graph lies in U_{iq}, whose Lipschitz constant does not exceed ρ_1 and $\|\varphi\| \leqq \rho_1 < \rho$.

3) The space $H_{ipq}^+(k)$ of surfaces $F_{iqp}^+(k) = F_{iq}^+ \cap \sigma_{ik}^0(p,q)$.

4) The space H_{iq}^- of surfaces F_{iq}^- whose equations are $y_1 = \psi(x_1)$, where ψ is a continuous function whose graph lies in U_{iq}, whose Lipschitz constant does not exceed $L_{iq}+1$, and $\|\psi\| \leqq \rho_1 < \rho$.

5) The space $H_{ip}^-(\rho_1)$ of surfaces $F_{ip}^-(\rho_1)$ whose equations are written in the form $y_0 = \psi(x_0)$, where ψ is a continuous function whose graph lies in U_{ip}, whose Lipschitz constant does not exceed ρ_1 and $\|\psi\| \leqq \rho_1 < \rho$.

6) The space $H_{iqp}^-(k)$ of surfaces $F_{iqp}^-(k) = F_{iq}^- \cap \sigma_{iq}^1(p,q)$.

We note that the elements of the spaces H_{ip}^+ and $H_{ip}^-(\rho_1)$, $H_{iq}^+(\rho_1)$ and H_{iq}^- intersect each other transversally for sufficiently small ρ_1. We note also that in all the spaces introduced one can introduce a C^0-metric, which, in order not to complicate the notation, will be denoted by the same letter d. The spaces with these metrics will be complete metric spaces, as is easy to see.

LEMMA 3.8. *For any sufficiently small $\rho > 0$ there exists a number $k_{pq}^i(\rho)$ such that for all $k \geqq k_{pq}^i(\rho)$ there are defined maps*

$$\mathcal{T}_{ipq}^+(k) : H_{ipq}^+(k) \to H_{iq}^+(\rho),$$

$$\mathcal{T}_{iqp}^-(k) : H_{iqp}^-(k) \to H_{ip}^-(\rho),$$

satisfying the conditions

$$d(\mathcal{T}^{+}_{ipq}(k)\,F^{+'}_{ipq}(k),\quad \mathcal{T}^{+}_{ipq}(k)\,F^{+''}_{ipq}(k)) < cq_0^k\cdot d(F^{+'}_{ipq}(k), F^{+''}_{ipq}(k)),$$

$$d(\mathcal{T}^{-}_{iqp}(k)\,F^{-1}_{iqp}(k),\quad \mathcal{T}^{-'}_{iqp}(k)\,F^{-''}_{iqp}(k)) < cq_0^k d(F^{-'}_{iqp}(k),\quad F^{-''}_{iqp}(k)),$$

where $0 < q_0 < 1$ *and* c *is a constant.*

The proof follows from (3.9), (3.10) and (3.4).

We assume now that the points P_{jp} and P_{iq} belong to some trajectory $\Gamma \subset \mathfrak{M}^{(\kappa)}_{m+1}$ with $\alpha(\Gamma) = L_i$ and $\omega(\Gamma) = L_j$. Hence between some neighborhood on S_i of the point P_{iq} and some neighborhood on S_j of the point P_{jp} one can establish a diffeomorphism along trajectories of the system (X, M), which we denote by $T^{+}_{(ij)(qp)}$. Hence for sufficiently small ρ_{iq} and ρ_{jp} we get

LEMMA 3.9. *There are defined maps*

$$\mathcal{T}^{+}_{(ij)(qp)}: H^{+}_{iq}(\rho_{iq}) \to H^{+}_{jp},$$

$$\mathcal{T}^{-}_{(ji)(pq)}: H^{-}_{jp}(\rho_{jp}) \to H^{-}_{iq}.$$

REMARK 3.1. *We denote by* $\Pi^{+}_{ipq}(k): H^{+}_{ip} \to H^{+}_{ipq}(k)$ *the map under which the surface* F^{+}_{ip} *corresponds to the surface* $F^{+}_{ipq}(k)$ *and by* $\Pi^{-}_{iqp}(k)$: $H^{-}_{iq} \to H^{-}_{iqp}(k)$ *the map under which the surface* F^{-}_{iq} *corresponds to the surface* $F^{-}_{iqp}(k)$. *From Lemmas* 3.8 *and* 3.9 *it follows that for all* $k \geqq k^{i}_{pq}$ *there are defined the compositions*

$$\mathcal{T}^{+}_{(ij)(qp_1)} \circ \mathcal{T}^{+}_{ipq}(k) \circ \Pi^{+}_{ipq}(k): H^{+}_{ip} \to H^{+}_{jp_1},$$

$$\mathcal{T}^{-}_{(ij)(pq_1)} \circ \mathcal{T}^{-}_{iqp}(k) \circ \Pi^{-}_{iqp}(k): H^{-}_{iq} \to H^{-}_{jq_1},$$

$$\mathcal{T}^{+}_{ipq}(k) \circ \Pi^{+}_{ipq}(k) \circ \mathcal{T}^{+}_{(ji)(q_1p)}: H^{+}_{jq_1}(\rho_{jq_1}) \to H^{+}_{iq}(\rho_{iq}),$$

$$\mathcal{T}^{-}_{iqp}(k) \circ \Pi_{iqp}(k) \circ \mathcal{T}^{-}_{(ji)(p_1q)}: H^{-}_{jp_1}(\rho_{jp_1}) \to H^{-}_{ip}(\rho_{ip}),$$

which by Lemma 3.8 *are contraction maps for sufficiently large numbers* k^{i}_{pq}.

§4. Preliminary classification of asymptotic trajectories of the critical set $\mathfrak{M}^{(\kappa)}_{m+1}$

For any $L_i \in \mathfrak{M}^{(\kappa)}_{m+1}$, by $\Gamma_{ip^0_\nu}$ and $\Gamma_{iq^0_\omega}$, $\nu = 1, \cdots, \nu^0_i$, $\omega = 1, \cdots, \omega^0_i$, we denote the trajectories of the zeroth order belonging respectively to the sets $\{\Gamma_{ip}\}$ and $\{\Gamma_{iq}\}$. We also choose neighborhoods $U_{ip^0_\nu}$ and $U_{iq^0_\omega}$ so that $U_{ip^0_{\nu_1}} \cap U_{ip^0_{\nu_2}} = \emptyset$ for $\nu_1 \neq \nu_2$ and $U_{iq^0_{\omega_1}} \cap U_{iq^0_{\omega_2}} = \emptyset$ for $\omega_1 \neq \omega_2$, so that they satisfy the assumptions of Lemma 3.3 if L_i is a periodic motion, and so that they are not both sections for the same trajectory.

Let $L_i \in \mathfrak{M}^{(\kappa)}_{m+1}$ be an arbitrary periodic motion, and let $P_{iq^0_{\omega_1}} \in \Gamma_{iq^0_{\omega_1}}$ and $P_{ip^0_{\nu_2}} \in \Gamma_{ip^0_{\nu_2}}$. Let $L_{j_1} = \omega(\Gamma_{iq^0_{\omega_1}})$ and $L_{j_2} = \alpha(\Gamma_{ip^0_{\nu_2}})$, and choose the points $P_{j_1p^0_{\nu_1}} \in \Gamma_{iq^0_{\omega_1}}$ and $P_{j_2q^0_{\omega_2}} \in \Gamma_{ip^0_{\nu_2}}$.

By Lemma 3.9 the maps

$$\mathcal{T}^{+}_{(ij_1)(q^0_{\omega_1}p^0_{\nu_1})}: H^{+}_{iq^0_{\omega_1}}(\rho_{iq^0_{\omega_1}}) \to H^{+}_{j_1p^0_{\nu_1}},$$

$$\mathcal{T}^{+}_{(j_2i)(q^0_{\omega_2}p^0_{\nu_2})}: H^{+}_{j_2q^0_{\omega_2}}(\rho_{j_2q^0_{\omega_2}}) \to H^{+}_{ip^0_{\nu_2}},$$

$$\mathcal{T}^{-}_{(j_1i)(p^0_{\nu_1}q^0_{\omega_1})}: H^{-}_{j_1p^0_{\nu_1}}(\rho_{j_1p^0_{\nu_1}}) \to H^{-}_{iq^0_{\omega_1}},$$

$$\mathcal{T}^{-}_{(ij_2)(p^0_{\nu_2}q^0_{\omega_2})}: H^{-}_{ip^0_{\nu_2}}(\rho_{ip^0_{\nu_2}}) \to H^{-}_{j_2q^0_{\omega_2}}$$

are defined. Taking into account Lemma 3.8 and Remark 3.1, we select the numbers $k^i_{p^0_{\nu_2}q^0_\omega}$ and $k^i_{p^0_\nu q^0_{\omega_1}}$ so that the composite maps

$$\mathcal{T}^{+}_{(ij_1)(q^0_{\omega_1}p^0_{\nu_1})} \circ \mathcal{T}^{+}_{ip^0_\nu q^0_{\omega_1}}(k) \circ \Pi^{+}_{ip^0_\nu q^0_{\omega_1}}(k): H^{+}_{ip^0_\nu} \to H^{+}_{j_1p^0_{\nu_1}},$$

$$\mathcal{T}^{+}_{ip^0_{\nu_2}q^0_\omega}(k) \circ \Pi^{+}_{ip^0_{\nu_2}q^0_\omega}(k) \circ \mathcal{T}^{+}_{(j_2i)(q^0_{\omega_2}p^0_{\nu_2})}: H^{+}_{j_2q^0_{\omega_2}}(\rho_{j_2q^0_{\omega_2}}) \to H^{+}_{iq^0_\omega}(\rho_{iq^0_\omega}),$$

$$\mathcal{T}^{-}_{ip^0_\nu q^0_{\omega_1}}(k) \circ \Pi^{-}_{iq^0_{\omega_1}p^0_\nu}(k) \circ \mathcal{T}^{-}_{(j_1i)(p^0_{\nu_1}q^0_{\omega_1})}: H^{-}_{j_1p^0_{\nu_1}}(\rho_{i_1p^0_{\nu_1}}) \to H^{-}_{ip^0_\nu}(\rho_{ip^0_\nu}),$$

$$\mathcal{T}^{-}_{(ij_2)(p^0_{\nu_2}q^0_{\omega_2})} \circ \mathcal{T}^{-}_{ip^0_{\nu_2}q^0_\omega}(k) \circ \Pi^{-}_{iq^0_\omega p^0_{\nu_2}}(k): H^{-}_{iq^0_\omega} \to H^{-}_{j_2q^0_{\omega_2}}$$

are defined for all k larger than $k^i_{p^0_\nu q^0_{\omega_1}}$ and $k^i_{p^0_{\nu_2}q^0_\omega}$ respectively. Now that we have thus defined the composite maps, we renumber the spaces from

$$\{H^{+}_{ip^0_\nu q^0_\omega}(k),\ k \geqslant k^i_{p^0_\nu q^0_\omega}\} \cup \{H^{+}_{iq^0_\omega}(\rho_{iq^0_\omega})\} \cup \{H^{+}_{ip^0_\nu}\}$$

(the union is taken over all indices considered), denoting them by H^+_{0s}, $s=1,2,\cdots$. One will write $H^+_{0s_1} \succ H^+_{0s_2}$ if either (i) $H^+_{0s_1} = H^+_{j_1q^0_\omega}(\rho_{j_1q^0_\omega})$, $H^+_{0s_2} = H^+_{j_2p^0_\nu}$ and the map $\mathcal{T}^+_{(j_1j_2)(q^0_\omega,p^0_\nu)}$ acts from $H^+_{j_1q^0_\omega}(\rho_{j_1q^0_\omega})$ into $H^+_{j_2p^0_\nu}$; or (ii) $H^+_{0s_1} = H^+_{ip^0_\nu q^0_\omega}(k)$ and $H^+_{0s_2} = H^+_{iq^0_\omega}(\rho_{iq^0_\omega})$; or (iii) $H^+_{0s_1} = H^+_{ip^0_\nu}$ and $H^+_{0s_2} = H^+_{ip^0_\nu q^0_\omega}(k)$.

DEFINITION 4.1. A sequence $H^+_{0s_1},\cdots,H^+_{0s_{n_0}}$ is called *admissible* if $H^+_{0s_{n_0}} = H^+_{ip^0_\nu}$, $H^+_{0s_1} = H^+_{jq^0_\omega}(\rho_{jq^0_\omega})$ and

$$H^+_{0s_1} \succ H^+_{0s_2} \succ \cdots \succ H^+_{0s_{n_0}}.$$

We shall show that to each trajectory $\Gamma \subset \mathfrak{M}^{(\kappa)}_{m+1}$ from a sufficiently small neighborhood V_0 of the set of trajectories of zeroth order and periodic motions there corresponds a unique admissible sequence. As follows from §2, the α-limit set of Γ will be some nonwandering motion L_i, and the ω-limit will be a nonwandering motion L_j.

Hence Γ has a point in common with some $U_{iq^0_\omega}$. As $H^+_{0s_1}$ we choose $H^+_{iq^0_\omega}(\rho_{iq^0_\omega})$. After $H^+_{0s_1}$ will follow a unique space $H^+_{0s_2}$, let us say $H^+_{i_1p^0_\nu}$. In this case the trajectory Γ intersects the neighborhood $U_{i_1p^0_\nu}$ on S_{i_1} at the point $P_{i_1p^0_\nu}$, and since Γ occurs in a sufficiently small neighborhood of a trajectory of zeroth order, it intersects also some neighborhood $U_{i_1q^0_\omega}$, and, as follows from §3, uniquely determines a number $k \geqq k^{i_1}_{p^0_\nu q^0_{\omega_1}}$. Hence $H^+_{0s_*} = H^+_{i_1p^0_\nu q^0_{\omega_1}}(k)$, and $H^+_{0s_4} = H^+_{i_1 q^0_{\omega_1}}(\rho_{i_1 q^0_{\omega_1}})$. After $H^+_{0s_4}$ there is a unique space $H^+_{0s_5}$ which can follow, etc. Since $\omega(\Gamma) = L_j$, Γ has a point in common with W_j^+ in some neighborhood $U_{jp^0_{\nu_1}}$. The last element of the sequence will be $H^+_{jp^0_{\nu_1}}$. The sequence thus constructed will obviously be admissible.

We now consider an arbitrary admissible sequence

$$H^+_{0s_1} \succ H^+_{0s_2} \succ \cdots \succ H^+_{0s_{n_0}}.$$

Let, for example, $H^+_{0s_1} = H^+_{iq^0_\omega}(\rho_{iq^0_\omega})$ and $H^+_{0s_{n_0}} = H^+_{jp^0_\nu}$. The intersection $W_i^- \cap U_{iq^0_\omega}$ obviously belongs to $H^+_{iq^0_\omega}(\rho_{iq^0_\omega})$. Its image under the composite map $H^+_{0s_1} \to H^+_{0s_{n_0}}$ belongs to $H^+_{jp^0_\nu}$, and so in $U_{jp^0_\nu}$ it intersects the surface $y_0 = 0$ transversally, thus defining a trajectory which by construction corresponds to the sequence $H^+_{0s_1}, \cdots, H^+_{0s_{n_0}}$.

Trajectories from $\mathfrak{M}^{(\kappa)}_{m+1}$ which correspond to admissible sequences from $\{H^+_{0s}\}$ will be called *coded*.

Analogously we renumber the spaces from

$$\{H^-_{iq^0_\omega p^0_\nu}(k),\ k \geqslant k^i_{p^0_\nu q^0_\omega}\} \cup \{H^-_{ip^0_\nu}(\rho_{ip^0_\nu})\} \cup \{H^-_{iq^0_\omega}\},$$

denoting them by H^-_{0r}, $r = 1, 2, \cdots$. The concept of admissible sequence was introduced relative to the maps $\mathcal{T}^+_{ip^0_\nu q^0_\omega}(k)$ and $\mathcal{T}^+_{(ij)(q^0_\omega p^0_\nu)}$, inducing maps $T^+_{ip^0_\nu}(k)$ and $T^+_{(ij)(q^0_\omega p^0_\nu)}$. If one reverses the direction of time, then by virtue of the complete time-symmetry of the constructions one can define the concept of admissible sequence from $\{H^-_{0r}\}$ with respect to the maps $\mathcal{T}^-_{iq^0_\omega p^0_\nu}(k)$ and $\mathcal{T}^-_{(ji)(p^0_\nu q^0_\omega)}$ and establish a one-to-one correspondence between the set of admissible sequences from $\{H^-_{0r}\}$ and the set of coded trajectories from $\mathfrak{M}^{(\kappa)}_{m+1}$.

Definition 4.2. A trajectory of the first order from $\mathfrak{M}^{(\kappa)}_{m+1}$ which does not correspond to an admissible sequence from $\{H^+_{0s}\}$, i.e. which is not coded, is called a *preferred trajectory of the first order*.

From Lemma 2.5 it follows that preferred trajectories of the first order are finite in number.

We denote by $\Gamma_{ip^1_\nu}$ and $\Gamma_{iq^1_\omega}$, $\nu = 1, \cdots, \nu^1_i$, $\omega = 1, \cdots, \omega^1_i$, the preferred trajectories of the first order belonging respectively to the sets $\{\Gamma_{ip}\}$ and $\{\Gamma_{iq}\}$. On the neighborhoods $U_{ip^1_\nu}$, $\nu = 1, \cdots, \nu^1_i$, and $U_{iq^1_\omega}$, $\omega = 1, \cdots, \omega^1_i$, satisfying the assumptions of Lemma 3.3, we place the following restrictions:

$$U_{ip^1_{\nu_1}} \cap U_{ip^1_{\nu_2}} = \varnothing,\ \nu_1 \neq \nu_2,\ U_{iq^1_{\omega_1}} \cap U_{iq^1_{\omega_2}} = \varnothing,\ \omega_1 \neq \omega_2,$$

and they are not simultaneously sections for the same trajectory. For each preferred trajectory of the first order to which the points $P_{iq^1_\omega}$ and $P_{jp^1_\nu}$ belong, by Lemma 3.9 we construct the map

$$\mathcal{T}^+_{(ij)(q^1_\omega p^1_\nu)} : H^+_{iq^1_\omega}(\rho_{iq^1_\omega}) \to H^+_{jp^1_\nu}$$

and we choose numbers $k^i_{p^1_\nu q^1_\omega}$, $k^i_{p^0_{\nu_1} q^1_\omega}$ and $k^i_{p^1_\nu q^0_{\omega_1}}$ sufficiently large that the compositions

$$\mathcal{T}^+_{(ij)(q^1_\omega p^1_\omega)} \circ \mathcal{T}^+_{ip^1_{\nu_1} q^1_\omega}(k) : H^+_{ip^1_{\nu_1} q^1_\omega}(k) \to H^+_{jp^1_\nu},$$

$$\mathcal{T}^+_{(ij)(q^0_{\omega_1} p^0_{\nu_2})} \circ \mathcal{T}^+_{ip^1_\nu q^0_{\omega_1}}(k) : H^+_{ip^1_{\nu_1} q^0_{\omega_1}}(k) \to H^+_{jp^0_{\nu_2}},$$

$$\mathcal{T}^+_{(ij)(q^1_\omega p^1_\nu)} \circ \mathcal{T}^+_{ip^0_{\nu_1} q^1_\omega}(k) : H^+_{ip^0_{\nu_1} q^1_\omega}(k) \to H^+_{jp^0_{\nu_2}}$$

are defined for all k larger than $k^i_{p^1_{\nu_1} q^1_\omega}$, $k^i_{p^1_\nu q^0_{\omega_1}}$ and $k^i_{p^0_{\nu_1} q^1_\omega}$ respectively.

We renumber the spaces from

$$\{H^+_{0s}\} \cup \{H^+_{ip^1_\nu q^1_\omega}(k),\ k \geqslant k^i_{p^1_\nu q^1_\omega}\} \cup \{H^+_{ip^0_\nu q^1_\omega}(k),$$

$$k \geqslant k^i_{p^0_\nu q^1_\omega}\} \cup \{H^+_{ip^1_\nu q^0_\omega}(k),\ k \geqslant k^i_{p^1_\nu q^0_\omega}\} \cup \{H^+_{iq^1_\omega}(\rho_{iq^1_\omega})\} \cup \{H^+_{ip^1_\nu}\}$$

(the union is taken over all the indices considered), denoting them by H^+_{1s}, $s = 1, 2, \cdots$. Just as for the spaces from $\{H^+_{0s}\}$, we introduce a relation $\succ$ between spaces from $\{H^+_{1s}\}$. One will write $H^+_{1s_1} \succ H^+_{1s_2}$ if either (i) $H^+_{1s_1} = H^+_{0s'_1}$, $H^+_{1s_2} = H^+_{0s'_2}$ and $H^+_{0s'_1} \succ H^+_{0s'_2}$; or (ii) $H^+_{1s_1} = H^+_{j_1 q^1_\omega}(\rho_{j_1 q^1_\omega})$, $H^+_{1s_2} = H^+_{j_2 p^1_\nu}$ and the map $\mathcal{T}^+_{(j_1 j_2)(q^1_\omega p^1_\nu)}$ acts from $H^+_{j_1 q^1_\omega}(\rho_{j_1 q^1_\omega})$ into $H^+_{j_2 p^1_\nu}$; or (iii) $H^+_{1s_1} = H^+_{ip^1_\nu}$, and $H^+_{1s_2} = H^+_{ip^1_\nu q^1_\omega}(k)$, $k \geqq k^i_{p^1_\nu q^1_\omega}$ (or $H^+_{1s_2} = H^+_{ip^1_\nu q^0_\omega}(k)$, $k \geqq k^i_{p^1_\nu q^0_\omega}$); or (iv) $H^+_{1s_1} = H^+_{ip^0_\nu}$, and $H^+_{1s_2} = H^+_{ip^0_\nu q^1_\omega}(k)$, $k \geqq k^i_{p^0_\nu q^1_\omega}$; or (v) $H^+_{1s_1}$ coincides with $H^+_{ip^0_\nu q^1_\omega}(k)$, $k \geqq k^i_{p^0_\nu q^1_\omega}$, or with $H^+_{ip^1_\nu q^0_\omega}(k)$, $k \geqq k^i_{p^1_\nu q^1_\omega}$, or with $H^+_{ip^1_\nu q^0_\omega}(k)$, $k \geqq k^i_{p^1_\nu q^0_\omega}$, and $H^+_{1s_2}$ coincides respectively with $H^+_{iq^1_\omega}(\rho_{iq^1_\omega})$, or with $H^+_{iq^0_\omega}(\rho_{iq^0_\omega})$.

The sequence $H^+_{1s_1}, \cdots, H^+_{1s_{n_0}}$ is called *admissible* if $H^+_{1s_{n_0}} = H^+_{ip^\gamma_\nu}$ and $H^+_{1s_1} = H^+_{jq^\delta_\omega}(\rho_{jq^\delta_\omega})$, $\gamma, \delta = 0, 1$ and $H^+_{1s_1} \succ H^+_{1s_2} \succ \cdots \succ H^+_{1s_{n_0}}$. Analogously one introduces the spaces $\{H^-_{1s}\}$ and admissible sequences of these spaces.

Just as above, one proves that to each admissible sequence from $\{H^+_{1s}\}$ there corresponds a trajectory from $\mathfrak{M}^{(\kappa)}_{m+1}$. Again, the proof that to each trajectory from $\mathfrak{M}^{(\kappa)}_{m+1}$ contained in a sufficiently small neighborhood of periodic motions, trajectories of zeroth order and preferred trajectories of the first order, which we denote by V_1, corresponds an admissible sequence from $\{H^+_{1s}\}$, differs from the proof of the analogous assertion of Lemma 4.1 only in that there as t increases each trajectory from the sufficiently small neighborhood intersects a uniquely determined sequence of neighborhoods $U_{iq^0_\omega}$, $U_{ip^0_\nu}$. Hence the admissible sequence from $\{H^+_{0s}\}$ is uniquely determined. But now a trajectory from V_1, generally speaking, has points in the intersections of neighborhoods $U_{iq^0_{\omega_1}} \cap U_{iq^1_{\omega_2}}$ and $U_{ip^0_{\nu_1}} \cap U_{ip^1_{\nu_2}}$. In this case preference is given to the indices $q^1_{\omega_2}$ and $p^1_{\nu_2}$. Analogously one carries out the construction of the metric spaces $\{H^+_{ks}\}$ and $\{H^-_{ks}\}$ and admissible sequences in them from the beginning for $k = 2$, then 3, etc.

For $k = \gamma_\kappa$ we arrive at the following theorem.

THEOREM 4.1. *The set of all trajectories from* $\mathfrak{M}^{(\kappa)}_{m+1}$, *with the exception of the nonwandering motions, is in one-to-one correspondence with the set of all admissible sequences from* $\{H^+_{\gamma_\kappa, s}\}$ (*and, what is the same, from* $\{H^-_{\gamma_\kappa, s}\}$).

REMARK 4.1. *Without loss of generality the number* $k^i_{p^\gamma_\nu q^\delta_\omega}$ *can be considered so large that for all* $k \geqq k^i_{p^\gamma_\nu q^\delta_\omega}$ *the maps*

$$\mathcal{T}^+_{ip^\gamma_\nu q^{\delta_1}_{\omega_1}}(k) \circ \Pi^+_{ip^\gamma_\nu q^{\delta_1}_{\omega}}(k) \circ \mathcal{T}^+_{(ij)(q^\delta_\omega p^\gamma_\nu)} : H^+_{jq^\delta_\omega}(\rho_{jq^\delta_\omega}) \to H^+_{iq^{\delta_1}_{\omega_1}}(\rho_{iq^{\delta_1}_{\omega_1}}),$$

$$\mathcal{T}^+_{(ij)(q^\delta_\omega p^{\gamma_1}_{\nu_1})} \circ \mathcal{T}^+_{ip^\gamma_\nu q^\delta_\omega}(k) \circ \Pi^+_{ip^\gamma_\nu q^\delta_\omega}(k) : H^+_{ip^\gamma_\nu} \to H^+_{jp^{\gamma_1}_{\nu_1}},$$

$$\mathcal{T}^-_{(ij)(q^\delta_\omega p^{\gamma_1}_{\nu_1})} \circ \mathcal{T}^-_{jq^\delta_\omega p^\gamma_\nu}(k) \circ \Pi^-_{jq^\delta_\omega p^\gamma_\nu}(k) : H^-_{jq^\delta_\omega} \to H^-_{iq^{\delta_1}_{\omega_1}},$$

$$\mathcal{T}^-_{ip^{\gamma_1}_{\nu_1} q^\delta_\omega}(k) \circ \Pi^-_{ip^{\gamma_1}_{\nu_1} q^\delta_\omega}(k) \circ \mathcal{T}^-_{(ji)(p^\gamma_\nu q^\delta_\omega)} : H^-_{jp^\gamma_\nu}(\rho_{jp^\gamma_\nu}) \to H^-_{ip^{\gamma_1}_{\nu_1}}(\rho_{ip^{\gamma_1}_{\nu_1}})$$

are contractions with coefficient $0 < q_0 < 1$ *for all indices considered* (*cf. Remark* 3.1 *and Lemma* 3.8).

REMARK 4.2. *To each asymptotic trajectory* Γ *from* $\mathfrak{M}^{(\kappa)}_{m+1}$ *by Theorem* 4.1 *there corresponds an admissible sequence* $H^+_{\gamma_\kappa, s_1} \succ \cdots \succ H^+_{\gamma_\kappa s_{n_0}}$ *from* $\{H^+_{\gamma_\kappa, s}\}$, *and also an admissible sequence* $H^-_{\gamma_\kappa, s_1} \prec \cdots \prec H^-_{\gamma_\kappa, s_{n_0}}$ *from* $\{H^-_{\gamma_\kappa s}\}$, *where by construction, if* $H^+_{\gamma_\kappa, s_i} = H^+_{jq^\delta_\omega}(\rho_{jq^\delta_\omega})$, *then* $H^-_{\gamma_\kappa s_i} = H^-_{jq^\delta_\omega}$, *and if* $H^+_{\gamma_\kappa, s_i} = H^+_{jp^\gamma_\nu}$, *then* $H^-_{\gamma_\kappa, s_i} = H^-_{jp^\gamma_\nu}(\rho_{jp^\gamma_\nu})$, *and elements of the spaces* $H^+_{\gamma_\kappa, s_i}$ *and* $H^-_{\gamma_\kappa, s_i}$ *meet each other transversally.*

§5. Suspension over a topological Markov chain

We consider an oriented graph G with a finite number of edges and vertices. In what follows we consider only connected graphs satisfying condition (*): *any vertex is the beginning of some edge and the end of some other* (*or the same*) *edge*. We make a one-to-one correspondence between the set of edges of G and the set of symbols $1, \cdots, n$, and we consider the $n \times n$ matrix $(G) = (a_{ij})$: $a_{ij} = 1$ if the end of edge i coincides with the beginning of edge j; otherwise $a_{ij} = 0$. We consider the set of sequences infinite in both directions with distinguished zeroth coordinate $\Omega = \{x = (\cdots i_{-1} i_0 \cdots i_k \cdots)\}$, consisting of the symbols $1, \cdots, n$, such that the symbol i_m can follow the symbol i_{m-1} if and only if $a_{i_{m-1} i_m} = 1$.

In the set Ω in the usual way one introduces a metric: for $x = (\cdots i_{-1} i_0 \cdots i_k \cdots)$ and $y = (\cdots j_{-1} j_0 \cdots j_k \cdots)$

$$d(x, y) = \sum_{k=-\infty}^{\infty} \frac{|i_k - j_k|}{2^{|k|}}.$$

After this the set Ω is turned into a zero-dimensional metric space. We consider the map $\sigma : \Omega \to \Omega$,

$$\sigma(\ldots i_{-1}i_0 \ldots i_k \ldots) = (\ldots j_{-1}j_0 \ldots j_k \ldots)$$

$j_k = i_{k+1}$, shift to the left by one symbol. It is easy to see that σ is a homeomorphism.

DEFINITION 5.1. The triple (G, Ω, σ) is called *the topological Markov chain corresponding to the graph G*. The edges of G will be called the *states* of the chain, the vertices will be called the *transitions*, and it will be said that *there is a transition from state i to state j* if $a_{ij} = 1$. The matrix (a_{ij}) will be called the *transition matrix* of the chain.

Obviously an edge j having beginning and end at the same vertex corresponds to the fixed point $(\cdots jj \cdots)$ of the shift σ.

Topological chains $(G_1, \Omega_1, \sigma_1)$ and $(G_2, \Omega_2, \sigma_2)$ are called *conjugate* if there exists a homeomorphism $g: \Omega_1 \to \Omega_2$ such that the diagram

$$\begin{array}{ccc} \Omega_1 & \xrightarrow{\sigma_1} & \Omega_1 \\ {\scriptstyle g}\downarrow & & \downarrow{\scriptstyle g} \\ \Omega_2 & \xrightarrow{\sigma_2} & \Omega_2 \end{array}$$

is commutative.

Let (G, Ω, σ) be a topological Markov chain. Let $f: \Omega \to R^1$ be a continuous nonnegative function which can vanish only at the fixed points of σ. With the help of the space Ω, the function f and the map σ we construct a continuous dynamical system.

We consider the set Σ_0 of points (x, s), $s \in \Omega$, $0 \leq s \leq f(x)$. In Σ_0 we identify the points $(x, f(x))$ and $(\sigma x, 0)$. We denote the resulting space by Σ. We define a flow on Σ in the following way: 1) if $f(x_0) = 0$, then $T_t(x_0, 0) = (x_0, 0)$ for all $t \in R^1$; 2) if $f(x) > 0$, then for $0 \leq t < 1$

$$T_t(x, s) = \begin{cases} (x, s + f(x)t), & \text{if} \quad t < 1 - \dfrac{s}{f(x)}, \\ \left(\sigma x, \left(t - 1 + \dfrac{s}{f(x)}\right) f(\sigma x)\right), & \text{if} \quad t \geqslant 1 - \dfrac{s}{f(x)}, \end{cases}$$

and for the remaining t the flow is uniquely given by the condition that $\{T_t\}$ is a one-parameter group of transformations. It is not hard to see that (T_t, Σ) is a continuous dynamical system.

DEFINITION 5.2. The flow $\{T_t\}$ on the space Σ will be called the *suspension* over the topological Markov chain (G, Ω, σ) and denoted by (G, Σ, f, T_t).

In the case when $f \equiv 1$, the suspension (G, Σ, f, T_t) will be denoted by (G, Σ, T_t).

Suspensions $(G_1, \Sigma_1, f_1, T_t^1)$ and $(G_2, \Sigma_2, f_2, T_t^2)$ are called *topologically equivalent* if there exists a homeomorphism $h: \Sigma_1 \to \Sigma_2$ under which trajectories of the flow $\{T_t^1\}$ are carried into trajectories of $\{T_t^2\}$.

Below we give a series of assertions about properties of suspensions over topological Markov chains which have independent interest.

LEMMA 5.1. *Let* (G, Ω, σ) *be a topological Markov chain, and let the functions* $f_1, f_2: \Omega \to R^1$ *vanish at the same fixed points of* σ. *Then the suspensions* $(G, \Sigma_1, f_1, T_t^1)$ *and* $(G, \Sigma_2, f_2, T_t^2)$ *are topologically equivalent.*

PROOF. To each point $(x_0, 0) \in \Sigma_1$ such that $f_1(x_0) = 0$ we associate the point $(x_0, 0) \in \Sigma_2$, and to each point $(x, s) \in \Sigma_1$ such that $f_1(x) \neq 0$ $(0 \leqq f_1(x))$ we associate the point $(x, (s/f_1(x))f_2(x))$. The map $\Sigma_1 \to \Sigma_2$ will obviously be the needed homeomorphism.

Let the homeomorphism $g: \Omega_1 \to \Omega_2$ realize the conjugacy of the topological Markov chains $(G_1, \Omega_1, \sigma_1)$ and $(G_2, \Omega_2, \sigma_2)$, and let $f_1: \Omega_1 \to R^1$ be an arbitrary function from the class considered by us. We denote by $f_2: \Omega_2 \to R^1$ the function whose value for any point $x \in \Omega_2$ is equal to $f_1(g^{-1}x)$. There is no difficulty in establishing the topological equivalence of the suspensions $(G_1, \Sigma_1, f_1, T_t^1)$ and $(G_2, \Sigma_2, f_2, T_t^2)$. The converse assertion, generally speaking, is not true, i.e. for two equivalent suspensions the original Markov chains, generally speaking, are not conjugate.

We consider a graph G_1 satisfying the condition (*). We assume that some vertex a of this graph is the initial point of only one edge A and is not its end. The end of A is at some other vertex, let us say b. We identify the vertices a and b, annihilating the edge A. We get some other graph G_2, also satisfying condition (*):

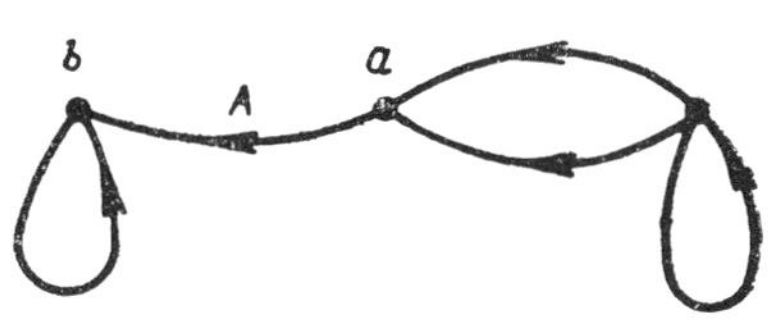

Graph G_1

Graph G_2

Under this operation to each edge of the graph G_2 there corresponds naturally an edge of G_1. We denote the edges of G_2, just like those of G_1 which correspond to them, by the symbols $1, \cdots, n$. The edge A of G_1 is denoted by the symbol $n+1$. Here to each fixed point of the shift σ_1 of the topological Markov chain $(G_1, \Omega_1, \sigma_1)$ there naturally corresponds a fixed point of the shift σ_2 of the chain $(G_2, \Omega_2, \sigma_2)$.

LEMMA 5.2. *For any two functions* $f_1: \Omega_1 \to R^1$ *and* $f_2: \Omega_2 \to R^1$ *of the class considered which vanish, if anywhere, only at corresponding fixed points, the suspensions* $(G_2, \Sigma_1, f_1, T_t^1)$ *and* $(G_2, \Sigma_2, f_2, T_t^2)$ *are topologically equivalent.*

PROOF. Let Ω_1^0 be the set of those points $x \in \Omega_1$ whose the zeroth coordinates are not equal to $n+1$, and let Ω_1^1 be the set of those points of Ω_1^0 whose first coordinates are equal to $n+1$. We consider an arbitrary point of Ω_1^0, which we write in the form

$$(\cdots i_{-1} i_0 \cdots i_k \cdots).$$

Removing the symbol $n+1$ everywhere that it is met in this sequence, and preserving the zeroth coordinate, we get a point of Ω_2 which is written in the form $(\cdots \bar{i}_{-1} i_0 \cdots \bar{i}_k \cdots)$. The map so defined of Ω_1^0 onto Ω_2, which we denote by g, will obviously be a homeomorphism (the continuity is obvious, the one-to-oneness between Ω_1^0 and $g(\Omega_1^0)$ is also obvious, and the fact that g is a map onto follows from the construction of G_2. We give a map $h: \Sigma_1 \to \Sigma_2$ in the following way: $h(x,0) = (gx, 0)$ if $f_1(x) = 0$. If $f_1(x) \neq 0$, then for $0 \leq t < f_1(x)$

$$h(x,t) = \begin{cases} \left(gx, \dfrac{t}{f_1(x)} \cdot f_2(gx)\right), & \text{if } x \in \Omega_1^0 \setminus \Omega_1^1, \\ \left(gx, \dfrac{t}{2f_1(x)} \cdot f_2(gx)\right), & \text{if } x \in \Omega_1^1, \\ \left(g\sigma_1^{-1}x, \dfrac{f_1(x)+t}{2f_1(x)} \cdot f_2(g\sigma_1^{-1}x)\right), & \text{if } x \in \Omega_1 \setminus \Omega_1^0. \end{cases} \tag{5.1}$$

We verify first that h is actually a map of Σ_1 into Σ_2. For this it suffices to show that $h(x, f_1(x)) = h(\sigma_1 x, 0)$. This is obvious for x such that $f_1(x) = 0$. Let $f_1(x) \neq 0$ and $x \in \Omega_1^0 \setminus \Omega_1^1$. Then

$$h(x, f_1(x)) = (gx, f_2(gx)) = (\sigma_2 gx, 0).$$

If $x \in \Omega_1^0 \setminus \Omega_1^1$, then obviously $\sigma_1 x \in \Omega_1^0$. Hence $h(\sigma_1 x, 0) = (g\sigma_1 x, 0)$, and from the definition of g is follows that $(\sigma_2 gx, 0) = (g\sigma_1 x, 0)$, i.e. $h(x, f_1(x)) = h(\sigma_1 x, 0)$. We assume now that $x \in \Omega_1^1$, so that $\sigma_1 x \in \Omega_1 \setminus \Omega_1^0$. Hence $h(x, f_1(x)) = (gx, \frac{1}{2} f_2(gx))$ and

$$h(\sigma_1 x, 0) = (g\sigma_1^{-1}\sigma_1 x, \frac{1}{2} f_2(g\sigma_1^{-1}\sigma_1 x)) = \left(gx, \frac{1}{2} f_2(gx)\right).$$

Let $x \in \Omega_1 \setminus \Omega_1^0$, so that $\sigma_1 x \in \Omega_1^0$ and $\sigma_1^{-1} x \in \Omega_1^0$. Hence

$$h(x, f_1(x)) = (g\sigma_1^{-1}x, f_2(g\sigma_1^{-1}x)) = (\sigma_2 g\sigma_1^{-1}x, 0)$$

and

$$h(\sigma_1 x, 0) = (g\sigma_1 x, 0).$$

The equation $\sigma_2 g \sigma_1^{-1} x = g\sigma_1 x$ for $x \in \Omega_1 \setminus \Omega_1^0$ follows from the definition of g. Thus h is a single-valued map of Σ_1 into Σ_2.

To each point $(y, 0) \in \Sigma_2$ such that $f_2(y) = 0$ we associate the point $(g^{-1}y, 0)$. Now let the point $(y,t) \in \Sigma_2$ and $f_2(y) \neq 0$. By construction $y \in g(\Omega_1^0)$ always. We set

$$h^{-1}(y,t) = \begin{cases} \left(g^{-1}y, \dfrac{t}{f_2(y)} \cdot f_1(g^{-1}y)\right), & \text{if } g^{-1}y \in \Omega_1^0 \setminus \Omega_1^1, \\ \left(g^{-1}y, \dfrac{2t}{f_2(y)} \cdot f_1(g^{-1}y)\right), & \text{if } g^{-1}y \in \Omega_1^1,\ 0 \leqslant t < \dfrac{1}{2} f_2(y), \\ \left(\sigma_1 g^{-1}y, \dfrac{2t - f_2(y)}{f_2(y)} \cdot f_1(\sigma_1 g^{-1}y)\right), & \text{if } g^{-1}y \in \Omega_1^1, \\ \qquad \dfrac{1}{2} f_2(y) \leqslant t < f_2(y). \end{cases} \tag{5.2}$$

h^{-1} is given on Σ_2. This assertion is verified in the same way as the preceding one, where one uses the relation $\sigma_1^2 g^{-1}y = g^{-1}\sigma_2 y$ for y such that $g^{-1}y \in \Omega_1^1$. It is easy to verify that h^{-1} is the inverse map for h. Thus h is one-to-one. The fact that h and h^{-1} are continuous and carry trajectories into trajectories follows from the properties of g and formulas (5.1) and (5.2). The lemma is proved.

As an example we consider the graphs G_1 and G_2 in the following figure.

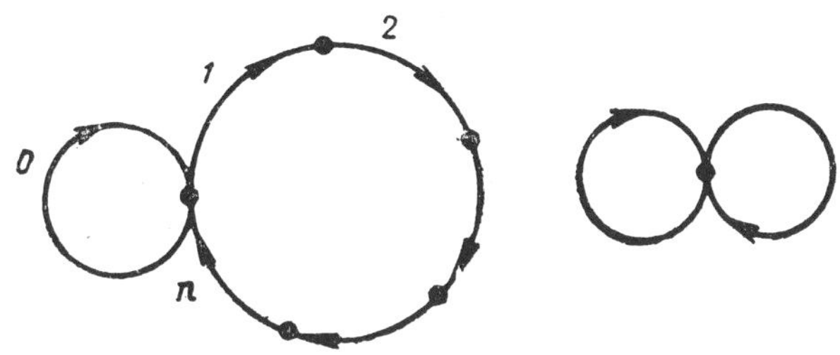

Graph G_1 Graph G_2

From Lemma 5.2 it follows that the suspensions (G_1, Σ_1, T_t^1) and (G_2, Σ_2, T_t^2) are topologically equivalent.

We now consider an arbitrary graph G_1 satisfying condition (*), and we assume that some vertex a is not simultaneously the beginning and end of one and the same edge. Let the symbols $1, \cdots, n$ denote the states of the topological Markov chain $(G_1, \Omega_1, \sigma_1)$ corresponding to the edges coming into the vertex a, let $k+1, \cdots, k+m$ denote those leaving it, and let $k+m+1, \cdots, k+m+n$, denote the remaining edges of G_1.

We replace each pair of edges i, j of the graph G_1, $i = 1, \cdots, k$, $j = k+1, \cdots, k+m$ by one edge (we denote it by (i,j)), whose beginning coincides with the beginning of the edge i and whose end coincides with the end of the edge j. We get a graph G_2 also satisfying condition (*). Fixed points of σ_1 in the chain $(G_1, \Omega_1, \sigma_1)$ correspond naturally to fixed points of σ_2 in the chain $(G_2, \Omega_2, \sigma_2)$.

LEMMA 5.3. *For any two functions $f_1 : \Omega_1 \to R^1$ and $f_2 : \Omega_2 \to R^1$ of the class considered which vanish, if anywhere, only at corresponding points, the suspensions $(G_1, \Sigma_1, f_1, T_t^1)$ and $(G_2, \Sigma_2, f_2, T_t^2)$ are topologically equivalent.*

PROOF. We divide each edge (i,j) of the graph G_2 into two edges $(i,j)_1$ and $(i,j)_2$ by inserting a vertex. We get a graph G_2^1. The topological Markov chains $(G_1, \Omega_1, \sigma_1)$ and $(G_2^1, \Omega_2^1, \sigma_2^1)$ are conjugate, which is proved by a simple recoding: the pair of symbols i, j in a sequence which is a point of Ω_1 is replaced by the pair of symbols $(i,j)_1$ and $(i,j)_2$ in a sequence from Ω_2^1. Consequently the suspensions $(G_1, \Sigma_1, f_1, T_t^1)$ and $(G_2^1, \Sigma_2^1, f_2^1, T_t^1)$ are topologically equivalent, where $f_2 : \Omega_2^1 \to R^1$ is the function whose value at the point $x \in \Omega_2^1$ is equal to $f_1(g^{-1}x)$, and g is the homeomorphism which establishes the conjugacy between the topological Markov chains $(G_1, \Omega_1, \sigma_1)$ and $(G_2^1, \Omega_2^1, \sigma_2^1)$. On the other hand, by Lemma 5.2 the suspensions $(G_2^1, \Sigma_2^1, f_2^1, T_t^1)$ and $(G_2, \Sigma_2, f_2, T_t^2)$ are topologically equivalent. Hence follows the assertion of the lemma.

We note that in Lemmas 5.2 and 5.3 the topological Markov chains, the suspensions over which are topologically equivalent, are not conjugate.

As an example using the lemmas proved we offer the following result. We consider a graph G_1:

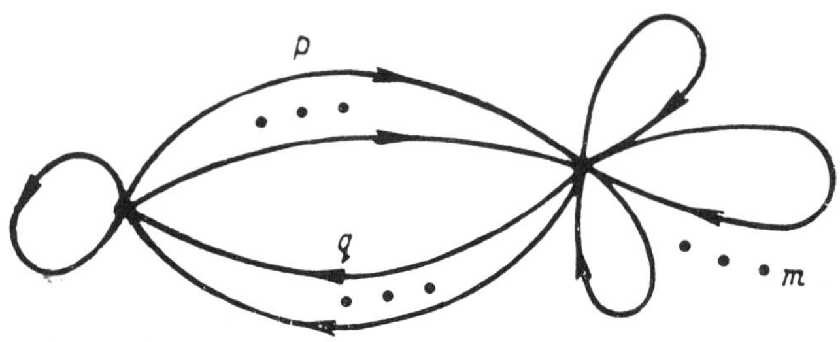

Let $r = \min(p, q)$. We consider also a graph G_2:

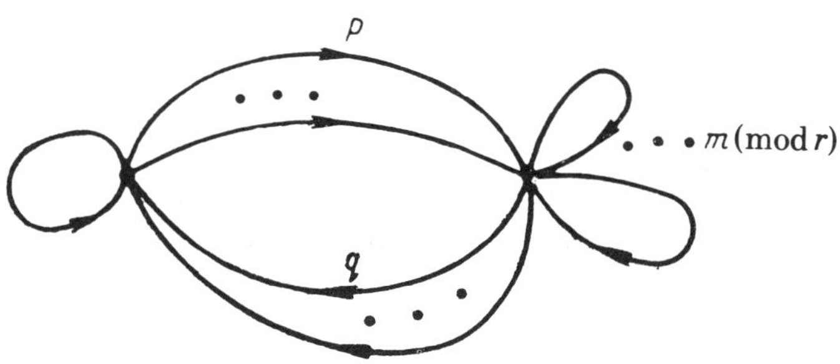

The following assertion is valid:

The suspensions (G_1, Σ_1, T_t^1) *and* (G_2, Σ_2, T_t^2) *are topologically equivalent.*

As is known, topological Bernoulli schemes with different numbers of symbols are not topologically conjugate.

CONJECTURE. *The suspensions with* $f \equiv 1$ *over Bernoulli schemes with different numbers of symbols are not topologically equivalent.*

§6. Proof of the basic results

In this section we shall construct a topological Markov chain, the suspension over which is homeomorphic to the subsystem $(T_t, \mathfrak{M}_{m+1}^{(\kappa)})$.

From Theorem 4.1 it follows that the behavior of each asymptotic trajectory Γ_ρ of $\mathfrak{M}_{m+1}^{(\kappa)}$ is determined in fact by its behavior only in a neighborhood of a periodic motion L_i, i.e. by the number $k^i \geq k^i_{p_\gamma^\nu q_\omega^\delta}$ which characterize the dwelling time of the trajectory in the neighborhood of the periodic motion L_i, i.e. the collection of trajectories of zeroth order and preferred trajectories of first through γ_κth order in the neighborhood of which it is contained. This fact lies at the base of the following construction.

We first construct a graph G_1. The set of edges $\{\hat{L}_j, \hat{\Gamma}_i\}$ of G_1 (or of states of the corresponding Markov chain) is in one-to-one correspondence with the set $\{L_j, \Gamma_i\}$, the elements of which are the following trajectories from $\mathfrak{M}_{m+1}^{(\kappa)}$: all periodic motions and saddles L_i, all trajectories of zeroth order and all preferred trajectories of first through γ_κth order. There is a transition from the state $\hat{\Gamma}_i$ into $\hat{\Gamma}_j$ if and only if $\Gamma_i \in W_{j_1}^+$ and $\Gamma_j \in W_{j_1}^-$; there is a transition from the state $\hat{\Gamma}_i$ into the state $\hat{L}_j$ if and only if $\Gamma_i \in W_j^+$; there is a transition from the state $\hat{L}_j$ into the state $\hat{\Gamma}_i$ if and only if $\Gamma_i \in W_j^-$; and there is a transition from the state $\hat{L}_{j_1}$ into the state $\hat{L}_{j_2}$ if and only if $L_{j_1} = L_{j_2}$.

We consider all states $\hat{L}_i$ corresponding to periodic motions or saddles L_i. The beginning and end of the edge $\hat{L}_i$ is some vertex a_i. Each edge corresponding to an asymptotic trajectory of the zeroth order, or preferred trajectory of first through γ_κth order, we divide into two by inserting a vertex. After this, the edges going into a_i will be denoted by $\hat{\Gamma}_{ip_\nu^\gamma}$ and those leaving a_i by $\hat{\Gamma}_{iq_\omega^\delta}$, $\nu = 1, \cdots, \nu_i^\gamma$, $\omega = 1, \cdots, \omega_i^\delta$; $\gamma, \delta = 0, 1, \cdots, \gamma_\kappa$, corresponding to the notation for the trajectories of the zeroth order and preferred trajectories of the first, second, $\cdots$ orders ω (α)-asymptotic to the trajectory L_i (cf. Figure 6.1). The rest of the construction is necessary only for the case when the length of $\mathfrak{M}_{m+1}^{(\kappa)}$ is positive.

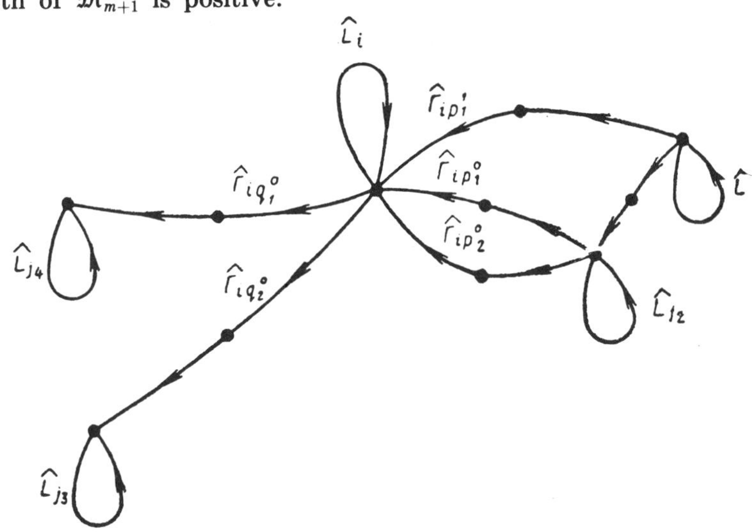

FIGURE 6.1

Let L_i be a periodic motion. We introduce the following notation: $k_{p_\nu^\gamma}^i = \max k_{p_\nu^\gamma q_\omega^\delta}^i$ (here $k_{p_\nu^\gamma q_\omega^\delta}^i$ is the number fixed in §4). Each edge $\hat{\Gamma}_{ip_\nu^\gamma}$ we divide into $k_{p_\nu^\gamma}^i$ new edges by $(k_{p_\nu^\gamma}^i - 1)$ vertices, so that each vertex appears as the beginning of one of the new edges and the end of a subsequent new edge. The

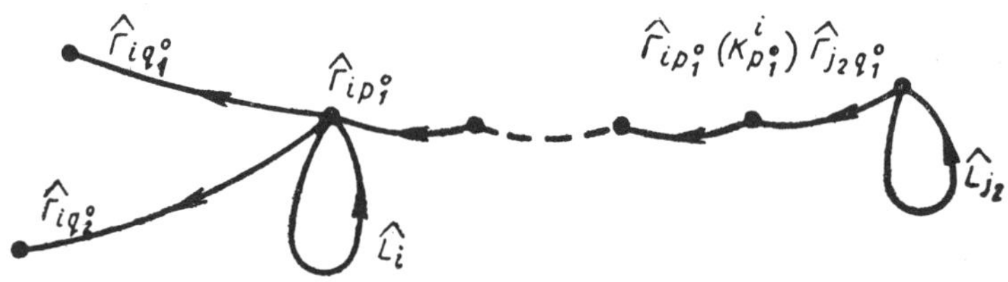

FIGURE 6.2

new edges we denote by $\hat{\Gamma}_{ip^\gamma_\nu}(s)$, $s=1,\cdots,k^i_{p^\gamma_\nu}$, and we suppose that from a state of the corresponding Markov chain $\hat{\Gamma}_{ip^\gamma_\nu}(s+1)$ there are transitions only into the states $\hat{\Gamma}_{ip^\gamma_\nu}(s)$, $s=1,\cdots,k^i_{p^\gamma_\nu}-1$. Here there will be a transition from the state $\hat{\Gamma}_{ip^\gamma_\nu}(1)$ into $\hat{L}_i$ and into all $\hat{\Gamma}_{iq^\delta_\omega}$. Thus in order to get from the state $\hat{\Gamma}_{ip^\gamma_\nu}(k^i_{p^\gamma_\nu})$ into any of the states $\hat{\Gamma}_{iq^\delta_\omega}$ it is necessary to pass through the states $\hat{\Gamma}_{ip^\gamma_\nu}(s)$, $s=1,\cdots,k^i_{p^\gamma_\nu}-1$. Further, we join the beginning of each edge $\hat{\Gamma}_{ip^\gamma_\nu}(1),\cdots,\hat{\Gamma}_{ip^\gamma_\nu}(k^i_{p^\gamma_\nu}-k^i_{p^\gamma_\nu q^\delta_\omega})$ with the end of the edge $\hat{\Gamma}_{iq^\delta_\omega}$ by some new edge whose beginning coincides with the beginning of the edge $\hat{\Gamma}_{ip^\gamma_\nu}(s)$, $s=1,\cdots,k^i_{p^\gamma_\nu}-k^i_{p^\gamma_\nu q^\delta_\omega}$, and its end with the end of the edge $\hat{\Gamma}_{iq^\delta_\omega}$. After this construction it is not hard to see that, first of all, the minimal number of states through which we must pass in order to get from the state $\hat{\Gamma}_{ip^\gamma_\nu}(k^i_{p^\delta_\omega})$ into a state following after $\hat{\Gamma}_{iq^\delta_\omega}$ (by construction from $\hat{\Gamma}_{iq^\delta_\omega}$ there is a transition into a unique state) is equal precisely to $k^i_{p^\gamma_\nu q^\delta_\omega}$, and, second, for any $k \geqq k^i_{p^\gamma_\nu q^\delta_\omega}$ a sequence of k states through which we must pass in order to get from the state $\hat{\Gamma}_{ip^\gamma_\nu}(k^i_{p^\gamma_\nu})$ into a state following $\hat{\Gamma}_{iq^\delta_\omega}$, is uniquely determined (cf. Figure 6.3).

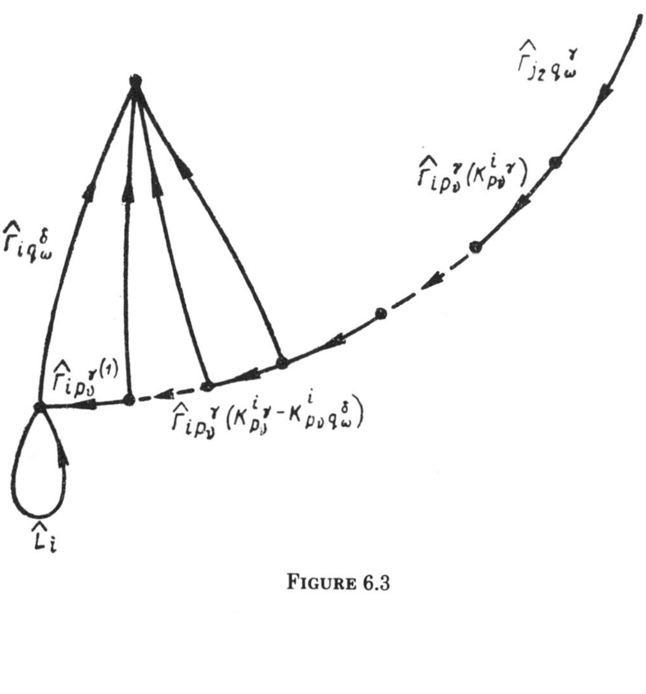

FIGURE 6.3

This construction on the graph G_1 of new edges and vertices is carried out for all indices i corresponding to periodic motions L_i. We denote the new graph by G.

REMARK 6.1. *From the construction it follows that for a suitable numbering of the edges of the graph G the transition matrix of the topological Markov chain will be triangular, i.e. one can renumber the states of the topological Markov chain (G, Ω, σ) so that from a state with lower number there are transitions only into states with higher numbers and, possibly, into the state itself. It will be assumed that such a numbering is used.*

One has

LEMMA 6.1. *The set of trajectories from $\mathfrak{M}_{m+1}^{(\kappa)}$ is in one-to-one correspondence with the set of trajectories of the topological Markov chain (G, Ω, σ)*

PROOF. By construction the periodic motions and saddles L_i are in one-to-one correspondence with the fixed points of the topological Markov chain (G, Ω, σ). Since by Theorem 4.1 all remaining trajectories from $\mathfrak{M}_{m+1}^{(\kappa)}$ are in one-to-one correspondence with the set of admissible sequences from $\{H_{\gamma_\kappa, s}^+\}$, for the proof of the lemma it suffices to put the remaining trajectories of the topological Markov chain (G, Ω, σ) into one-to-one correspondence with the set of admissible sequences from $\{H_{\gamma_\kappa, s}^+\}$.

We consider an arbitrary asymptotic trajectory of the topological Markov chain (G, Ω, σ). Let the α-limit set of this trajectory be a fixed point corresponding to a nonwandering motion L_i, and let the ω-limit set be a point corresponding to L_j. Then, in the sequence of symbols corresponding to the given trajectory, on the right will be in succession an infinite repetition of the symbol $\hat{L}_j$ corresponding to L_j, on the left an infinite repetition of the symbol $\hat{L}_i$ corresponding to L_i, and in between will be some segment, for example $[A_1, \cdots, A_{k_0}]$. Obviously the symbol A_1 must denote one of the states $\hat{\Gamma}_{iq_\omega^\delta}$, since from $\hat{L}_i$ there are transitions only into $\hat{L}_i$ and $\hat{\Gamma}_{iq_\omega^\delta}$. Hence as the first element of the admissible sequence we take $H_{iq_\omega^\delta}^+(\rho_{iq_\omega^\delta})$. From $\hat{\Gamma}_{iq_\omega^\delta}$ by construction there is a transition only into some state $\hat{\Gamma}_{i_1 p_\nu^\delta}(k_{p_\nu^\delta}^i)$, then necessarily followed by the symbols $\hat{\Gamma}_{i_1 p_\nu^\delta}(s), \cdots$, and, finally, if L_{i_1} does not coincide with L_{j_0} the symbol denoting one of the states $\hat{\Gamma}_{i_1 q_{\omega_1}^{\delta_1}}$. The number of symbols between $\hat{\Gamma}_{iq_\omega^\delta}$ and $\hat{\Gamma}_{i_1 q_{\omega_1}^{\delta_1}}$ is equal to some number k_1, while, by construction, $k_1 \geqq k_{p_\nu^\delta q_{\omega_1}^\delta}^i$. As the second element of the admissible sequence we take $H_{i_1 p_\nu^\delta}^+$, as the third we take $H_{i_1 p_\nu^\delta q_{\omega_1}^{\delta_1}}^+(k_1)$, and as the fourth we take $H_{i_1 q_{\omega_1}^{\delta_1}}^+(\rho_{i_1 q_{\omega_1}^{\delta_1}})$. Further, we construct three more elements of the admissible sequence, etc. At the very end we make the symbol A_{k_0}, which denotes some state $\hat{\Gamma}_{ip_{\nu_0}^{\gamma_0}}$, correspond to the space $H_{jp_{\nu_0}^{\gamma_0}}^+$, which will be the last element of the admissible sequence.

Now suppose an arbitrary admissible sequence from $\{H_{\gamma_\kappa, s}^+\}$ is given and let the trajectory from $\mathfrak{M}_{m+1}^{(\kappa)}$ corresponding to this sequence by Theorem

4.1 be α-asymptotic to the nonwandering motion L_i and ω-asymptotic to L_j. Hence the first element of the admissible sequence is some space $H^+_{iq^\delta_\omega}(\rho_{iq^\delta_\omega})$, and the last is the space $H^+_{jp^\gamma_\nu}$.

This sequence will be put in correspondence with a trajectory of the topological Markov chain (G, Ω, σ) α-asymptotic to a fixed point corresponding to L_i and ω-asymptotic to a fixed point corresponding to L_j, with an intermediate segment $[A_1, \cdots, A_{k_0}]$. We choose the symbol A_1 to correspond to the state $\hat{\Gamma}_{iq^{\delta_1}_{\omega_1}}$. Let the second, third and fourth elements of the admissible sequence be respectively $H^+_{i_1p^\gamma_\nu}$, $H^+_{i_1p^\gamma_\nu q^\delta_\omega}(k_1)$ and $H^+_{i_1q^\delta_\omega}(\rho_{i_1q^\delta_\omega})$. The symbol A_{k_1+1} we let correspond to the state $\hat{\Gamma}_{i_1q^\delta_\omega}$. But since by construction one can get from the state $\hat{\Gamma}_{iq^{\delta_1}_{\omega_1}}$ to the state $\hat{\Gamma}_{i_1q^\delta_\omega}$ in k steps in a unique way, the symbols $A_2, \cdots, A_{k_1}$ are uniquely determined. Then all the symbols of the segment $[A_1, \cdots, A_{k_0}]$ are determined in exactly the same way, and consequently all of the corresponding trajectory of the topological Markov chain (G, Ω, σ). In correspondence with the above construction, this trajectory will be mapped onto the given admissible sequence. The lemma is proved.

Before defining the suspension over the topological Markov chain (G, Ω, σ), we introduce some preliminary constructions.

We recall that for each saddle $L_j \in \mathfrak{M}^{(\kappa)}_{m+1}$ in §3 we constructed a sphere without contact D_j for trajectories belonging to W^+_j if $\dim W^+_j = m+1$, or belonging to W^-_j if $\dim W^-_j = n+1$.

In some small neighborhood of each saddle L_j, $\dim W^+_j = m+1$, we introduce euclidean coordinates (x, y), where x is an $(m+1)$-vector and y is an n-vector, in which the equation of W^+_j will be $y = 0$. By $\theta_j(x)$ we denote a Ljapunov function for trajectories belonging to W^+_j. Without loss of generality one can assume that the equation of D_j has the form $\theta_j(x) = c^j_0$ $(c^j_0 > 0)$. We fix the sequence $\{c^j_k\}^\infty_{k=1}$, $c^j_k \to 0$ as $k \to \infty$, where $0 < c^j_k < c^j_{k+1}$. Each equation $\theta_j(x) = c^j_k$ determines a sphere without contact for trajectories belonging to W^+_j, which we denote by D^k_j. Obviously the topological limit of D^k_j as $k \to \infty$ is the saddle L_j.

Analogously, in some small neighborhood of each saddle $L_i \in \mathfrak{M}^{(\kappa)}_{m+1}$, such that $\dim W^-_i = n+1$, we introduce euclidean coordinates (x, y), where x is an m-vector and y is an $(n+1)$-vector, in which the equation of W^-_i will be $x = 0$; and we construct a Ljapunov function $\theta_i(y)$ for trajectories belonging to W^-_i and a sequence of spheres without contact D^k_i, $k = -1, -2, \cdots$, the topological limit of which as $k \to -\infty$ is the saddle L_i.

We associate with each asymptotic trajectory $\Gamma \in \mathfrak{M}^{(\kappa)}_{m+1}$ the two-sidedly infinite sequence of its points of intersection with sections S_i, and, possibly, with D^k_j in the following way. Let $\alpha(\Gamma) = L_{i_0}$ and $\omega(\Gamma) = L_{j_0}$. Hence, in the admissible sequence from $\{H^+_{\gamma_\kappa, s}\}$ corresponding to Γ by Theorem 4.1, the first element will be $H^+_{i_0q^{\delta_0}_{\omega_0}}(\rho_{i_0q^{\delta_0}_{\omega_0}})$, and the last will be $H^+_{j_0p^{\gamma_0}_{\nu_0}}$. We fix a point of intersection of Γ with $U_{i_0q^{\delta_0}_{\omega_0}}$ and we denote it by $P_0(\Gamma)$.

In case L_{i_0} is a periodic motion, as $P_{-1}(\Gamma), \cdots, P_{-k}(\Gamma), \cdots$ we take the se-

financial position of DVC changed from red to black i.e. from a loss of Rs. 2 crores during the six months preceding the launching of the new style of management, the Corporation earned a profit of Rs. 7 crores during the succeeding six months. The following two years yielded a profit of Rs. 57 crores and Rs. 72 crores respectively.

The Neyveli Lignite Corporation (NLC) is an another example of public sector undertaking which was in red with accumulated losses of Rs. 840 millions by the financial year 1975-76, was turned around in 1976-77 by Mr. G.L. Tandon, Chairman and Managing Director and maintained by his successor Mr. P. Narayanan, NLC earned a profit of Rs. 120 millions on a turnover of Rs. 790 millions in 1976-77 and thereafter continued to grow steadily. And the accumulated losses were wiped out by the year 1981-82 having a net credit balance of Rs. 30 millions. The Corporation was ranked second in terms of absolute profits among all the public sector corporations in 1984-85. The credit of such a remarkable turnaround goes mainly to a determined management and a dedicated workforce.

The Anglo French Textiles (a subsidiary of the Pondicherry Textiles Corporation) the 90 Year old mill spreaded over 56 acres of land, which was closed in July 1983 despite of revival efforts at various levels for many years, was turned around by P.M. Nair who took over as Chairman and Managing Director in 1985. P.M. Nair and Joint Managing Director, P. Somasundaram, set up a task-force with an action plan and turned around it within six months. The mill achieved a turnover of Rs. 16 crores in the first year, making a profit of Rs. 1.2 lakhs. The turnover touched Rs. 31 crores in the second year and expected to be in the region of Rs. 35 crores for the current year. The Management of AFT succeeded in its revival through its dedicated efforts and style of management i.e. an open door approach, which motivated the workforce and inculcated the sense of belongingness among them.

Before 1981, Hindustan Latex was gravely sick. The newly appointed Chairman introduced a new style of management in which workers' suggestions were invited, considered and implemented. The result of this style of management was better capacity utilisation and within two years the organisation not only became solvent but earned several crores as profit. The strikes which were endemic in 1980-81 virtually disappeared through workers' participation and placement of better personnel in key areas. The workers were also given a fair share of increase in productivity and profit. They also got a good

quence of points of intersection as $t \to -\infty$ of the semitrajectory of the trajectory Γ with initial point $P_0(\Gamma)$ with the section S_{i_0}. If L_{i_0} is a saddle, then as $P_k(\Gamma)$ we take the point of intersection of Γ with $D^k_{i_0}$, $k = -1, -2, \cdots$.

In the admissible sequence from $\{H^+_{\gamma_\kappa, s}\}$ corresponding to Γ, the second element will be some space $H^+_{ip^{\delta_0}_{\nu_1}}$. As $P_1(\Gamma)$ we take the point of intersection of Γ with $U_{i_1 p^{\delta_0}_{\nu_1}}$. The third element of the admissible sequence will be $H^+_{i_1 p^{\delta_0}_{\nu_1} q^{\delta_1}_{\omega_1}}(k)$; the fourth, $H^+_{i_1 q^{\delta_1}_{\omega_1}}(\rho_{i_1 q^{\delta_1}_{\omega_1}})$, if $L_{i_1} \neq L_{j_0}$. As $P_2(\Gamma), P_3(\Gamma), \cdots, P_{k_1+1}(\Gamma)$ we take the sequence of points of intersection of the semitrajectory of the trajectory Γ with initial point $P_1(\Gamma)$ with the section S_i. Obviously $P_{k_1+1}(\Gamma) \in U_{i_1 q^{\delta_1}_{\omega_1}}$. The fifth element of the admissible sequence will be some space $H^+_{i_2 p^{\delta_1}_{\nu_2}}$. As $P_{k_1+2}(\Gamma)$ we take the point of intersection of Γ with $U_{i_2 p^{\delta_1}_{\nu_2}}$, etc. Finally, let $P_{k_0}(\Gamma)$ be the point of intersection of Γ with $U_{j_0 p^{\gamma_0}_{\nu_0}}$. If L_{j_0} is a periodic motion, as $P_{k_0+1}(\Gamma), P_{k_0+2}(\Gamma), \cdots$ we take the sequence of points of intersection of the semitrajectory with initial point $P_{k_0}(\Gamma)$ of the trajectory Γ as $t \to \infty$ with the section S_{j_0}. Now if L_{j_0} is a saddle, then as $P_{k_0+s}(\Gamma)$ we take the point of intersection of Γ with $D^s_{j_0}$, $s = 1, 2, \cdots$.

Each nonwandering trajectory $L_i \in \mathfrak{M}^{(\kappa)}_{m+1}$ we also make correspond to a sequence of points $\{P_k(L_i)\}^{\infty}_{k=-\infty}$ by the following rule: each $P_k(L_i)$ is the point of intersection of L_i with S_i if L_i is a periodic motion, and is L_i if L_i is a saddle.

We denote the set

$$\bigcup_{L_i, \Gamma_j \in \mathfrak{M}^{(\kappa)}_{m+1}} \left(\bigcup_k P_k(\Gamma_j) \cup P_k(L_i) \right),$$

considered as a topological space with the relative topology, by $\Omega^{(\kappa)}_{m+1}$, and we denote the map $\Omega^{(\kappa)}_{m+1} \to \Omega^{(\kappa)}_{m+1}$, under which the point $P_{k+1}(\Gamma)$ corresponds to the point $P_k(\Gamma)$, and $P_{k+1}(L_i)$ to $P_k(L_i)$, by T.

It is easy to see that T is a *homeomorphism*.

Lemma 6.2. *There exists a homeomorphism $g: \Omega^{(\kappa)}_{m+1} \to \Omega$ such that the diagram*

$$\begin{array}{ccc} \Omega^{(\kappa)}_{m+1} & \xrightarrow{T} & \Omega^{(\kappa)}_{m+1} \\ {\scriptstyle g}\downarrow & & \downarrow{\scriptstyle g} \\ \Omega & \xrightarrow{\sigma} & \Omega \end{array}$$

is commutative, where (G, Ω, σ) is the topological Markov chain constructed at the beginning of the section.

Proof. By construction each nonwandering point $P_0 \in \Omega^{(\kappa)}_{m+1}$ is a fixed point of the map T. Trajectories which belong to P_0 correspond by Lemma 6.1 to trajectories of the topological Markov chain (G, Ω, σ) consisting of one point x_0, a fixed point of the shift σ. We make the point $x_0 \in \Omega$ correspond to the point $P_0 \in \mathfrak{M}^{(\kappa)}_{m+1}$.

We now consider an arbitrary asymptotic trajectory $\Gamma \in \mathfrak{M}^{(\kappa)}_{m+1}$. Let $\alpha(\Gamma) = L_i$ and $\omega(\Gamma) = L_j$. By Lemma 6.1, to it there corresponds a trajectory of the topological Markov chain (G, Ω, σ). We fix on this trajectory a point x_0 with zeroth coordinate denoting the state $\hat{\Gamma}_{iq^\delta_\omega}$. From the construction of the chain (G, Ω, σ) it follows that the symbol denoting $\hat{\Gamma}_{iq^\delta_\omega}$ occurs only once. We make the point $P_0(\Gamma)$ correspond to the point x_0, and the point $P_k(\Gamma) = T^k P_0(\Gamma)$ to the point $\sigma^k x_0$, $k \in Z$.

Thus each point of $\Omega^{(\kappa)}_{m+1}$ corresponds to a point of Ω. We denote the map $\Omega^{(\kappa)}_{m+1} \to \Omega$ thus constructed by g. From Lemma 6.1 it follows that g is a bijective map, and from the definition of g it follows that the diagram

$$\begin{array}{ccc} \Omega^{(\varkappa)}_{m+1} & \xrightarrow{T} & \Omega^{(\varkappa)}_{m+1} \\ g\downarrow & & \downarrow g \\ \Omega & \xrightarrow{\sigma} & \Omega \end{array}$$

is commutative. It remains to show that g is a homeomorphism. For this it suffices to show that for any sequence $\{P^s\}_1^\infty$, $P^s \to P^0$, we have $gP^s \to gP^0$. It is easy to show that this is satisfied if all points P^s beginning with some number belong to one and the same trajectory Γ. Hence the proof is required only in the case when not all points belong to one trajectory. Then without loss of generality one can assume, first, that all points P^s, $s = 1, 2, \cdots$, belong to different trajectories $\Gamma_s \subset \mathfrak{M}^{(\kappa)}_{m+1}$, and, second, that all of them belong to the same surface S_i or D^k_j.

We denote by Γ_0 the trajectory containing P^0. Let $\alpha(\Gamma_0) = L_{i_1}$ and $\omega(\Gamma_0) = L_{j_1}$. The trajectory Γ_0 by Theorem 4.1 corresponds to a sequence from $\{H^+_{\gamma_\kappa, k}\}$, say, $H^+_{\gamma_\kappa, k_1}, \cdots, H^+_{\gamma_\kappa, k_r}$, the first element of which is some space $H^+_{i_1 q^{\delta_0}_{\omega_0}}(\rho_{i_1 q^{\delta_0}_{\omega_0}})$, and the last is the space $H^+_{j_1 p \gamma_1}$. Hence the point $P_0(\Gamma_0) \in U_{i_1 q^{\delta_0}_{\omega_0}}$, introduced in the construction of the set $\Omega^{(\kappa)}_{m+1}$, and for some $k \in Z_+$ the point $P_k(\Gamma_0) \in U_{j_1 p \gamma_1}$. Since the transition time from P^0 to $P_0(\Gamma_0)$ and from P^0 to $P_k(\Gamma_0)$ is finite, the points $P_0(\Gamma_0)$ and $P_k(\Gamma_0)$ are limits for the set of trajectories $\{\Gamma_s\}$, $s = 1, 2, \cdots$.

Since the segment of Γ_0 included between the points $P_0(\Gamma_0)$ and $P_k(\Gamma_0)$ is a limit for the set $\{\Gamma_s\}$, there exists an $s_0 > 0$ such that for any $s > s_0$ in the admissible sequence from $\{H^+_{\gamma_\kappa, k}\}$ corresponding to Γ_s there will be contained a segment

$$H^+_{\gamma_\varkappa, k_0},\ H^+_{\gamma_\varkappa, k_1},\ \ldots,\ H^+_{\gamma_\varkappa, k_r},\ H^+_{\gamma_\varkappa, k_{r+1}},$$

where

$$H^+_{\gamma_\varkappa, k_0} = H^+_{i_0 p^{\gamma_0}_{\gamma_0} q^{\delta_0}_{\omega_0}}(\bar{k}^{i_0}_s),\quad H^+_{\gamma_\varkappa, k_{r+1}} = H^+_{j_1 p^{\gamma_1}_{\gamma_1} q^{\delta_1}_{\omega_1}}(\bar{k}^{j_1}_s),$$

$$\bar{k}^{i_0}_s,\ \bar{k}^{j_1}_s \to \infty,\ s \to \infty,$$

and $H^+_{\gamma_\kappa, k_1}, \cdots, H^+_{\gamma_\kappa, k_r}$ coincides with the admissible sequence for Γ_0.

Let A_{i_1} denote the state of the chain (G, Ω, σ) corresponding to L_{i_1}, and A_{j_1} the state corresponding to L_{j_1}. Then by construction the point $P_0(\Gamma_0)$ corresponds under the map g to the point

$$x^0 = (\cdots A_{i_1} A_{i_1} A_0 A_1 \cdots A_k A_{j_1} A_{j_1} \cdots),$$

where A_0 denotes the symbol $\hat{\Gamma}_{i_1 q_{\omega_0}^{\delta_0}}$, and A_k the symbol $\hat{\Gamma}_{j_1 p_{\nu_1}^{\gamma_1}}(1)$. We denote by $P(\Gamma_s)$ the point of intersection of the trajectory Γ_s and $U_{i_1 q_{\omega_0}^{\delta_0}}$.

From the above argument it follows that the point $P(\Gamma_s)$ corresponds under the map g to the point

$$x^s = (\ldots \underbrace{A_{t_0} A_{i_0} \ldots A_{i_0}}_{\bar{\bar{k}}_s^{i_0}} A_0 \ldots A_k \underbrace{A_{j_1} \; A_{j_1} \ldots A_{j_1}}_{\bar{\bar{k}}_s^{j_1}} \ldots),$$

where

$$\bar{\bar{k}}_s^{i_0} = \bar{k}_s^{i_0} - k^{i_0}_{p_{\nu_0}^{\gamma_0} q_{\omega_0}^{\delta_0}}, \qquad \bar{\bar{k}}_s^{j_1} = \bar{k}_s^{j_1} - k^{j_1}_{p_{\nu_1}^{\gamma_1} q_{\omega_1}^{\delta_1}}.$$

Consequently $\bar{\bar{k}}_s^{i_0}, \bar{\bar{k}}_s^{j_1} \to \infty$ as $s \to \infty$, i.e. $x^s \to x^0$. And since by construction $gP^s = \sigma^l x^s$, $s = 0, 1, 2, \cdots$, for some l, one also has $gP^s \to gP^0$.

The lemma is proved.

We now construct the suspension over the topological Markov chain (G, Ω, σ). For this we equip the manifold M with a riemannian metric.

We give a function $f_0 : \Omega_{m+1}^{(\kappa)} \to R^1$ in the following way: $f_0(L_i) = 0$ if $L_i \in \Omega_{m+1}^{(\kappa)}$ is a saddle; otherwise $f_0(P_k(\Gamma))$ is equal to the length of the segment of the trajectory Γ included between the points $P_k(\Gamma)$ and $P_{k+1}(\Gamma)$; and, for a periodic motion L_j, $f_0(P_k(L_j))$ is equal to the length of L_j. It is easy to see that f_0 is a nonnegative continuous function. For any $x \in \Omega$ we set $f(x) = f_0(g^{-1}x)$, where g is the homeomorphism of Lemma 6.2. From Lemma 6.2 and the definition of f_0 it follows that $f : \Omega \to R^1$ is a continuous nonnegative function which vanishes only at the fixed points of σ corresponding to saddles $L_i \in \mathfrak{M}_{m+1}^{(\kappa)}$.

Theorem 6.1. *The subsystem* $(T_g, \mathfrak{M}_{m+1}^{(\kappa)})$ *is topologically equivalent to the suspension* (G, Σ, f, T_t).

Proof. Each point $P \in \Gamma \subset \mathfrak{M}_{m+1}^{(\kappa)}$ is contained in a half-closed segment $[P_k, P_{k+1})$ of the trajectory Γ and hence uniquely determines a pair $(P_k(\Gamma), s)$, where s is the length of the segment of the trajectory Γ included between the points $P_k(\Gamma)$ and P. From the definition it follows that $0 \leqq s < f_0(P_k(\Gamma))$. Conversely, each pair $(P_k(\Gamma), s)$, $0 \leqq s < f_0(P_k)$, uniquely determines a point $P \in \Gamma$ which lies at distance s along the trajectory Γ from $P_k(\Gamma)$ between $P_k(\Gamma)$ and $P_{k+1}(\Gamma)$.

We set $hP = (gP_k, s)$. From Lemma 6.2 it follows that $h : \mathfrak{M}_{m+1}^{(\kappa)} \to \Sigma$ is a homeomorphism carrying trajectories of the system $(\mathfrak{M}_{m+1}^{(\kappa)}, T_t)$ into trajectories of the system (G, Σ, f, T_t).

The theorem is proved.

Remark 6.2. *From Lemmas* 3.5, 3.6 *and* 3.7, *the concept of continuity and Theorem* 6.1 *it follows that the subsystem* $(T_t, \mathfrak{M}_{m+1}^{(\kappa)})$ *is structurally stable.*

Received 13/SEPT/71

Bibliography

[1] A. A. Andronov and L. S. Pontrjagin, *Systèmes grossiers,* C. R. (Dokl.) Acad. Sci. URSS **14** (1937), 247-250.

[2] E. Leontovič and A. G. Maĭer, *On a scheme determining the topological structure of a separation of trajectories,* Dokl. Akad. Nauk SSSR **103** (1955), 557-560. (Russian) MR **17**, 262.

[3] L. Markus, *Global structure of ordinary differential equations in the plane,* Trans. Amer. Math. Soc. **76** (1954), 127-148. MR **15**, 704.

[4] I. N. Vrublevskaja, *On geometric equivalence of the trajectories of dynamical systems,* Mat. Sb. **42** **(84)** (1957), 361-424. (Russian) MR **20** # 3534.

[5] S. Smale, *Morse inequalities for a dynamical system,* Bull. Amer. Math. Soc. **66** (1960), 43-49. MR **22** # 8519.

[6] ———, *Differentiable dynamical systems,* Bull. Amer. Math. Soc. **73** (1967), 747-817. MR **37** # 3598.

[7] J. Palis and S. Smale, *Structural stability theorems,* Proc. Sympos. Pure Math., vol. 14, Amer. Math. Soc., Providence, R. I., 1970, pp. 223-231. MR **42** # 2505.

[8] Ja. G. Sinaĭ, *Markov partitions and Y-diffeomorphisms,* Funkcional. Anal. i Priložen. **2** (1968), no. 1, 64-89 = Functional Anal. Appl. **2** (1968), 61-82. MR **38** # 1361.

[9] V. M. Alekseev, *Quasirandom dynamical systems.* I. *Quasirandom diffeomorphisms,* Mat. Sb. **76** **(118)** (1968), 72-134 = Math. USSR Sb. **5** (1968), 73-128. MR **43** # 2687b.

[10] V. V. Nemyckiĭ, *Topological problems of the theory of dynamical systems,* Uspehi Mat. Nauk **4** (1949), no. 6 (34), 91-153; English transl., Amer. Math. Soc. Transl. (1) **5** (1962), 414-497. MR **11**, 526.

[11] D. V. Anosov, *Geodesic flows on closed Riemannian manifolds of negative curvature,* Trudy Mat. Inst. Steklov. **90** (1967) = Proc. Steklov Inst. Math. **90** (1967). MR **36** # 7157; **39** # 3527.

[12] J. Palis, *On Morse-Smale dynamical systems,* Topology **8** (1968), 385-404. MR **39** # 7620.

[13] L. P. Šil′nikov, *On a Poincaré-Birkhoff problem,* Mat. Sb. **74** **(116)** (1967), 378-397 = Math. USSR Sb. **3** (1967), 353-372. MR **38** # 1322.

[14] ———, *Structure of the neighborhood of a homoclinic tube of an invariant torus,* Dokl. Akad. Nauk SSSR **180** (1968), 286-289 = Soviet Math. Dokl. **9** (1968), 624-628. MR **40** # 2111.

Translated by
N. STEIN

Trudy Moskov. Mat. Obšč.
Tom 28 (1973)

Trans. Moscow Math. Soc.
Vol. 28 (1973)

LIMITING EQUIVALENCE OF VARIOUS ENSEMBLES FOR ONE-DIMENSIONAL STATISTICAL SYSTEMS

A. HAITOV

Contents

Introduction

It is usual to assume in statistical physics that for those values of the thermodynamic parameters for which the system consists of a single phase, the three fundamental ensembles (microcanonical, canonical and grand canonical) are thermodynamically (that is, for a large system) equivalent. In its most complete and rigorous form this means that the limit Gibbs distributions (see below) obtained by means of each of these ensembles coincide if there are certain relations among the thermodynamic parameters which define the various ensembles.

Till now this equivalence has been proven for the canonical and grand canonical ensembles when the particle density is small for both continuous and lattice systems of any dimension. One method for proving this equivalence, consisting in the direct investigation of the limit correlation functions, is contained in the work of Bogoljubov, Petrina and Hacet [**22**] (see also the note [**24**] by Simjatickiĭ). The method for demonstrating the equivalence adopted here is contained in the paper [**17**] by Halfina (this procedure is based on certain results from the work of Minlos and Sinaĭ [**23**]; see also the paper [**7**] of Minlos and Halfina which is close to the spirit of the present work). The basis of Halfina's proof [**17**] is the derivation of the local limit theorem for the particle number in the grand canonical ensemble, using as a technical device the Kirkwood-Salzburg correlation equations, which up

---

AMS (MOS) subject classifications (1970). Primary 82A30, 60F05; Secondary 60J10.

to the time had been used effectively only in studying systems having low particle densities (and at high temperatures in the case of lattice systems).

Since one-dimensional systems, which we shall be concerned with in the following, do not undergo phase transitions (when the interactions decay sufficiently rapidly with interparticle distance), it has been assumed that the three ensembles are thermodynamically equivalent for all values of the thermodynamic parameters. The present paper is concerned with the proof of this statement. We shall devote most of our attention to proving the equivalence of the microcanonical and grand canonical ensembles for continuous systems obeying classical mechanics. The equivalence of these ensembles to the canonical ensemble will be obtained as a simple corollary.

In order to prove the equivalence of the microcanonical and grand canonical ensembles, following the outline of [**17**], we shall first obtain the two-dimensional local limit theorem for the distribution of energy and particle number in the grand canonical ensemble. As to the technical aspects of the demonstration, since, as it was already stated, we are unable to make use of the Kirkwood-Salzburg system of correlation equations for all values of the thermodynamic parameters, we shall obtain all the necessary estimates relating to the statistical properties of the Gibbs distribution by making various modifications of the methods in the work of van Hove [**19**], Dobrušin [**1,2**], Gallavotti and Miracle-Sole [**20**] and Suhov [**15,16**] on one-dimensional systems. The basis of all these procedures is the consideration of the Gibbs distribution for one-dimensional systems as a Markov process. In this regard we point out that in their probability-theoretic aspects the results obtained here are related to work concerning limit theorems for the sums of weakly interdependent stochastic quantities and, in particular, to quantities which depend on the values of a Markov chain (see the work of Statulevičius [**14**], Siraždinov [**13**], Nagaev [**9**], and Ibragimov and Linnik [**4**]).

In this paper the equivalence of ensembles is developed for one-dimensional systems for the case of a finite potential having a "hard core". The equivalence of the ensembles for potentials of more general sort will compose the subject matter of a subsequent publication. A general theorem concerning the equivalence of ensembles for the case of a general potential has been published (without proof) in a joint paper by Minlos and the author [**8**].

In many respects this paper has been inspired by Ja. G. Sinaĭ in connection with questions relating to his work on the dynamics of one-dimensional systems [**12**].

Briefly, the plan of this work is as follows: In the first section we formulate the basic concepts and the problem itself. In the second, we discuss briefly the necessary technical approximations (their detailed justification will compose the second part of this publication, which we shall henceforth refer to as II). The third section is devoted to the central point of the whole construction: the proof: the proof of the local limit theorem for the particle number and energy in the grand canonical ensemble. In the fourth section, based on this theorem, we shall derive the basic result concerning the equivalence of the limit Gibbs distributions for the microcanonical and grand canonical ensembles, and thence as a corollary we shall prove the equivalence of the canonical and grand canonical ensembles.

I take this occasion to express my deep thanks to my scientific director R. A. Minlos for posing this problem and for his constant attention to this work. I also express my sincere gratitude to Ja. G. Sinaĭ, who is one of those responsible for the appearance of this paper.

§1. Basic definitions and statement of the problem

I. *The space of states of the system.* Although the concepts introduced here are meaningful for systems having any number of dimensions, in what follows we shall limit consideration to the subject of immediate concern, one-dimensional systems.

Let $\Omega \subset R^1$ be a bounded measurable set on the real axis. Denote by C_Ω the collection of all finite subsets of Ω and by $C_{\Omega,N}$ the collection of all N-point subsets of Ω ($C_{\Omega,0} = \{\emptyset\}$). We have $C_\Omega = \bigcup_{N \geq 0} C_{\Omega,N}$. We shall call a subset $c \subset \Omega$ an *arrangement* (or *configuration*) in Ω; $N(c)$ will denote the number of points in c.

We shall call the pair $a = \{c, p_c(\cdot)\}$, where $c \in C_\Omega$ and $p_c(\cdot)$ is a numerical function defined on the elements of c and which takes on real values, a *state of the one-component mechanical system enclosed in the set* Ω. The quantity $p_c(x)$ is called the *momentum* of the particle located at the point $x \in c$.

We shall denote the space of all states by $\mathfrak{B}_\Omega$, and the state $a = \{c, p_c(\cdot)\}$ for which $c \in C_{\Omega,N}$ by $\mathfrak{B}_{\Omega,N}$. It is clear that

$$\mathfrak{B}_\Omega = \bigcup_{N \geqslant 0} \mathfrak{B}_{\Omega,N}\ . \tag{1}$$

The space $\mathfrak{B}_{\Omega,N}$ can be expressed as a factor space of a reduced N-fold cartesian power $(\Omega \otimes R^1)^{\otimes N}$, factored by the permutation group S_N, i.e.

$$\mathfrak{B}_{\Omega,N} = ((\Omega \otimes R^1)^{\otimes N})' / S_N \tag{1.2}$$

(the prime denotes the reduction which consists in eliminating the sets $(x_1, p_1; \cdots; x_N, p_N) \in (\Omega \otimes R^1)^{\otimes N}$ for which at least two x_i are the same).

With the help of the representation (1.2) we can introduce into $\mathfrak{B}_{\Omega,N}$ a Lebesgue volume by defining for the set $A \subset \mathfrak{B}_{\Omega,N}$

$$\lambda_\Omega(A) = \frac{1}{N!} \tilde{\lambda}_\Omega(\tilde{A}),$$

where $\tilde{\lambda}_\Omega(\cdot)$ is the Lebesgue measure in $((\Omega \otimes R^1)^{\otimes N})'$, and $\tilde{A}$ is the complete preimage of A under the factorization (1.2). By introducing the measure into each $\mathfrak{B}_{\Omega,N}$ we extend it to $\mathfrak{B}_\Omega$ (assuming that $\lambda_\Omega(\mathfrak{B}_{\Omega,0}) = 1$). In what follows, the integral of a measurable function on $\mathfrak{B}_\Omega$ with respect to this measure will be denoted by $\int_{\mathfrak{B}_\Omega} f(a)\, da$.

In the same way $C_{\Omega,N}$ can be represented as a factor space of a reduced N-fold Cartesian power $\Omega^{\otimes N}$ factored by the permutation group S_N, i.e.

$$C_{\Omega,N} = (\Omega^{\otimes N})'/S_N. \tag{1.2'}$$

As before, with the help of (1.2′) a Lebesgue measure $\mu_\Omega(\cdot)$ can be introduced into $C_{\Omega,N}$, and thence into C_Ω; the integral of a measurable function $f(c)$ with respect to this measure will be denoted by $\int_{C_\Omega} f(c)\,dc$.

In a canonical way the space $\mathfrak{B}_{\Omega,N}$ can be identified with the Cartesian product of the space $C_{\Omega,N}$ (configuration space) and the space R^N (the space of the values of the momenta of all the particles):

$$\mathfrak{B}_{\Omega,N} = C_{\Omega,N} \otimes R^N, \tag{1.3}$$

in such a way that the Lebesgue measure in $\mathfrak{B}_{\Omega,N}$ is the product of the Lebesgue measures of $C_{\Omega,N}$ and R^N.

By C_∞ we denote the totality of all finite or countable locally finite subsets $c \subset R^1$ (that is, such that only a finite number of points of the subset c occur on any bounded part of the line R^1). The elements of the space C_∞ are called *configurations of particles in* R^1.

By the *state of an infinite mechanical system* we shall understand a pair $a = \{c, p_c(\cdot)\}$, where $c \in C_\infty$ and $p_c(\cdot)$ is a real function defined over the elements $x \in c$. We shall denote the space of these states by $\mathfrak{B}_\infty$.

Next, for any two bounded measurable sets U and $V \subset R^1$, $U \subset V$, we define the restriction mapping $\varphi_{U,V}: \mathfrak{B}_V \to \mathfrak{B}_U$, under which to each state $a_2 = \{c_2, p_{c_2}(\cdot)\} \in \mathfrak{B}_V$ we associate the state $\varphi_{U,V} a_2 = a_1 = \{c_1, p_{c_1}(\cdot)\} \in \mathfrak{B}_U$, where $c_1 = c_2 \cap U$ and $p_{c_1}(x) = p_{c_2}(x)$ for $x \in c_1$.

Similarly we define the mapping $\varphi_U: \mathfrak{B}_\infty \to \mathfrak{B}_U$, where U is a bounded measurable set:

$$\varphi_U(\{c, p_c(\cdot)\}) = \{c \cap U, p_{c\cap U}(\cdot)\} \in \mathfrak{B}_U,\ \{c, p_c(\cdot)\} \in \mathfrak{B}_\infty.$$

A set $S = (S_{U,A}) \subset \mathfrak{B}_\infty$ is called *cylindrical* if there exists a measurable subset $U \subset R^1$ and a measurable set $A \subset \mathfrak{B}_U$ such that S is the complete preimage of A under the mapping $\varphi_U: S = \varphi_U^{-1}(A)$.

II. *The energy of a system*. We shall consider the function $E(a)$ defined on the space $\mathfrak{B}_\Omega$ (or $\mathfrak{B}_{\Omega,N}$), called the *energy* and defined, as usual, by the formula

$$E(a) = \frac{1}{2} \sum_{x\in c} p_c^2(x) + \sum_{x\in c,\, y\in c} U(|x-y|),\ a = \{c, p_c(\cdot)\},$$

where $U(r)$, $r > 0$, is a real function called the *pairwise interaction potential*.

In the following we shall consider the class of potentials obeying the following conditions:

I. $U(r)$ has compact support, i.e. $U(r) = 0$ for $r > R_0$.

The quantity R_0 is called the *range* of the potential $U(r)$.

II. The potential $U(r)$ has a "hard core", i.e. $U(r) = +\infty$ for $0 \leqq r \leqq \delta_0$.

III. There exists $r_0 > \delta_0$ such that $U(r)$ decreases monotonically for $\delta_0 < r < r_0$. Moreover, the limit $\lim_{r\to\delta_0} U(r) \leqq \infty$ exists. The interaction $U(r)$ will be called *finite at zero* if this limit is finite, and *infinite at zero* otherwise.

IV. Derivatives through second order exist for all $\delta_0 < r \leqq R_0$. In addition we shall require that for any $\beta > 0$

$$\int_{\delta_0}^{R_0} | U''(r) | \exp\{-\beta U(r)\} \, dr < \infty.$$

V. Constants $a > 0$ and $\zeta_0 > 0$ exist such that $U'(r) < -\zeta_0$ for $\delta_0 < r < \delta_0 + a$.

A number of consequences follow from these conditions.

1. *It follows from* IV *that* $U(r)$ *is bounded from below, i.e.* $U(r) > A > -\infty$.

2. *It follows from* I, II *and* 1 *that a constant* $B > 0$ *exists such that for any finite set* $c \subset R^1$ *and any* $x \in R^1$

(1.4)
$$\sum_{y \in c} U(|x-y|) > -B.$$

This is called the *stability property* of $U(r)$.

3. *It follows from* V *that for all* $\beta > 0$ *the integral*

$$\int_{\delta_0}^{R_0} | U'(r) | \exp\{-\beta U(r)\} \, dr < \infty$$

exists.

III. *Basic statistical ensembles.* Now we turn to the definition of the three basic probability ensembles used in statistical mechanics.

I. The probability distribution on $\mathfrak{B}_\Omega$ given by the density (with respect to the measure $\lambda_\Omega(\cdot)$)

(1.5)
$$p(a \mid \beta, \mu, \Omega) = \Xi^{-1}(\beta, \mu, \Omega) \, e^{-\beta E(a) - \mu N(a)},$$

is called the *Gibbs distribution.* Here $\beta > 0$ and μ is a real parameter; the normalizing factor

(1.5′)
$$\Xi(\beta, \mu, \Omega) = \int_{\mathfrak{B}_\Omega} e^{-\beta E(a) - \mu N(a)} \, da$$

is called the *grand partition function.*

It is customary to call the space $\mathfrak{B}_\Omega$ together with the distribution (1.5) the *grand canonical ensemble.*

II. Under the condition that the number of particles N is fixed, the conditional distribution generated by the Gibbs distribution (1.5) gives in the space $\mathfrak{B}_{\Omega,N}$, with respect to the measure $\lambda_\Omega(\cdot)$, the density

(1.6)
$$p(a \mid \beta, N, \Omega) = Z^{-1}(\beta, N, \Omega) \, e^{-\beta E(a)},$$

where the normalizing factor

$$Z(\beta, N, \Omega) = \int_{\mathfrak{B}_{\Omega,N}} e^{-\beta E(a)}\, da$$

is called the *canonical partition function*. The space $\mathfrak{B}_{\Omega,N}$ together with the distribution (1.6) is called the *canonical ensemble*.

III. Finally, if along with fixing the number of particles, we also fix the energy E and consider the surface Γ_E defined in the space $\mathfrak{B}_{\Omega,N}$ by the condition $E(a) = E$, it is possible to consider the conditional probability distribution generated on this surface by (1.6). It coincides with the uniform distribution on Γ_E, i.e. the distribution given by the generalized density (with respect to the Lebesgue measure on $\mathfrak{B}_{\Omega,N}$)

(1.7) $$p(a \mid E, N, \Omega) = \sum\nolimits^{-1}(E, N, \Omega)\, \delta(E(a) - E),$$

where the normalizing factor is equal to

$$\sum(E, N, \Omega) = \int_{\mathfrak{B}_{\Omega,N}} \delta(E(a) - E)\, da,$$

and $\delta(x)$ denotes the one-dimensional Dirac delta function.

The distribution (1.7) is called the *microcanonical distribution* (a more detailed description of the microcanonical distribution is given, for example, in Hinčin's article [**18**] or Ruelle's book [**10**]).

It is easy to see that in the canonical ensemble the conditional distribution of the values $\{p_c(\cdot)\}$ for a given configuration c does not depend on c. If we make use of the representation (1.6), the canonical ensemble can be viewed as the product of the two distributions:

1) The distribution induced into the configuration space $C_{\Omega,N}$ by the density (with respect to the measure $\mu_\Omega(\cdot)$)

(1.8) $$p_{\text{conf.}}(c) = p(c, \beta, N, \Omega) = \tilde{Z}^{-1}(\beta, N, \Omega)\, e^{-\beta H(c)},$$

where $H(c)$ denotes the *potential energy* of the system:

$$H(c) = \sum_{x \in c,\, y \in c} U(|x - y|),$$

and $\tilde{Z}(\beta, N, \Omega)$ denotes the integral $\tilde{Z}(\beta, N, \Omega) = \int_{C_{\Omega,N}} e^{-\beta H(c)} dc$. Note that $\tilde{Z}(\beta, N, \Omega) = (\pi/\beta)^{N/2} \tilde{Z}(\beta, N, \Omega)$. This ensemble will be called the *configurational canonical ensemble*.

2) The momentum distribution induced on the space R^N by the density (with respect to ordinary Lebesgue measure)

(1.9) $$\pi(p_1, p_2, \ldots p_N) = \left(\frac{\beta}{\pi}\right)^{\frac{N}{2}} \exp\left\{-\frac{1}{2} \sum_{i=1}^{N} p_i^2\right\}.$$

In particular, it follows from this that in the canonical ensemble the potential energy $H=\sum U(|x-y|)$ and the kinetic energy $T=\frac{1}{2}\sum_1^N p_i^2$ are statistically independent, and therefore the total energy $E(a)$ is the sum of two independent terms.

In other words, if we denote by $P_{\text{tot}}(E|\Omega,N)$ the probability density distribution for the total energy in the canonical ensemble, by $E_{\text{pot}}(H|\Omega,N)$ the probability distribution function for the potential energy and by $p_{\text{kin}}(T|N)$ the density distribution for the kinetic energy in this ensemble, we can write

(1.10)
$$p_{\text{tot}}(E\mid\Omega,N)=\int_{-\infty}^{E} p_{\text{kin}}(E-H\mid N)\,dF_{\text{pot}}(H\mid\Omega,N)$$
$$(p_{\text{kin}}(T\mid N)=0 \text{ for } T<0).$$

The existence of $p_{\text{tot}}(E|\Omega,N)$ will follow from (1.10) and the obvious existence of $p_{\text{kin}}(T|N)$.

We turn now to the grand canonical ensemble and introduce the quantity

$$p_{\text{tot}}(E,\ N\mid\Omega)=\frac{d}{dE}\{P_r(E(a)<E,\ N(a)=N)\}$$

(where the probability standing to the right refers to the grand ensemble). This is the probability density for the values of E and the probability of the values of N in the joint distribution for the energy and particle number in the grand ensemble.

It is clear that

$$p_{\text{tot}}(E,\ N\mid\Omega)=p_{\text{tot}}(E\mid\Omega,\ N)\cdot P_r(N(a)=N)$$
$$=p_{\text{tot}}(E\mid\Omega,\ N)\frac{Z(\beta,\ N,\ \Omega)}{\Xi(\beta,\ \mu,\ \Omega)}e^{-\mu N}=p_{\text{tot}}(E\mid\Omega,\ N)\frac{\widetilde{Z}(\beta,\ N,\ \Omega)}{\widetilde{\Xi}(\beta,\ \widetilde{\mu},\ \Omega)}e^{-\widetilde{\mu}N},$$

where $\tilde{\mu}=\mu-\frac{1}{2}\ln(\pi/\beta)$ and $\tilde{\Xi}(\beta,\tilde{\mu},\Omega)=\Xi(\beta,\mu,\Omega)$.

From (1.10) we obtain

$$p_{\text{tot}}(E,\ N\mid\Omega)=\int_{-\infty}^{E} p_{\text{kin}}(E-H\mid N)\,dF_{\text{pot}}(H,\ N\mid\Omega),$$

where $F_{\text{pot}}(H,N|\Omega)=P_r(H(c)<H,N(c)=N)$ is calculated in the configurational grand canonical ensemble with the chemical potential equal to $\tilde{\mu}$, i.e. for the distribution in C_Ω with the density

$$p(c,\ \widetilde{\mu},\ \beta,\ \Omega)=\widetilde{\Xi}^{-1}(\beta,\ \widetilde{\mu},\ \Omega)\,e^{-\beta H(c)-\widetilde{\mu}N(c)}$$

where

(1.11)
$$\widetilde{\Xi}(\beta,\ \widetilde{\mu},\ \Omega)=\int_{C_\Omega} e^{-\beta H(c)-\widetilde{\mu}N(c)}\,dc.$$

(In what follows, if no confusion will result, we will simply write μ in place of $\tilde{\mu}$).

Thus, in order to study $p_{\text{tot}}(E,N|\Omega)$, in view of the simplicity of the momentum distribution, in the following it will only be necessary to examine in detail the joint distribution of energy and particle number in the configurational grand ensemble C_Ω with probability density distribution (1.11).

IV. *Limit Gibbs distribution*. Let a segment $\Omega \subset R^1$ and a measurable set $U \subset \Omega$ be given. For each $A \subset \mathfrak{B}_U$ let $\widetilde{A}_\Omega^U$ denote its complete preimage under the mapping $\varphi_{\Omega,U}: \widetilde{A}_\Omega^U = \overset{-1}{\varphi}_{\Omega,U}(A) \subset \mathfrak{B}_\Omega$. For the cylindrical set $S_{U,A}$ we define the probability

$$P_{\Omega,\beta,\mu}(S_{U,A}) = P_{\Omega,\beta,\mu}(\widetilde{A}_\Omega^U) \tag{1.12}$$

and

$$P_{\Omega,E,N}(S_{U,A}) = P_{\Omega,E,N}(\widetilde{A}_\Omega^U \cap \mathfrak{B}_{\Omega,N}), \tag{1.13}$$

where $P_{\Omega,\beta,\mu}(\cdot)$ and $P_{\Omega,E,N}(\cdot)$ are the probabilities of the sets $\widetilde{A}_\Omega^U$ and $\widetilde{A}_{\Omega,N}^U$ for the grand canonical and microcanonical ensembles respectively. The formulas (1.12) and (1.13) give probability distributions over the set $\mathfrak{B}_U$ which we denote by $P_{\Omega,U,\beta,\mu}(\cdot)$ and $P_{\Omega,U,E,N}(\cdot)$ respectively.

Now consider any expanding sequence of line segments $\Omega_1 \subset \Omega_2 \subset \cdots$ such that $\bigcup_{N \geq 1} \Omega_N = R^1$. For each cylindrical set $S_{U,A}$ let the limit

$$\lim_{n\to\infty} P_{\Omega_n,U,\beta,\mu}(A) = P_{\beta,\mu}(S_{U,A}) \tag{1.14}$$

exist and for fixed U converge uniformly with respect to all $A \subset \mathfrak{B}_U$ independently of the choice of the sequence $\{\Omega_n\}$. Then the limit probabilities $P_{\beta,\mu}(S_{U,A})$ generate a limit distribution on the σ-algebra $\mathfrak{M}$ generated by the cylindrical sets (in this regard see Minlos [**6**]). This is called the *limit Gibbs distribution on* $\mathfrak{B}_\infty$.

Similarly to (1.14) we can consider the limit

$$\lim_{n\to\infty} P_{\Omega_n,U,\beta,\mu}(A) = P_{\beta,\mu}(S_{U,A}) \tag{1.14$'$}$$

in the configurational ensemble $\{C_\Omega, H(c), \beta, \mu\}$. The existence of (1.14′) under the conditions formulated above concerning the potential and for any values of the parameters β and μ follows from the work of Dobrušin [**1, 2**], Gallavotti and Miracle-Sole [**20**] and Suhov [**15, 16**]. Again for each fixed U the convergence of (1.14′) is uniform with respect to all $A \subset C_U$, and the limit is independent of the choice of the sequence $\{\Omega_n\}$. The limit probabilities (1.14′) generate a limit (configurational) distribution on the σ-algebra $\mathfrak{M}_{\text{conf}}$, called the *limit configurational Gibbs distribution on* C_∞. The existence of the limit (1.14) follows easily from the existence of (1.14′), i.e. the existence of the limit Gibbs distribution for the whole mechanical system follows from the existence of the configurational limit distribution.

Similar to the preceding, if for a sequence of ensembles $\{\mathfrak{B}_{\Omega_n}, E_n, N_n\}$ $(\Omega_n \subset \Omega_{n+1}, \bigcup_{n\geq 0} \Omega_n = R^1)$, such that $N_n/|\Omega_n| \to \rho$ and $E_n/|\Omega_n| \to h$, $-\infty < h < \infty$, as $n \to \infty$, for all cylindrical sets $S_{U,A}$, the limits

$$\lim_{n\to\infty} P_{\Omega_n,E_n,N_n,U}(A) = P_{\rho,h}(S_{U,A}), \tag{1.15}$$

exist, and if for fixed U the convergence is uniform with respect to all $A \subset \mathfrak{B}_U$ and the limits (1.15) do not depend on the choice of the sequence $\{\Omega_n, E_n, N_n\}$, then the limits $P_{\rho,h}(\cdot)$ generate a probability distribution on $\mathfrak{B}_\infty$ which, as before, we shall call the limit Gibbs distribution $P_{\rho,h}(\cdot)$ on $\mathfrak{B}_\infty$.

The proof that $P_{\rho,h}(\cdot)$ exists and coincides with the limit Gibbs distribution $P_{\beta,\mu}(\cdot)$ when there is a certain relation between the parameters (β,μ) and (ρ,h) is one of the problems considered here.

Since in what follows we shall consider the distributions $P_{\Omega,U,\beta,\mu}(\cdot)$ and $P_{\Omega,U,E,N}(\cdot)$, we shall introduce some useful formulas for their densities here.

Note that if $\Omega = U \cup V$ and $U \cap V = \emptyset$, then $\mathfrak{B}_\Omega = \mathfrak{B}_U \otimes \mathfrak{B}_V$ and the Lebesgue measure on $\mathfrak{B}_\Omega$ is the product of the Lebesgue measures on $\mathfrak{B}_U$ and $\mathfrak{B}_V$. Thus for any $A \subset \mathfrak{B}_U$

$$P_{\Omega,U,\beta,\mu}(A) = \Xi^{-1}(\beta, \mu, \Omega) \int_A da_1 \int_{\mathfrak{B}_V} e^{-\beta E(a_1 \cup a_2) - \mu(N(a_1)+N(a_2))} da_2$$

$$= \int_A \pi_U(a_1, \beta, \mu, \Omega)\, da_1,$$

where

$$\pi_U(a, \beta, \mu, \Omega) = \Xi^{-1}(\beta, \mu, \Omega)\, e^{-\beta E(a) - \mu N(a)} \int_{\mathfrak{B}_V} e^{-\beta E(a_1 \mid a) - \mu N(a_1)} da_1.$$

Here

$$E(a_1 \mid a) = E(a_1) + \sum_{x \in c \subset a,\, y \in c_1 \subset a_1} U(|x-y|).$$

By introducing the notation

$$\xi_a(x) = \exp\Big\{-\beta \sum_{y \in c \subset a} U(|x-y|)\Big\}, \quad z = e^{-\mu} \text{ и } V = \Omega \setminus U,$$

we obtain

$$\pi_U(a, \beta, \mu, \Omega) = z^{N(a)} e^{-\beta E(a)} \frac{\Xi(\beta, \mu, \xi_a, \Omega \setminus U)}{\Xi(\beta, \mu, \Omega)}, \tag{1.16}$$

where

$$\Xi(\beta, \mu, \xi_a, \Omega \setminus U) = \int_{\mathfrak{B}_{\Omega \setminus U}} e^{-\beta E(a_1)} \prod_{x \in c_1 \subset a_1} (z\xi_a(x))\, da_1.$$

In the same way the distribution $P_{\Omega,U,E,N}(\cdot)$ on $\mathfrak{B}_U$ is given by the density

$$\begin{aligned} \pi_U(a, E, N, \Omega) &= \Sigma^{-1}(E, N, \Omega) \int_{\mathfrak{B}_{\Omega \setminus U, N-N(a)}} \delta(E(a \cup a_1) - E)\, da_1 \\ &= \Sigma^{-1}(E, N, \Omega) \int \delta(E(a) - E')\, \Sigma_a(E - E', N, \Omega)\, dE', \end{aligned} \tag{1.17}$$

where

$$\Sigma_a(E, N, \Omega) = \int_{\mathfrak{B}_{\Omega \setminus U, N-N(a)}} \delta(E(a_1 \mid a) - E)\, da_1.$$

Next, it is easily seen that the density $\pi_U(a,\beta,\mu,\Omega)$ of the distribution in the ensemble $\{\mathfrak{B}_\Omega, \beta, \mu, E(a)\}$ is related to the density $\pi_U(c, \beta, \tilde{\mu}\ \Omega)$ of the distribution in the configurational ensemble $\{C_\Omega, \beta, \mu, H(c)\}$ by the following formula:

$$\pi_U(a, \beta, \mu, \Omega) = \left(\frac{\beta}{\pi}\right)^{\frac{N(a)}{2}} e^{-\frac{\beta}{2}\sum_{x\in c} p_c^2(x)} \pi_U(c, \beta, \tilde{\mu}, \Omega). \tag{1.18}$$

Here

$$\pi_U(c, \beta, \tilde{\mu}, \Omega) = e^{-\beta H(c) - \tilde{\mu} N(c)} \frac{\tilde{\Xi}(\beta, \tilde{\mu}, \xi_c, \Omega\setminus U)}{\tilde{\Xi}(\beta, \tilde{\mu}, \Omega)},$$

where

$$\tilde{\mu} = \mu - \frac{1}{2}\ln\frac{\pi}{\beta} \quad \text{and} \quad \xi_c(x) = \exp\Big\{-\beta\sum_{y\in c} U(|x-y|)\Big\}.$$

Here $\tilde{\Xi}(\beta, \tilde{\mu}, \xi_c, \Omega\setminus U)$ denotes the partition function

$$\tilde{\Xi}(\beta, \tilde{\mu}, \xi_c, \Omega\setminus U) = \int\limits_{C_{\Omega\setminus U}} e^{-\beta H(c_1)} \prod_{x\in c_1} (e^{-\tilde{\mu}}\xi_c(x))\, dc_1.$$

It follows from the results of the papers [**1, 2, 20, 15, 16**] that in the thermodynamic limit $|\Omega| \to \infty$ the density $\pi_U(c, \beta, \tilde{\mu}, \Omega)$ evaluated in the configurational ensemble converges uniformly with respect to $c \in C_U$ to the limit density $\pi_U(a, \beta, \tilde{\mu}, \infty)$ which generates the limit distribution on C_U. Hence it follows by (1.17) that in particular the density $\pi_U(a, \beta, \mu, \Omega)$ of the total distribution also converges uniformly with respect to $a \in \mathfrak{B}_U$ to the limit density $\pi_U(a, \beta, \mu, \infty)$, which generates the limit distribution on $\mathfrak{B}_U$.

We shall prove below that in the limit as $E/|\Omega| \to h$ and $N/|\Omega| \to \rho$ the density $\pi_U(a, E, N, \Omega)$ also converges to the same limit $\pi_U(a, \beta, \mu, \infty)$, where β and μ depend in a certain way on ρ and h.

§2. Technical procedures and basic approximations

In this section we introduce the estimates we shall require later.

In the space C_Ω, where Ω is a line segment, we consider the probability distribution defined by the density (with respect to the measure $\mu_\Omega(\cdot)$)

$$p(c, \xi, \psi, \Omega) = \Xi^{-1}(\xi, \psi, \Omega) \prod_{x\in c} \xi(x) \prod_{\substack{x\in c,\, y\in c\\ x\neq y}} e^{\psi(|x-y|)}, \tag{2.1}$$

where $\xi(x)$, $x \in R^1$, and $\psi(s)$, $s > 0$, are any two functions which satisfy the following conditions:

(2.2)
1) $\xi(x)$ is positive and bounded, $0 < \xi(x) < D$, where D is a constant,
2) $\psi(s) = -\infty$ if $0 \leqq s \leqq \delta_0$,
3) $\psi(s) = 0$ if $s \geqq R_0$,
4) $\psi(s) < A$ if $\delta_0 < s < R_0$ (A is a constant).

It follows from $2-4$ that an estimate similar to (1.4) for $U(r)$ is also satisfied by $\psi(s)$. Hence

$$\prod_{y \in c} e^{\psi(|x-y|)} < e^{B_0} \tag{2.3}$$

for all $x \in \Omega$ and $B_0 > 0$. The normalization factor in (2.1) is equal to

$$\Xi(\xi, \psi, \Omega) = \int_{C_\Omega} \prod_{x \in c} \xi(x) \prod_{\substack{x \in c,\, y \in c \\ x \neq y}} e^{\psi(|x-y|)} dc$$

(the convergence of the integral follows from (2.2) and (2.3)). In the case $\xi(x) = e^{-\mu}$ and $\psi(s) = -\beta U(s)$ the distribution (2.1) coincides with the Gibbs distribution (1.11).

If we fix some arrangement $c_0 \in C_\Omega$, then in the usual manner we can define a conditional Gibbs distribution $p(\cdot | c_0 \subset c)$ on the space C_Ω. Its density has the form

$$p(c, \xi, \psi, \Omega | c_0) = \Xi^{-1}(\xi, \psi, \Omega | c_0) \prod_{x c \in} \xi(x) \prod_{\substack{x \in c,\, y \in c \cup c_0 \\ x \neq y}} e^{\psi(|x-y|)}, \tag{2.4}$$

where

$$\Xi(\xi, \psi, \Omega | c_0) = \int_{C_\Omega} \prod_{x \in c} \xi(x) \prod_{\substack{x \in c,\, y \in c \cup c_0 \\ x \neq y}} e^{\psi(|x-y|)} dc.$$

We note that the conditional distribution (2.4) is of the type (2.1), where in place of $\xi(x)$ we have the function

$$\xi_{c_0}(x) = \xi(x) \prod_{y \in c_0} e^{\psi(|x-y|)}.$$

The function $r(c, \xi, \psi, \Omega)$ defined by the equation

$$r(c, \xi, \psi, \Omega) = \int_{C_\Omega} p(c \cup \tilde{c}, \xi, \psi, \Omega) d\tilde{c} \tag{2.5}$$

is called the *correlation function* of the distribution (2.1). The restriction of the correlation function (2.5) to the space $C_{\Omega,k}$ is called the *kth order correlation function,* and we shall frequently write it in the form $r(c, \xi, \psi, \Omega) = r_k(x_1, \cdots, x_k, \xi, \psi, \Omega)$, where $c = \{x_i\}$, $i = 1, \cdots, k$.

As follows from the results of [**1, 2, 15, 16, 20**], in the thermodynamic limit $|\Omega| \to \infty$ the distribution (2.1) converges to the limit distribution $p(\cdot, \xi, \psi, \infty)$ defined on the space C_∞. This distribution generates a corresponding family of densities $\pi_U(\cdot, \xi, \psi, \infty)$ on each C_U (see the preceding section), where U is any line segment. In addition the correlation function $r(c, \xi, \psi, \Omega)$ of the distribution (2.1) converges to the correlation function $r(c, \xi, \psi, \infty)$ of the limit distribution $p(\cdot, \xi, \psi, \infty)$ defined by the formula

$$r(c, \xi, \psi, \infty) = \int_{C_U} \pi_U(c \cup \tilde{c}, \xi, \psi, \infty) d\tilde{c}, \tag{2.6}$$

where $c \subset U$ (it is easily verified that the left-hand side of (2.6) does not actually depend on U). In what follows we shall call the function $r(c,\xi,\psi,\infty)$ the *limit correlation function*. The restriction of $r(c,\xi,\psi,\infty)$ to a configuration c having $N(c)=k$ points is called the *kth order correlation function* and will be denoted $r_k(x_1,\cdots,x_k,\xi,\psi,\infty)$.

At this point we present a number of estimates and relations for the correlation functions $r(c,\xi,\psi,\Omega)$ and $r(c,\xi,\psi,\infty)$, putting off the proofs until II.

1. There exist constants A_1 and q such that for any Ω and c

$$r(c,\ \xi,\ \psi,\ \Omega)<A_1q^{N(c)}\prod_{x\in c}\xi(x)\prod_{\substack{x\in c,\ y\in c\\ x\neq y}}e^{\psi(|x-y|)}. \tag{2.7}$$

It is clear that the relation (2.7) remains valid for the limit correlation function as well.

2. The deviation of the correlation function $r(c,\xi,\psi,\Omega)$ from the limit $r(c,\xi,\psi,\infty)$ obeys the inequality

$$\begin{aligned}&|r(c,\ \xi,\ \psi,\ \Omega)-r(c,\ \xi,\ \psi,\ \infty)|\\ &<A_2q^{N(c)}e^{-bd(c,\Gamma(\Omega))}\prod_{x\in c}\xi(x)\prod_{\substack{x\in c,\ y\in c\\ x\neq y}}e^{\psi(|x-y|)},\end{aligned} \tag{2.8}$$

where A_2 is an absolute constant, $d(c,\Gamma(\Omega))$ is the distance from the set c to the boundary $\Gamma(\Omega)$ of the segment Ω, and $b>0$ is a constant independent of c and Ω.

3. If the two functions $\xi_1(x)$ and $\xi_2(x)$ coincide outside of some set S, i.e. $\xi_1(x)=\xi_2(x)$ if $x\notin S$, then for any $c\in C_\Omega$

$$\begin{aligned}&|r(c,\ \xi_1,\ \psi,\ \Omega)-r(c,\ \xi_2,\ \psi,\ \Omega)|\\ &<A_3q^{N(c)}e^{-bd(c,S)}\prod_{x\in c}\xi(x)\prod_{\substack{x\in c,\ y\in c\\ x\neq y}}e^{\psi(|x-y|)},\end{aligned} \tag{2.9}$$

where A_3 is an absolute constant, $d(c,S)$ is the distance between the sets c and S (here and in the following inequalities the constant b is the same as in (2.8)), and $\xi(x)=\max(\xi_1(x),\xi_2(x))$; the limit correlation function obeys a similar inequality.

4. If for $c_0\in C_\Omega$ we denote by $\delta(c,c_0,\xi,\psi,\Omega)$ the difference between the conditional and unconditional correlation functions, i.e. $\delta(c,c_0,\xi,\psi,\Omega)=r(c,\xi,\psi,\Omega|c_0)-r(c,\xi,\psi,\Omega)$, then for any $c\in C_\Omega$

$$\begin{aligned}&|\delta(c,\ c_0,\ \xi,\ \psi,\ \Omega)|\\ &<A_4q^{N(c)}e^{-bd(c,c_0)}\prod_{x\in c}\xi(x)\prod_{\substack{x\in c,\ y\in c\\ x\neq y}}e^{\psi(|x-y|)},\end{aligned} \tag{2.10}$$

where A_4 is an absolute constant and $d(c,c_0)$ is the distance between c and c_0. A similar estimate is also valid for the limit correlation functions.

5. The difference $\delta(c,c_0,\xi_1,\psi,\Omega)-\delta(c,c_0,\xi_2,\psi,\Omega)$ for two functions $\xi_1(x)$ and $\xi_2(x)$ which differ only on some set S satisfies the inequality

$$\begin{aligned}&|\delta(c, c_0, \xi_1, \psi, \Omega) - \delta(c, c_0, \xi_2, \psi, \Omega)| \\ (2.11)\qquad &< A_5 q^{N(c)} e^{-\frac{1}{2} b(d(c_0,c)+d(c,S))} \prod_{x\in c} \xi(x) \prod_{\substack{x\in c,\, y\in c\\ x\neq y}} e^{\psi(|x-y|)},\end{aligned}$$

where A_5 is an absolute constant, $d(c_0,c)$ and $d(c,S)$ are the distances between c and c_0 and c and S, respectively, and $\xi(x) = \max(\xi_1(x), \xi_2(x))$.

By making use of these inequalities we can prove a number of properties for the quantity $\ln \Xi(\xi, \psi, \Omega)$ considered as a functional of $\xi(x)$ and $\psi(s)$. The equation

$$\xi(x_0) \frac{\delta \ln \Xi(\xi, \psi, \Omega)}{\delta \xi(x_0)} = r_1(x_0, \xi, \psi, \Omega).$$

is easily derived.[1] From it the expression

$$(2.12)\qquad -\frac{\partial \ln \Xi(\beta, \mu, \Omega)}{\partial \mu} = \bar{N} = \int_\Omega r_1(x, \beta, \mu, \Omega)\, dx.$$

for the average particle number in the Gibbs ensemble is easily obtained. Next,

$$\begin{aligned}\xi(x_1) \frac{\delta}{\delta\xi(x_1)} \xi(x_2) \frac{\delta}{\delta\xi(x_2)} \ln \Xi(\xi, \psi, \Omega) &= r_2(x_1, x_2, \xi, \psi, \Omega) \\ - r_1(x_1, \xi, \psi, \Omega) r_1(x_2, \xi, \psi, \Omega) &+ r_1(x_1, \xi, \psi, \Omega)\, \delta(x_1 - x_2).\end{aligned}$$

Hence for the dispersion of the particle number in the Gibbs ensemble we find

$$\begin{aligned}\frac{\partial^2 \ln \Xi(\beta, \mu, \Omega)}{\partial \mu^2} &= M_{22}^\Omega = M(N - \bar{N})^2 = \int_\Omega\int_\Omega r_2(x, y, \beta, \mu, \Omega)\, dx dy \\ (2.13)\qquad &- \int_\Omega r_1(x, \beta, \mu, \Omega)\, dx \int_\Omega r_1(y, \beta, \mu, \Omega)\, dy + \int_\Omega r_1(x, \beta, \mu, \Omega)\, dx.\end{aligned}$$

Now by using (2.8) we obtain

$$(2.14)\qquad |\bar{N}(\beta, \mu, \Omega) - |\Omega| \rho(\beta, \mu)| < A_6,$$

where $\rho(\beta, \mu)$ is the limit particle density in the grand canonical ensemble (identical to the correlation function $r_1(x, \beta, \mu, \infty)$), $|\Omega|$ is the length of the interval Ω, and A_6 is an absolute constant. In exactly the same way, by using (2.8) and (2.10) we find from (2.13) that

$$(2.15)\qquad |M_{22}^\Omega - |\Omega| g| < A_7,$$

[1] Let $\Phi(\xi)$ be a functional depending on the functions $\xi(x)$, $x \in R^1$. If the increment $\Delta\Phi = \Phi(\xi) - \Phi(\xi_0)$ can be represented in the form $\Delta\Phi = \int_{R^1} K(x_0; \xi_0) \Delta\xi(x_0)\, dx_0 + O(\|\Delta\xi\|)$, where $K(x_0; \xi_0)$ is some continuous function (which depends on x_0 and $\xi_0(x_0)$), $\Delta\xi(x_0) = \xi_0(x_0) - \xi_0(x_0)$ and $\|\Delta\xi\| = \max_x |\Delta\xi(x)|$, then the quantity $\delta\Phi(\xi)/\delta\xi(x)|_{\xi=\xi_0} = K(x_0; \xi_0)$ is called the *variational derivative* of the functional $\Phi(\xi)$ at the point x_0 for $\xi(x) = \xi_0(x)$.

where A_7 is an absolute constant and

$$g \equiv g(\beta, \mu) = \rho(\beta, \mu) + \int_{-\infty}^{\infty} [r_2(0, y, \beta, \mu, \infty) - \rho^2(\beta, \mu)]\, dy. \tag{2.16}$$

Here $r_2(x, y, \beta, \mu, \infty)$ is the second limit correlation function in the grand ensemble (it depends only on the distance $x - y$:

$$r_2(x, y, \beta, \mu, \infty) = \bar{r}_2(x - y, \beta, \mu, \Omega)).$$

The convergence of the integral in (2.16) is ensured by (2.10). It follows from the inequality $M_{22}^{\Omega} > 0$ that $g \geqq 0$.

As we have pointed out, in what follows it is necessary to examine the probability distribution for the values of the potential energy in the configurational grand canonical ensemble. In this case we find

$$\overline{H(c)} = -\frac{\partial \ln \Xi(\beta, \mu, \Omega)}{\partial \beta} = \iint_{|x-y|>\delta_0} U(|x-y|)\, r_2(x, y, \beta, \mu, \Omega)\, dx dy, \tag{2.17}$$

$$\begin{aligned} \tilde{M}_{11}^{\Omega} &= \frac{\partial^2 \ln \Xi(\beta, \mu, \Omega)}{\partial \beta^2} = \overline{H^2(c)} - \overline{H(c)}^2 \\ &= \iiiint_{\substack{|x-y|>\delta_0 \\ |z-t|>\delta_0}} U(|x-y|) \cdot U(|z-t|) \{r_4(x, y, z, t, \beta, \mu, \Omega) \\ &\quad - r_2(x, y, \beta, \mu, \Omega)\, r_2(z, t, \beta, \mu, \Omega)\}\, dx\, dy\, dz\, dt \\ &+ \iiint_{\substack{|x-y|>\delta_0 \\ |z-t|>\delta_0}} U(|x-y|)\, U(|x-t|)\, r_3(x, y, t, \beta, \mu, \Omega)\, dx\, dy\, dt \\ &+ \iint_{|x-y|>\delta_0} U^2(|x-y|\, r_2(x, y, \beta, \mu, \Omega)\, dx dy. \end{aligned} \tag{2.18}$$

for the mean energy $\overline{H(c)}$ and its dispersion $DH(c) = \tilde{M}_{11}^{\Omega}$, respectively.

If we now write

$$\tilde{\xi}(\beta, \mu) = \int_{\delta_0}^{\infty} U(t)\, r_2(0, t, \beta, \mu, \infty)\, dt$$

(the convergence of the integral follows from (2.7) and condition III with respect to the potential $U(r)$), then from (2.8) we find

$$|\overline{H(c)} - |\Omega|\, \tilde{\xi}(\beta, \mu)| < A_8, \tag{2.19}$$

where A_8 is an absolute constant. In the same way we find from (2.8), (2.10), (2.11) and (2.18) that

$$|\tilde{M}_{11}^{\Omega} - |\Omega|\, \tilde{d}_{11}(\beta, \mu)| < A_9. \tag{2.20}$$

Here A_9 is an absolute constant and

$$\tilde{d}_{11}(\beta, \mu) = 2\int_{-\infty}^{\infty} d\tau \int_{\delta_0}^{\infty} U(t_1)\,dt_1 \int_{\delta_0}^{\infty} U(t_2)\{r_4(0, t_1, \tau, \tau+t_2, \beta, \mu, \infty)$$

$$-r_2(0, t_1, \beta, \mu, \infty)\,r_2(0, t_2, \beta, \mu, \infty)\}\,dt_2$$

$$+2\int_{\delta_0}^{\infty} U(t_1)\,dt_1 \int_{\delta_0}^{\infty} U(t_2)\,r_3(0, t_1, t_2, \beta, \mu, \infty)\,dt_2 \tag{2.21}$$

$$+\int_{\delta_0}^{\infty} U^2(t)\,r_2(0, t, \beta, \mu, \infty)\,dt,$$

where $r_4(0,t_1,\tau,\tau+t_2,\beta,\mu,\infty)$, $r_3(0,t_1,t_2,\beta,\mu,\infty)$ and $r_2(0,t,\beta,\mu,\infty)$ are the limit correlation functions of the fourth, third and second order, respectively. The convergence of the integral in (2.21) is ensured by the inequality (2.10).

Next, in the same way we find

$$\tilde{M}_{12}^{\Omega} = \overline{N(c)\,H(c)} - \overline{N(c)}\cdot\overline{H(c)} = \frac{\partial^2 \ln \Xi(\beta, \mu, \Omega)}{\partial\beta\,\partial\mu}$$

$$= \iiint\limits_{\substack{|y-t|>\delta_0\\ x\in\Omega}} U(|y-t|)\{r_3(x, y, t, \beta, \mu, \Omega) - r_1(x, \beta, \mu, \Omega)\cdot r_2(y, t, \beta, \mu, \Omega)\}$$

$$\times\, dx\,dy\,dt + 2\iint\limits_{|x-y|>\delta_0} U(|x-y|)(1-r_1(x, \beta, \mu, \Omega))\,r_2(x, y, \beta, \mu, \Omega)\,dx\,dy. \tag{2.22}$$

Hence, by making use of (2.8), (2.10) and (2.11), we get

$$|\tilde{M}_{12}^{\Omega} - |\Omega|\,\tilde{d}_{12}(\beta, \mu)| < A_{10}, \tag{2.23}$$

where A_{10} is an absolute constant and

$$d_{12}(\beta,\mu)$$

$$= \int_{-\infty}^{\infty} dt_1 \int_{\delta_0}^{\infty} U(t_2)\{r_3(0, t_1, t_2, \beta, \mu, \infty) - \rho(\beta, \mu)\,r_2(0, t_2, \beta, \mu, \infty)\}\,dt_2$$

$$+2\int_{\delta_0}^{\infty} U(t)\,(1-\rho(\beta, \mu))\,r_2(0, t, \beta, \mu, \infty)\,dt.$$

Besides studying the joint probability distribution for the values of the potential energy $H(c)$ and particle number $N(c)$ in the Gibbs grand canonical ensemble, we shall have to examine the joint probability distribution for the total energy $E(a)$ and particle number $N(a)$ in the Gibbs ensemble for the total mechanical system. For the mean value $\overline{E(a)}$ and the dispersion $DE(a) = M_{11}^{\Omega}$ we find

$$\overline{E(a)} = -\frac{\partial \ln \Xi(\beta, \mu, \Omega)}{\partial\beta} = \frac{1}{2}\overline{N(a)}\,\beta^{-1} + \overline{H(c)}, \tag{2.17'}$$

$$M_{11}^{\Omega} = \frac{\partial^2 \ln\Xi(\beta, \mu, \Omega)}{\partial\beta^2} = \frac{1}{2}\overline{N(a)}\,\beta^{-2} + \tilde{M}_{11}^{\Omega}. \tag{2.18'}$$

Moreover, estimates similar to (2.19) and (2.20) are valid in this case as well, i.e.

$$(2.19') \qquad |\overline{E(a)} - |\Omega|\xi(\beta,\mu)| < A_8^1$$

and

$$(2.20') \qquad |M_{11}^{\Omega} - |\Omega| d_{11}(\beta,\mu)| < A_9^1,$$

where A_8^1 and A_9^1 are absolute constants which do not depend on Ω, and

$$\xi(\beta,\mu) = \tfrac{1}{2}\beta^{-1}\rho(\beta,\mu) + \tilde{\xi}(\beta,\mu)$$

and

$$d_{11}(\beta,\mu) = \tfrac{1}{2}\beta^{-2}\rho(\beta,\mu) + \tilde{d}_{11}(\beta,\mu).$$

Similarly we find

$$M_{12}^{\Omega} = \overline{E(a)\,N(a)} - \overline{E(a)}\cdot\overline{N(a)} = \frac{\partial^2 \ln \Xi(\beta,\mu,\Omega)}{\partial\beta\,\partial\mu}$$
$$= \overline{H(c)\,N(c)} - \overline{H(c)}\cdot\overline{N(c)} = \widetilde{M}_{12}^{\Omega},$$

i.e. $M_{12}^{\Omega} = \widetilde{M}_{12}^{\Omega}$. Consequently (2.23) also holds for the quantity M_{12}^{Ω}.

We consider now the quadratic forms

$$(2.24) \qquad \begin{aligned} D(H(c)\,t_1 + N(c)\,t_2) &= \widetilde{M}_{11}^{\Omega}\,t_1^2 + 2\widetilde{M}_{12}^{\Omega}\,t_1t_2 + M_{22}^{\Omega}\,t_2^2, \\ D(E(a)\,t_1 + N(a)\,t_2) &= M_{11}^{\Omega}\,t_1^2 + 2M_{12}^{\Omega}\,t_1t_2 + M_{22}^{\Omega}\,t_2^2. \end{aligned}$$

By repeating the argument in the proof of Lemma 2 from the paper of Dobrušin and Minlos [3], we conclude that there exists a positive definite quadratic form $A(t_1,t_2)$, which does not depend on Ω, such that

$$D(H(c)\,t_1 + N(c)\,t_2) \geqslant A(t_1,\ t_2)\,|\Omega|$$

and

$$D(E(a)\,t_1 + N(a)\,t_2) \geqslant A(t_1,\ t_2)\,|\Omega|.$$

By taking the limit as $|\Omega| \to \infty$, we have

$$\lim_{|\Omega|\to\infty} \frac{D(H(c)\,t_1 + N(c)\,t_2)}{|\Omega|} \geqslant A(t_1,\ t_2) \quad \text{and} \quad \lim_{|\Omega|\to\infty} \frac{D(E(a)\,t_1 + N(a)\,t_2)}{|\Omega|} \geqslant A(t_1,\ t_2).$$

whence we find that the quadratic forms

$$(2.24) \qquad \begin{aligned} &\tilde{d}_{11}(\beta,\ \mu)\,t_1^2 + 2\tilde{d}_{12}(\beta,\ \mu)\,t_1t_2 + \tilde{d}_{22}(\beta,\ \mu)\,t_2^2 \quad (\tilde{d}_{22}(\beta,\ \mu) = g), \\ &d_{11}(\beta,\ \mu)\,t_1^2 + 2d_{12}(\beta,\ \mu)\,t_1t_2 + d_{22}(\beta,\ \mu)\,t_2^2 \quad (d_{12} = \tilde{d}_{12},\ d_{22} = g) \end{aligned}$$

are positive definite.

Next, for the third derivatives

$$\frac{\partial^3 \ln \Xi(\beta,\ \mu,\ \Omega)}{\partial\beta^3}, \quad \frac{\partial^3 \ln \Xi(\beta,\ \mu,\ \Omega)}{\partial\beta^2\partial\mu}, \quad \frac{\partial^3 \ln \Xi(\beta,\ \mu,\ \Omega)}{\partial\beta\partial\mu^2},$$
$$\frac{\partial^3 \ln \Xi(\beta,\ \mu,\ \Omega)}{\partial\mu^3}$$

we can in the same way arrive at several expressions composed of correlation functions (quite complicated) and from these prove with the help of (2.7) – (2.11) that all these derivatives satisfy the inequalities

$$\left| \frac{\partial^3 \ln \Xi(\beta, \mu, \Omega)}{\partial \beta^3} \right| < A_{11} |\Omega| \tag{2.25}$$

($A_{11} = A_{11}(\beta, \mu)$ is an absolute constant independent of Ω), and similarly for the remaining ones. In particular it follows that the third derivative of $\ln \Xi(\beta, \mu, \Omega)$ obeys the inequality

$$\begin{aligned} \Bigg| \frac{\partial^3 \ln \Xi(\beta, \mu, \Omega)}{\partial \beta^3} t_1^3 + 3 \frac{\partial^3 \ln \Xi(\beta, \mu, \Omega)}{\partial \beta^2 \partial \mu} t_1^2 t_2 + 3 \frac{\partial^3 \ln \Xi(\beta, \mu, \Omega)}{\partial \beta \partial \mu^2} t_1 t_2^2 \\ + \frac{\partial^3 \ln \Xi(\beta, \mu, \Omega)}{\partial \mu^3} t_2^3 \Bigg| < A_{12} |\Omega| (|t_1|^3 + |t_2|^3), \end{aligned} \tag{2.26}$$

where $A_{12} = A_{12}(\beta, \mu)$ is an absolute constant which does not depend on Ω.

We now make the following important remark. In the initial definition of the Gibbs distribution the parameters β and μ were assumed to be always real, and all the estimates relating to the correlation function and the derivatives of the logarithm of the partition function were derived on the basis of this assumption. However, we can formally consider the partition function for complex values of β and μ as well. It appears that if the imaginary parts of β and μ are sufficiently small, all the estimates derived above for $\ln \Xi(\beta, \mu, \Omega)$ and its derivatives with respect to β and μ remain valid. More precisely, we have

PROPOSITION 2.1. *Let*

$$\Xi(\beta + z_1, \mu + z_2, \Omega | \tilde{c}) = \int_{C_\Omega} e^{-(\beta + z_1) H(c \tilde{c}) - (\mu + z_2) N(c)} \, dc,$$

where β and μ are real, z_1 and z_2 are complex numbers, and c is an arrangement outside of Ω. Then there exist $\delta = \delta(\beta, \mu)$ and a constant $A = A(\beta, \mu)$ satisfying the following conditions:

1) *The function $\Omega^{-1} \ln \Xi(\beta + z_1, \mu + z_2, \Omega | \tilde{c})$ is analytic in the region $|z_1|$, $|z_2| < \delta$ for all Ω and any $\tilde{c}$.*

2) $|\ln \Xi(\beta + z_1, \mu + z_2, \Omega | \tilde{c})| < A |\Omega|$.

This assertion will be proved in II. From it we have the following

COROLLARY. *For the partial derivatives of any order of the function $\ln \Xi(\beta, \mu, \Omega | \tilde{c})$ with respect to β and μ (β and μ are complex; $\tilde{c}$ is arbitrary) we have the estimate*

$$\left| \frac{\partial^{k_1 + k_2} \ln \Xi(\beta, \mu, \Omega | \tilde{c})}{\partial \beta^{k_1} \partial \mu^{k_2}} \right| < A_{k_1 k_2}(\beta, \mu) |\Omega|,$$

where $A_{k_1, k_2}(\beta, \mu) > 0$ is an absolute constant which does not depend on Ω and $\tilde{c}$.

§3. Local limit theorem for the particle number and energy in the grand canonical ensemble (for a one-dimensional mechanical system)

Let $\{\mathfrak{B}_\Omega, \beta, \mu\}$ be the grand canonical ensemble for the total mechanical system and consider the joint distribution for $E(a)$ and $N(a)$ generated by this ensemble. In the first section we introduced the quantity $p_{\text{tot}}(E, N)$ given by

$$p_{\text{tot}}(E, N) = \frac{d}{dE}\{P_r(E(a) < E,\ N(a) = N)\}, \tag{3.1}$$

and showed that

$$p_{\text{tot}}(E, N) = \int_{-\infty}^{\infty} p_{\text{kin}}(E - H \mid N)\, dF_{\text{pot}}(H, N \mid \Omega), \tag{3.2}$$

where $p_{\text{kin}}(T|N)$ is the density for the distribution of values of the kinetic energy evaluated by means of (1.9) and $F_{\text{pot}}(H, N|\Omega)$ is the joint distribution function for the potential energy H and the particle number N in the configurational grand canonical ensemble $\{C_\Omega, \beta, \mu\}$.

The purpose of this section is to establish the asymptotic behavior of $p_{\text{tot}}(E, N)$ as a local limit theorem.

Theorem 3.1. *Consider the grand canonical ensemble for a one-dimensional system of particles enclosed on an interval Ω. Let the numbers E and N be such that*

$$|E - \overline{E(\beta,\mu)}| \le d\sqrt{|\Omega|}, \qquad |N - \overline{N(\beta,\mu)}| \le d\sqrt{|\Omega|}, \tag{3.3}$$

where $d > 0$ is any fixed constant, and $\overline{E(\beta,\mu)}$ and $\overline{N(\beta,\mu)}$ are the mean value of the energy and the particle number, respectively, in the ensemble $\{\mathfrak{B}_\Omega, \beta, \mu\}$. Under these conditions the asymptotic expressions

$$p_{\text{tot}}(E, N) = \frac{1}{\pi |\Omega| \sqrt{D}} \exp\left\{-\frac{1}{2}\alpha_{11}\frac{(E-\bar{E})^2}{|\Omega|} - \alpha_{12}\frac{(E-\bar{E})(N-\bar{N})}{|\Omega|} - \frac{1}{2}\alpha_{22}\frac{(N-\bar{N})^2}{|\Omega|}\right\} + O(|\Omega|^{-\frac{6}{5}}) \tag{3.4}$$

hold uniformly with respect to all E and N which satisfy (3.3). *Here the matrix $\|\alpha_{ik}\| = \|d_{ik}\|^{-1}$, and $D = \det\|d_{ik}\|$, where $\|d_{ik}\|$ is the covariance matrix for the distribution of E and N.*

The proof of this theorem can be broken down into a series of lemmas. Thus we use formula (3.2) and derive (3.4) for $p_{\text{tot}}(E, N)$ with the help of the corresponding asymptotic expressions for $p_{\text{kin}}(T|N)$ and $F_{\text{pot}}(H, N|\Omega)$. Unfortunately the distribution function $F_{\text{pot}}(H, N|\Omega)$ is not differentiable with respect to H (because of the fact that, for example, the set of configurations $c \in C_\Omega$ such that $|x - y| > R_0$ for all pairs $(x, y) \in c$ has positive probability in the ensemble $\{C_\Omega, \beta, \mu\}$ and thus the probability $P_r(H(c) = 0, N(c) = N)$ is positive). However, if from $\{C_\Omega, \beta, \mu\}$ we exclude the set of such arrangements (in fact it will be easier for us to exclude an even larger set), then on the remaining part of the ensemble the joint distribution $\widetilde{F}_{\text{pot}}(H, N|\Omega)$ of potential energy H and particle number N will possess the derivative $\widetilde{p}_{\text{pot}}(H,N) = (d/dH)\widetilde{F}_{\text{pot}}(H,N|\Omega)$, and for it we have an asymptotic estimate in the form of a local limit theorem. The excluded set of arrangements has an exponentially small (with respect to $|\Omega|$) probability. Thus we have the following lemma.

LEMMA 3.1. *There exists a set of configurations $\mathfrak{M} \subset C_\Omega$ (which will be described below) satisfying the following conditions:*

I. *If we denote $\overline{\mathfrak{M}} = C_\Omega \backslash \mathfrak{M}$, then*

$$P_r(\overline{\mathfrak{M}}) < d_0 |\Omega|^{3/2} e^{-d_1|\Omega|} \tag{3.5}$$

where d_0 and d_1 are absolute constants which do not depend on Ω.

II. *If we consider the conditional Gibbs distribution on $\mathfrak{M}$ and the conditional joint distribution function $\widetilde{F}_{\text{pot}}(H,N|\Omega)$ for the values of H and N which it generates, then the quantity*

$$\widetilde{p}_{\text{pot}}(H,N) = (d/dH)\widetilde{F}_{\text{pot}}(H,N|\Omega)$$

exists. If N obeys the estimate

$$|N - \overline{N}| \leqq d\sqrt{|\Omega|} \tag{3.3'}$$

($d > 0$ is any given constant), then

$$\widetilde{p}_{\text{pot}}(H, N) = \frac{1}{\pi|\Omega|\sqrt{D'}} \exp\left\{-\frac{1}{2} h_{11} \frac{(H-\overline{H})^2}{|\Omega|} - h_{12}\frac{(H-\overline{H})(N-\overline{N})}{|\Omega|} - \frac{1}{2} h_{22} \frac{(N-\widetilde{N})^2}{|\Omega|}\right\} + O(|\Omega|^{-\frac{6}{5}}), \tag{3.6}$$

uniformly with respect to all N satisfying (3.3′) and all $-\infty < H < +\infty$. Here the matrix $\|h_{ik}\| = \|\widetilde{d}_{ik}\|^{-1}$, and $D' = \det\|\widetilde{d}_{ik}\|$, where the matrix $\|\widetilde{d}_{ik}\|$ is the covariance matrix for the distribution of the quantities H and N in the ensemble $\{C_\Omega, \beta, \mu\}$, and $\overline{N}$ and $\overline{H}$ are the mean values of N and N, respectively, in this ensemble.

LEMMA 3.2. *If N obeys (3.3′), the following estimates hold:*

$$\text{I.} \quad \left\| p_{\text{kin}}(T|N) - \frac{2\beta}{\sqrt{N\pi}} \exp\left\{-\frac{2\beta^2(T-\overline{T})^2}{N}\right\}\right\| \leqslant C_1(\beta) N^{-\frac{3}{4}}, \tag{3.7}$$

$$\text{II.} \quad |p_{\text{kin}}(T|N)| < C_2(\beta) \tag{3.8}$$

where $C_1(\beta)$ and $C_2(\beta)$ are absolute constants which do not depend on N, and T is the mean value of the kinetic energy T in the ensemble $\{\mathfrak{B}_\Omega, \beta, \mu\}$. Here the norm $\|\cdot\|$ is the norm in the space $L_2(-\infty, \infty)$.

Since the distribution $p_{\text{kin}}(\cdot\,|N)$ has a simple form, the proof of Theorem 3.2 is obtained by means of an elementary investigation of the characteristic function $f_N(t)$.

Before proving Lemma 3.1, we derive our theorem from Lemmas 3.1 and 3.2. Assume that (3.6), (3.7) and (3.8) are satisfied. We write

$$p_{\text{kin}}(T\,|\,N)=\frac{2\beta}{\sqrt{N\pi}}\exp\left\{-\frac{2\beta^2(T-\bar{T})^2}{N}\right\}+q_N(T), \tag{3.9}$$

where, in view of (3.7), we have for the function $q_N(T)$ the estimate $\|q_N(t)\| < C_1(\beta)N^{-3/4}$. Next, it is not hard to see that the function $F_{\text{pot}}(H,N|\Omega)$ of the joint distribution of the quantities H and N in the ensemble $\{C_\Omega,\beta,\mu\}$ can be represented in the form

$$F_{\text{pot}}(H,\,N\,|\,\Omega)=\tilde{F}_{\text{pot}}(H,\,N)\,|\,\Omega)\,P_r(\mathfrak{M})+\tilde{\tilde{F}}_{\text{pot}}(H,\,N\,|\,\Omega)\,P_r(\overline{\mathfrak{M}}), \tag{3.10}$$

where $\tilde{F}_{\text{pot}}(H,N|\Omega)$ and $\tilde{\tilde{F}}_{\text{pot}}(H,N|\Omega)$ are conditional distribution functions on the sets $\mathfrak{M}$ and $\overline{\mathfrak{M}}$, respectively, and $P_r(\mathfrak{M})$ and $P_r(\overline{\mathfrak{M}})$ are the probabilities of these sets in the ensemble $\{C_\Omega,\beta,\mu\}$. From (3.2) and with the aid of (3.9) and (3.10) we obtain

$$\begin{aligned}
p_{\text{tot}}(E,\,N) &= \int_{-\infty}^{\infty} p_{\text{kin}}(E-H\,|\,N)\,dF_{\text{pot}}(H,\,N\,|\,\Omega)\\
&= P_r(\mathfrak{M})\int_{-\infty}^{\infty} p_{\text{kin}}(E-H\,|\,N)\,d\tilde{F}_{\text{pot}}(H,\,N\,|\,\Omega)\\
&\quad + P_r(\overline{\mathfrak{M}})\int_{-\infty}^{\infty} p_{\text{kin}}(E-H\,|\,N)\,d\tilde{\tilde{F}}_{\text{pot}}(H,\,N\,|\,\Omega)\\
&= \frac{1}{\pi\,|\,\Omega\,|\sqrt{D'}}\int_{-\infty}^{\infty} p_{\text{kin}}(E-H\,|\,N)\exp\left\{-\frac{1}{2}h_{11}\frac{(H-\bar{H})^2}{|\,\Omega\,|}\right.\\
&\quad\left. -h_{12}\frac{(H-\bar{H})(N-\bar{N})}{|\,\Omega\,|}-\frac{1}{2}h_{22}\frac{(N-\tilde{N})^2}{|\,\Omega\,|}\right\}+O(|\,\Omega\,|^{-\frac{6}{5}})\\
&\quad + P_r(\overline{\mathfrak{M}})\left\{\int_{-\infty}^{\infty} P_{\text{kin}}(E-H\,|\,N)\cdot d\tilde{\tilde{F}}_{\text{pot}}(H,\,N\,|\,\Omega)\right.\\
&\quad\left. -\int_{-\infty}^{\infty} p_{\text{kin}}(E-H\,|\,N)\,d\tilde{F}_{\text{pot}}(H,\,N\,|\,\Omega)\right\}.
\end{aligned} \tag{3.11}$$

It is not difficult to see that the third term in (3.11), due to (3.5) and (3.8), makes an exponentially small contribution. By taking the expansion (3.9) into account, we find for the second term

$$\frac{1}{\pi|\Omega|\sqrt{D'}}\int_{-\infty}^{\infty} p_{\text{kin}}(E-H|N)\exp\Big\{-\frac{1}{2}h_{11}\frac{(H-\bar{H})^2}{|\Omega|} - h_{12}\frac{(H-\bar{H})(N-\bar{N})}{|\Omega|}-\frac{1}{2}h_{22}\frac{(N-\bar{N})^2}{|\Omega|}\Big\}dH$$

$$(3.12)\quad = \frac{2\beta}{\pi^{3/2}|\Omega|\sqrt{D'N}}\int_{-\infty}^{\infty}\exp\Big\{-\frac{2\beta^2(E-H-\bar{T})^2}{N}\Big\}\cdot\exp\Big\{-\frac{1}{2}h_{11} \times\frac{(H-\bar{H})^2}{|\Omega|} - h_{12}\frac{(H-\bar{H})(N-\bar{N})}{|\Omega|}-\frac{1}{2}h_{22}\frac{(N-\bar{N})^2}{|\Omega|}\Big\}dH$$

$$+\frac{1}{\pi|\Omega|\sqrt{D'}}\int_{-\infty}^{\infty} q_N(E-H)\exp\Big\{-\frac{1}{2}h_{11}\frac{(H-\bar{H})^2}{|\Omega|} - h_{12}\frac{(H-\bar{H})(N-\bar{N})}{|\Omega|}-\frac{1}{2}h_{22}\frac{(N-\bar{N})^2}{|\Omega|}\Big\}dH.$$

The first term in this expression is easily evaluated. It is equal to

$$(3.13)\quad \frac{1}{\pi|\Omega|\sqrt{D}}\exp\Big\{-\frac{1}{2}\alpha_{11}\frac{(F-\bar{E})^2}{|\Omega|} - \alpha_{12}\frac{(E-\bar{E})(N-\bar{N})}{|\Omega|}-\frac{1}{2}\frac{(N-\bar{N})^2}{|\Omega|}\Big\}.$$

Here $\bar{E}$ and $\bar{N}$ are the mean values of the quantities $E(a)$ and $N(a)$ in the ensemble $\{\mathfrak{B}_\Omega,\beta,\mu\}$ and $\|\alpha_{ik}\|=\|d_{ik}\|^{-1}$, where $\|d_{ik}\|$ is the limit covariance matrix of the quantities E and N; $D=\det\|d_{ik}\|$. Next we estimate the second term in (3.12). By applying the Cauchy-Bunjakovskiĭ inequality we obtain

$$(3.14)\quad \frac{1}{\pi|\Omega|\sqrt{D'}}\Big|\int_{-\infty}^{\infty} q_N(E-H)\exp\Big\{-\frac{1}{2}h_{11}\frac{(H-\bar{H})^2}{|\Omega|} - h_{12}\frac{(H-\bar{H})(N-\bar{N})}{|\Omega|} -\frac{1}{2}h_{22}\frac{(N-\bar{N})^2}{|\Omega|}\Big\}DH\Big| \leqslant \frac{1}{\pi|\Omega|\sqrt{D'}}\Big[\int_{-\infty}^{\infty}|q_N(t)|^2\,dt\Big]^{\frac{1}{2}}$$

$$\times\Big[\int_{-\infty}^{\infty}\exp\Big\{-h_{11}\frac{t^2}{|\Omega|}-2h_{12}\frac{t(N-\bar{N})}{|\Omega|}-h_{22}\frac{(N-\bar{N})^2}{|\Omega|}\Big\}dt\Big]^{\frac{1}{2}}.$$

It is easily seen that

$$
(3.15)\qquad \begin{aligned}
&\frac{1}{\pi\,|\Omega|\,\sqrt{D'}}\int_{-\infty}^{\infty}\exp\left\{-h_{11}\frac{t^2}{|\Omega|}-2h_{12}\frac{t\,(N-\overline{N})}{|\Omega|}-h_{22}\frac{(N-\overline{N})^2}{|\Omega|}\right\}dt\\
&=\frac{1}{\sqrt{\pi h_{11}D'}}\exp\left\{-\frac{1}{h_{11}D'}\frac{(N-\overline{N})^2}{|\Omega|}\right\}<A,
\end{aligned}
$$

where $A>0$ is some absolute constant which does not depend on Ω and N. Consequently, by using (3.7) and also the fact that $N\sim\Omega$, as follows from (3.3′), we have

$$
(3.16)\qquad \begin{aligned}
&\frac{1}{\pi\,|\Omega|\,\sqrt{D'}}\int_{-\infty}^{\infty}q_N(E-H)\exp\left\{-\frac{1}{2}h_{11}\frac{(H-\overline{H})^2}{|\Omega|}-h_{12}\frac{(H-\overline{H})(N-\overline{N})}{|\Omega|}\right.\\
&\qquad\left.-\frac{1}{2}h_{22}\frac{(N-\overline{N})^2}{|\Omega|}\right\}dH=O(|\Omega|^{-\frac{5}{4}}).
\end{aligned}
$$

Finally, with the help of (3.13) and (3.16) we finally obtain from (3.11)

$$
(3.17)\qquad \begin{aligned}
p_{\mathrm{tot}}\,(E,N)&=\frac{1}{\pi\,|\Omega|\,\sqrt{D}}\exp\left\{-\frac{1}{2}\alpha_{11}\frac{(E-\overline{E})^2}{|\Omega|}\right.\\
&\quad\left.-\alpha_{12}\frac{(E-\overline{E})(N-\overline{N})}{|\Omega|}-\frac{1}{2}\alpha_{22}\frac{(N-\overline{N})^2}{|\Omega|}\right\}+O(|\Omega|^{-6/5}).
\end{aligned}
$$

Thus it remains to prove Lemma 3.1.

PROOF OF LEMMA 3.1. We begin by describing the set $\mathfrak{M}$. We break up the line segment Ω into segments $l_1,\cdots,l_k$ of lengths $|l_i|=l=2R_0+\delta_0+a$ (where the constant a is defined in §1 in the list of conditions imposed on the potential $U(r)$ (condition V)), $i=1,\cdots,k,$ and such that the distance between these segments equals R_0. It is clear that $k=[|\Omega|/(3R_0+\delta_0)]$. We denote $\widetilde{S}=\bigcup_1^k l_i$ and $\widetilde{\Omega}=\Omega\backslash\widetilde{S}$. For each arrangement $c\in C_\Omega$ we set $\widetilde{c}=c\cap\widetilde{\Omega}$ and $c_i=c\cap l_i$, $i=1,\cdots,k$. Next, in each segment l_i we select an interior segment $\widetilde{l}_i$ of length $|\widetilde{l}_i|=\delta_0+a$ at a distance R_0 from the end points of l_i.

The arrangement $c_i\subset l_i$ for $i<[k/2]$ and the segment l_i itself will be called *distinguished from the left* if

$$
(*)\qquad c_i\cap(l_i\backslash\widetilde{l}_i)=\emptyset\quad\text{and}\quad N(c_i)=2\quad\text{for}\quad i<[k/2].
$$

If (*) is not satisfied, i.e. either $c_i\cap(l_i\backslash\widetilde{l}_i)\neq\emptyset$ or $N(c_i)\neq 2$, $i<[k/2]$, the arrangement c_i (and the segment l_i) will be called *undistinguished from the left*. Similarly the arrangement $c_j\subset l_j$ and the segment l_j itself, for $j>[k/2]$, will be called *distinguished from the right* if

$$
(**)\qquad c_j\cap(l_j\backslash\widetilde{l}_j)=\emptyset\quad\text{and}\quad N(c_j)<2\quad\text{for}\quad j>[k/2].
$$

If the condition (**) is not satisfied, then the arrangement c_j and the segment l_j is called *undistinguished from the right*.

Now let $0<\gamma<1$ be some given number, and set $n=n(\gamma)=[\gamma k/2]$. We define $\mathfrak{M}_\gamma^{(1)}$ to be the totality of all arrangements $c\in C_\Omega$ for which there are n segments $\widetilde{l}_{i_1},\cdots,\widetilde{l}_{i_n}$, $i_m<[k/2]$, $m=1,\cdots,n$, distinguished from the left. In the same way we define $\mathfrak{M}_\gamma^{(2)}$ as the totality of those arrangements $c\in C_\Omega$ for which there are n segments $\widetilde{l}_{j_1},\cdots,\widetilde{l}_{j_n}$, $j_m>[k/2]$, $m=1,\cdots,n$, distinguished from the right. We set $\mathfrak{M}_\gamma=\mathfrak{M}_\gamma^{(1)}\cap\mathfrak{M}_\gamma^{(2)}$ and shall show that for sufficiently small γ the probability of the set $\overline{\mathfrak{M}}_\gamma=C_\Omega\backslash\mathfrak{M}_\gamma$ is exponentially small (with respect to $|\Omega|$). This follows from the following lemma.

Lemma 3.4. *There exist γ_0 and absolute constants d_0 and d_1 such that*

$$P_r(\overline{\mathfrak{M}}_{\gamma_0}^{(i)})<d_0\,|\,\Omega\,|^{\frac{3}{2}}\,e^{-d_1|\Omega|},\quad i=1,\,2, \tag{3.18}$$

where $\overline{\mathfrak{M}}_\gamma^{(i)}=C_\Omega\backslash\mathfrak{M}_\gamma^{(i)}$, $i=1,2$.

Proof. Having fixed an arrangement $\widetilde{c}_0\in C_{\widetilde{\Omega}}$, we consider the conditional joint distribution for the configuration $\{c_1,\cdots,c_k\}$ generated by the Gibbs distribution in the configurational grand canonical ensemble on the segment Ω. Note that the density of this distribution has the form

$$\begin{aligned} &p\,(c_1,\,c_2,\,\ldots\,,\,c_k|\widetilde{c}_0)\\ &=\prod_{s=1}^{k}\Xi^{-1}\,(\beta,\,\mu,\,l_s|\widetilde{c}_0)\exp\{-\beta H\,(c_s|\widetilde{c}_0)-\mu N\,(c_s)\},\end{aligned} \tag{3.19}$$

where

$$\Xi\,(\beta,\,\mu,\,l|\widetilde{c}_0)=\int_{C_l}\exp\{-\beta H\,(c|\widetilde{c}_0)-\mu\,N\,(c)\}\,dc,$$

$$H\,(c|\widetilde{c}_0)=H\,(c)+H\,(c,\,\widetilde{c}_0),\;\;H\,(c,\,\widetilde{c}_0)=\sum_{x\in c,y\in\widetilde{c}_0}U\,(|\,x-y\,|).$$

From this it is clear that in the conditional distribution (3.19) the configurations c_s, $s=1,\cdots,k$, are statistically independent and each has the density

$$p_l\,(c|\widetilde{c}_0)=\Xi^{-1}\,(\beta,\;\mu,\;l|\widetilde{c}_0)\,e^{-\beta H(c|\widetilde{c}_0)-\mu N(c)}\,,\quad c\in C_l. \tag{3.20}$$

We denote by $\mathfrak{M}_\gamma^{(1)}(\widetilde{c}_0)\subset\mathfrak{M}_\gamma^{(1)}$ the collection of all those arrangements $c\in\mathfrak{M}_\gamma^{(1)}$ for which $c\cap\widetilde{\Omega}=\widetilde{c}_0$. It is clear that $\mathfrak{M}_\gamma^{(1)}(\widetilde{c}_0)$ consists of those arrangements for which among the segments $l_1,\cdots,l_{[k/2]}$ there are at least $t_0=[k/2]-[\gamma k/2]+1$ segments $l_{i_1},\cdots,l_{i_t}$, $t>t_0$, undistinguished from the left. It is not difficult to see that for fixed t there exist $C_{[k/2]}^t$ different sets of t segments undistinguished from the left. Next we can show that for any $j=1,\cdots,k$ and any $\widetilde{c}_0$ the probability $P_r(\cdot)$ that the segment l_j is undistinguished is

$$P_r\,(c_j:c_j\cap(l_j|\widetilde{l}_j)\neq\varnothing\;\;\text{or}\;\;N\,(c_j)\neq 2|\widetilde{c}_0)<1-\delta, \tag{3.21}$$

where $\delta>0$ is an absolute constant which does not depend on $\widetilde{c}_0$.

We note that (3.21) follows from the inequalities proven below for the probability that the arrangement $c_j \subset l_j$ will be distinguished from the left, under the condition that the arrangement $\tilde{c}_0$ is fixed:

$$(\theta_1) \qquad P_r(c_j \subset l_j : c_j \cap (l_j | \tilde{l}_j) = \varnothing,\ N(c_j) = 2 | \tilde{c}_0) > \delta$$

and a similar inequality for the probability that $c_j \subset l_j$ will be distinguished from the right.

From the definition of the segments l_j we have

$$(\theta_2) \qquad \begin{aligned} &P_r(c_j \subset l_j : c_j \cap (l_j | \tilde{l}_j) = \varnothing,\ N(c_j) = 2 | \tilde{c}_0) \\ &= e^{-2\mu}\, \Xi^{-1}(\beta, \mu, l_j | \tilde{c}_0) \iint_{(x,y)\in \tilde{l}_j} e^{-\beta U(|x-y|)}\, dx\, dy. \end{aligned}$$

It is easily seen that

$$\Xi(\beta, \mu, l_j | \tilde{c}_0) < \exp\{(2R_0 + \delta_0 + a)\, e^{-\mu}\, e^{\frac{2R_0+\delta_0+a}{\delta_0}\beta B}\}.$$

Consequently, by denoting

$$I = \iint_{(x,y)\in \tilde{l}_j} \exp\{-\beta U(|x-y|)\}\, dx\, dy,$$

we obtain from (θ_2)

$$\begin{aligned} &P_r(c_j \subset l_j : c_j \cap (l_j | \tilde{l}_j) = \varnothing,\ N(c_j) = 2 | \tilde{c}_0) \\ &> e^{-2\mu} \cdot I \cdot \exp\{-(2R_0 + \delta_0 + a)\, e^{-\mu}\, e^{\frac{2R_0+\delta_0+a}{\delta_0}\beta B}\}. \end{aligned}$$

Now setting

$$\delta = \delta(\beta, \mu) = I \cdot \exp\{-2\mu - (2R_0 + \delta_0 + a)\, e^{-\mu}\, e^{\frac{2R_0+\delta_0+a}{\delta_0}\beta B}\},$$

we arrive at the inequality (θ_1). In a similar way we can prove the inequality

$$P_r(c_j \subset l_j : c_j \cap (l_j | \tilde{l}_j) = \varnothing,\ N(c_j) < 2 | \tilde{c}_0) > \delta$$

for the conditional probability that the arrangement c_j is distinguished from the right. Hence it follows easily that the probability of the set $\mathfrak{M}_\gamma^{(1)}(\tilde{c}_0)$, evaluated in the conditional distribution (3.19), obeys the estimate

$$(3.22) \qquad P_r(\overline{\mathfrak{M}}_\gamma^{(1)}(\tilde{c}_0) | \tilde{c}_0) < \sum_{t=t_0}^{\left[\frac{k}{2}\right]} C^t_{\left[\frac{k}{2}\right]} (1-\delta)^t.$$

We note that if $t_0 > [k/4] + 1$, then the terms in the sum (3.22) decay monotonically with increasing t. Consequently

$$P_r(\overline{\mathfrak{M}}_\gamma^{(1)}(\tilde{c}_0) | \tilde{c}_0) < \left[\frac{k}{2}\right] C^{t_0}_{\left[\frac{k}{2}\right]} (1-\delta)^{t_0} < \left[\frac{k}{2}\right] C^{\left[(1-\gamma)\frac{k}{2}\right]}_{\left[\frac{k}{2}\right]} (1-\delta)^{\left[(1-\gamma)\frac{k}{2}\right]}$$

$$< k^{\frac{3}{2}} \widetilde{d}_0(\gamma) \exp\left\{-(\gamma \ln \gamma + (1-\gamma)\ln(1-\gamma) + (1-\gamma)\ln\frac{1}{1-\delta} \Big/ \left[\frac{k}{2}\right]\right\}.$$

Next we select $\gamma_0 = \gamma_0(\delta)$ small enough that

$$\widetilde{d}_1 \equiv \widetilde{d}_1(\delta) = \left\{\gamma_0 \ln \gamma_0 + (1-\gamma_0)\ln(1-\gamma_0) + (1-\gamma_0)\ln\frac{1}{1-\delta}\right\} > 0.$$

Thus

$$P_r(\overline{\mathfrak{M}}_{\gamma_0}^{(1)}(\widetilde{c}_0)|\widetilde{c}_0) < \widetilde{d}_0 k^{\frac{3}{2}} e^{-\widetilde{d}_1\frac{k}{2}}. \tag{3.23}$$

By using the equation

$$P_r(\overline{\mathfrak{M}}_{\gamma_0}^{(1)}) = \int_{C_{\widetilde{\Omega}}} P_r(\overline{\mathfrak{M}}_{\gamma_0}^{(1)}(\widetilde{c}_0)|\widetilde{c}_0)\, p(\widetilde{c}_0)\, d\widetilde{c}_0$$

(where $p(\widetilde{c}_0)$ is the density for the probability distribution for the configuration $\widetilde{c}_0$, evaluated by means of the Gibbs distribution), the estimate (3.23), and also the fact that $k \sim |\Omega|$, we obtain (3.19).

The inequality (3.15) for the case $\overline{\mathfrak{M}}_{\gamma_0}^{(2)}$ is proved similarly.

II. Now let $\varphi(\lambda,\kappa)$ be the characteristic function of the joint distribution for $H(c)$ and $N(c)$ in the configurational grand canonical ensemble in Ω. It is evident that $\varphi(\lambda,\kappa)$ is equal to

$$\varphi(\lambda,\varkappa) = \int_{C_\Omega} p(c)\, e^{i\lambda H(c)+i\varkappa N(c)}\, dc = \Xi^{-1}(\beta,\mu,\Omega) \times \int_{C_\Omega} e^{-(\beta-i\lambda)H(c)-(\mu-i\varkappa)N(c)}\, dc = \frac{\Xi(\beta-i\lambda,\mu-i\varkappa,\Omega)}{\Xi(\beta,\mu,\Omega)}, \tag{3.24}$$

where $\Xi(\beta-i\lambda,\mu-i\kappa,\Omega)$ is the partition function for complex values of the parameters β and μ. On the other hand

$$\varphi(\lambda,\varkappa) = \widetilde{\varphi}(\lambda,\varkappa|\mathfrak{M}_\gamma)\, P_r(\mathfrak{M}_\gamma) + \widetilde{\widetilde{\varphi}}(\lambda,\varkappa|\overline{\mathfrak{M}}_\gamma)\, P_r(\overline{\mathfrak{M}}_\gamma), \tag{3.25}$$

where $\widetilde{\varphi}(\lambda,\kappa|\mathfrak{M}_\gamma)$ and $\widetilde{\widetilde{\varphi}}(\lambda,\kappa|\overline{\mathfrak{M}}_\gamma)$ are the conditional characteristic functions for the joint distribution of the values of H and N, respectively, on the sets $\mathfrak{M}_\gamma$ and $\overline{\mathfrak{M}}_\gamma$. From (3.5) and the fact that the characteristic functions are bounded in modulus by unity, we find from (3.25)

$$|\varphi(\lambda,\varkappa) - \widetilde{\varphi}(\lambda,\varkappa|\mathfrak{M}_\gamma)| < D\,|\Omega|^{\frac{3}{2}} e^{-d_1|\Omega|}, \tag{3.26}$$

where D is some absolute constant which does not depend on Ω.

Next, the quantity which concerns us, $\widetilde{p}_{\text{pot}}(H,N|\Omega)$, if it exists, can be represented in the form

$$\widetilde{p}_{\text{pot}}(H,N|\Omega) = \frac{1}{\pi^2}\int_{-\infty}^{\infty} d\lambda \int_{-\pi}^{\pi} \widetilde{\varphi}(\lambda,\varkappa|\mathfrak{M}_\gamma)\, e^{-i\lambda H - i\varkappa N}\, d\varkappa, \tag{3.27}$$

and, conversely, if the characteristic function $\tilde{\varphi}(\lambda,\kappa|\mathfrak{M}_\gamma)$ dies off sufficiently rapidly with increasing λ, then $\tilde{p}_{\text{pot}}(H,N|\Omega)$ exists and (3.27) is valid. The rapid decrease of $\tilde{\varphi}(\lambda,\kappa|\mathfrak{M}_\gamma)$ will follow from the estimates we shall derive below.

Thus we shall now obtain the asymptotic behavior of the integral (3.27) for large Ω and N which satisfy (3.3′).

As will be seen, the basic contribution to this integral comes from the neighborhood of the point $\lambda=\kappa=0$. This will be shown later by estimating the values of the integrand outside of this neighborhood.

We let Q denote the strip $-\pi\leqq\kappa\leqq\pi$, $-\infty<\lambda<\infty$ on the plane (λ,κ); $Q=Q(-\pi\leqq\kappa\leqq\pi,\ -\infty<\lambda<\infty)$, and let S denote the subset of this strip

$$S_\Omega=S_\Omega\{(\lambda,\ \varkappa)\in Q:|\lambda|<|\Omega|^{-\alpha},\ |\varkappa|<|\Omega|^{-\alpha}\},$$

where $\alpha<\frac{1}{2}$ is an absolute constant which will be chosen below.

We now write the integral in (3.27) in the form

$$\tag{3.28}\begin{aligned}&\iint_Q\tilde{\varphi}(\lambda,\ \varkappa\,|\,\mathfrak{M}_\gamma)\,e^{-i\lambda H-i\varkappa N}\,d\lambda\,d\varkappa\\ &=\iint_{S_\Omega}\tilde{\varphi}(\lambda,\ \varkappa\,|\,\mathfrak{M}_\gamma)\,e^{-i\lambda H-i\varkappa N}\,d\lambda\,d\varkappa+\iint_{Q\setminus S_\Omega}\tilde{\varphi}(\lambda,\ \varkappa\,|\,\mathfrak{M}_\gamma)\,e^{-i\lambda H-i\varkappa N}\,d\lambda\,d\varkappa.\end{aligned}$$

Consider the first term in (3.28). From (3.26) we find

$$\tag{3.29}\begin{aligned}&\iint_{S_\Omega}\tilde{\varphi}(\lambda,\ \varkappa\,|\,\mathfrak{M}_\gamma)\,e^{-i\lambda H-i\varkappa N}\,dx\,d\varkappa\\ &=\iint_{S_\Omega}\varphi(\lambda,\ \varkappa)\,e^{-i\lambda H-i\varkappa N}\,dx\,d\varkappa+O(|\Omega|^{\frac{3}{2}-2\alpha}\,e^{-d_1|\Omega|}).\end{aligned}$$

In order to find the asymptotic behavior of the integral on the right-hand side of (3.29) we expand the logarithm of the integrand in powers of λ and κ in the neighborhood S_Ω up to terms of the third order, taking advantage for this purpose of the representation (3.24), Proposition 2.1 from §2 and the fact that $(\lambda,\kappa)\in S_\Omega$ are small for sufficiently large $|\Omega|$. We obtain

$$\tag{3.30}\begin{aligned}\varphi(\lambda,\ \varkappa)\,e^{-i\lambda H-i\varkappa N}=\exp\Big\{&i(\overline{H}-H)\,\lambda+i(\overline{N}-N)\,\varkappa\\ &-\frac{1}{2}\widetilde{M}_{11}^{\Omega}\lambda^2-\widetilde{M}_{12}^{\Omega}\lambda\varkappa-\frac{1}{2}M_{22}^{\Omega}\varkappa^2+R(\lambda,\ \varkappa,\ \Omega)\Big\},\end{aligned}$$

where $\widetilde{M}_{11}^{\Omega}$, $\widetilde{M}_{12}^{\Omega}$ and M_{22}^{Ω} denote the covariance coefficients for the distribu- of the quantities H and N, and $R(\lambda,\kappa,\Omega)$, for brevity, denotes the third derivative of the logarithm of the partition function

$$R(\lambda, \varkappa, \Omega) = \frac{1}{6}\left\{ \frac{\partial^3 \ln \Xi(\beta - i\lambda,\ \mu - i\varkappa,\ \Omega)}{\partial\lambda^3}\lambda^3 + 3\frac{\partial^3 \ln \Xi(\beta - i\lambda,\ \mu - i\varkappa,\ \Omega)}{\partial\lambda^2\,\partial\varkappa}\lambda^2\varkappa + 3\frac{\partial^3 \ln \Xi(\beta - i\lambda,\ \mu - i\varkappa,\ \Omega)}{\partial\lambda\,\partial\varkappa^2}\lambda\varkappa^2 + \frac{\partial^3 \ln \Xi(\beta - i\lambda,\ \mu - i\varkappa,\ \Omega)}{\partial\varkappa^3}\varkappa^3 \right\}_{\substack{\tilde\lambda=\theta\lambda \\ \tilde\varkappa=\theta\varkappa}} \quad 0<\theta<1,$$

evaluated at some intermediate point $\{\beta - i\tilde{\lambda}, \mu - i\tilde{\kappa}\}$. Since, for $|\lambda| < |\Omega|^{-\alpha}$ and $|\kappa| < |\Omega|^{-\alpha}$, $\tilde{\lambda}$ and $\tilde{\kappa}$ also satisfy these respective inequalities, we have the estimate for the remainder term (see (2.27))

$$(3.31) \qquad |R(\lambda, \varkappa, \Omega)| < \tilde{A}\cdot|\Omega|\cdot(|\lambda|^3 + |\varkappa|^3),$$

where $\tilde{A} > 0$ is an absolute constant which does not depend on Ω. Next we can replace the quadratic form in the exponent of (3.17) by the quadratic form

$$(3.31) \qquad \frac{1}{2}|\Omega|\{\tilde{d}_{11}(\beta, \mu)\lambda^2 + 2\,\tilde{d}_{12}(\beta, \mu)\lambda\varkappa + \tilde{d}_{22}(\beta, \mu)\varkappa^2\}$$

(where $\tilde{d}_{11}(\beta,\mu)$, $\tilde{d}_{12}(\beta,\mu)$ and $\tilde{d}_{22}(\beta,\mu)$ are defined in §2), which by (2.15), (2.20) and (2.23) does not change the remainder by more than $A_1(\lambda^2 + \kappa^2)$, $\tilde{A}_1 > 0$ being an absolute constant which does not depend on Ω. Thus for the new remainder term $\tilde{R}(\lambda, \kappa, \Omega)$ we obtain

$$|\tilde{R}(\lambda, \varkappa, \Omega)| < \tilde{A}|\Omega|(|\lambda|^3 + |\varkappa|^3) + \tilde{A}_1(\lambda^2 + \varkappa^2).$$

Since $|\lambda| < |\Omega|^{-\alpha}$ and $|\kappa| < |\Omega|^{-\alpha}$, we have

$$|\tilde{R}(\lambda, \varkappa, \Omega)| < 2\tilde{A}|\Omega|^{1-3\alpha} + 2\tilde{A}_1|\Omega|^{-2\alpha}.$$

Now if we set $\alpha = 2/5$ and note that

$$\exp\{\tilde{R}(\lambda, \varkappa, \Omega)\} = 1 + O(|\Omega|^{-\frac{1}{5}}),$$

for the integral

$$I_1 = \iint\limits_{S_\Omega} \varphi(\lambda, \varkappa)\, e^{-i\lambda H - i\varkappa N}\, d\lambda\, d\varkappa$$

$$= \iint\limits_{|\lambda|<|\Omega|^{-\frac{2}{5}},\ |\varkappa|<|\Omega|^{-\frac{2}{5}}} (1 + O(|\Omega|^{-\frac{1}{5}}))\exp\{i(\bar{H} - H)\lambda + i(\bar{N} - N)\varkappa - \frac{1}{2}|\Omega|(\tilde{d}_{11}\lambda^2 + 2\tilde{d}_{12}\lambda\varkappa + \tilde{d}_{22}\varkappa^2)\}\, d\lambda\, d\varkappa$$

we obtain the expression

$$I = \frac{\pi}{|\Omega| \sqrt{D'}} \exp\left\{-\frac{1}{2} h_{11} \frac{(H-\bar{H})^2}{|\Omega|} - h_{12} \frac{(H-\bar{H})(N-\bar{N})}{|\Omega|} - \frac{1}{2} h_{22} \frac{(N-\bar{N})^2}{|\Omega|}\right\} + O(|\Omega|^{-\frac{6}{5}}). \tag{3.33}$$

Here

$$\begin{Vmatrix} h_{11} & h_{12} \\ h_{12} & h_{22} \end{Vmatrix} = \begin{Vmatrix} \tilde{d}_{11} & \tilde{d}_{12} \\ \tilde{d}_{12} & \tilde{d}_{22} \end{Vmatrix}^{-1}, \quad D = \det \begin{Vmatrix} \tilde{d}_{11} & \tilde{d}_{12} \\ \tilde{d}_{12} & \tilde{d}_{22} \end{Vmatrix}.$$

Consequently in view of (3.33) and (3.29) we find

$$\frac{1}{\pi^2} \iint_{S_\Omega} \tilde{\varphi}(\lambda, \varkappa \mid \mathfrak{M}_\gamma) e^{-i\lambda H - i\varkappa N} \, d\lambda \, d\varkappa$$

$$= \frac{1}{\pi |\Omega| \sqrt{D'}} \exp\left\{-\frac{1}{2} h_{11} \frac{(H-\bar{H})^2}{|\Omega|} - h_{12} \frac{(H-\bar{H})(N-\bar{N})}{|\Omega|} - \frac{1}{2} h_{22} \frac{(N-\bar{N})^2}{|\Omega|}\right\} + O(|\Omega|^{-\frac{6}{5}}).$$

Hence it remains to evaluate the integral

$$I_2 = \iint_{Q \setminus S_\Omega} \tilde{\varphi}(\lambda, \varkappa \mid \mathfrak{M}_\gamma) e^{-i\lambda H - i\varkappa N} \, d\lambda \, d\varkappa.$$

III. In the region $Q \setminus S_\Omega$ it is expedient to estimate the characteristic function $\tilde{\varphi}(\lambda, \kappa \mid \mathfrak{M}_\gamma)$ directly; we have the following lemma.

Lemma 3.5. *The characteristic function $\tilde{\varphi}(\lambda, \kappa \mid \mathfrak{M}_\gamma)$ obeys the following inequalities in the region $Q \setminus S_\Omega$:*

1. *There exists a λ_0 (which does not depend on Ω) such that for $|\Omega|^{-2/5} < |\lambda| \leqq \lambda_0$ and arbitrary κ, or for $|\lambda| < |\Omega|^{-2/5}$ and $|\Omega|^{-2/5} < |\kappa| \leqq \pi$*

$$|\tilde{\varphi}(\lambda, \varkappa \mid \mathfrak{M}_\gamma)| < a_0 e^{-a_1 |\Omega|^{\frac{1}{5}}}, \tag{3.34}$$

where $a_0 > 0$ and $a_1 > 0$ are absolute constants which do not depend on Ω.

2. *For $|\lambda| > \lambda_0$ and arbitrary κ*

$$|\tilde{\varphi}(\lambda, \varkappa \mid \mathfrak{M}_\gamma)| < \frac{\sigma_0}{(\lambda^2 + \beta^2)^2} e^{-b_0 |\Omega|}, \tag{3.35}$$

where $\sigma_0 > 0$ and $b_0 > 0$ are absolute constants which do not depend on Ω.

Proof. At this point we turn to the decomposition of the segment Ω introduced at the start of the proof of Lemma 3.1 and consisting of the segments $l_1, \cdots, l_k$ of length $|l_i| = l = 2R_0 + \delta_0 + a$, $i = 1, \cdots, k$, which are a distance R_0 apart. Next if for each arrangement $c \subset C_\Omega$ we let $\tilde{c}$ denote the portion of this arrangement which falls outside the set $\tilde{S} = \bigcup_1^k l_j$, and let

$\mathfrak{M}_\gamma(\tilde{c}_0)$ denote the subset of those arrangements in $\mathfrak{M}_\gamma(\tilde{c})$ for which $\tilde{c} = \tilde{c}_0$ is fixed, and then consider the characteristic function $\tilde{\varphi}(\lambda, \kappa \mid \mathfrak{M}_\gamma(\tilde{c}_0) \mid \tilde{c}_0)$ for the conditional joint distribution of the quantities $H(c' \mid \tilde{c}_0)$ and $N(c')$, $c' = c \mid \tilde{c}_0$, in the conditional ensemble on the set $\mathfrak{M}_\gamma(\tilde{c}_0)$, then we have the obvious relation

$$\tilde{\varphi}(\lambda, \varkappa \mid \mathfrak{M}_\gamma) = \int_{C_{\tilde{\Omega}}} \tilde{\varphi}(\lambda, \varkappa \mid \mathfrak{M}_\gamma(\tilde{c}_0) \mid \tilde{c}_0) \times e^{i\lambda H(\tilde{c}_0) + i\varkappa N(\tilde{c}_0)} \pi(\tilde{c}_0, \beta, \mu, \tilde{\Omega})\, d\tilde{c}_0, \tag{3.36}$$

where $\pi(\tilde{c}_0, \beta, \mu, \tilde{\Omega})$ $(\tilde{\Omega} = \Omega \backslash \tilde{S})$ denote the density of the probability distribution on C_Ω for the arrangement $\tilde{c}_0$ within $\tilde{\Omega}$ which is generated by the Gibbs distribution in the configurational ensemble $\{C_\Omega, \beta, \mu\}$.

We recall that the set $\mathfrak{M}_\gamma$ consists of those collections of arrangements $c' = \{c_1, \cdots, c_k\} \in C_{\tilde{S}}$, $c_i \subset l_i$, for which there are at least n_0 $(= [\gamma_0 k/2])$ segments $l_{i_1}, \cdots, l_{i_{n_0}}$, $i_s < [k/2]$, $s = 1, \cdots, n_0$, distinguished from the left (i.e. such that (*) of page 234 is satisfied) and at least n_0 segments $l_{j_1}, \cdots, l_{j_{n_0}}$, $j_s > [k/2]$, $s = 1, \cdots, n_0$, distinguished from the right (i.e. such that (**) of page 234 is satisfied). For each of the arrangements $c \in \mathfrak{M}_\gamma(\tilde{c}_0)$ we denote by

$$i_1(c) < i_2(c) < \cdots < i_{\left[\gamma \frac{k}{2}\right]}(c) \leqslant \left[\frac{k}{2}\right] < j_1(c) < \cdots < j_{\left[\gamma \frac{k}{2}\right]} \leqslant k,$$

the n leftmost segments distinguished from the left and right, respectively, and by $\mathfrak{M}_\gamma(\tilde{c}_0, \{i_s\}, \{j_s\}, s = 1, \cdots, [\gamma k/2])$ the totality of all configurations $c' = \{c_1, \cdots, c_k\} \in \mathfrak{M}_\gamma(\tilde{c}_0)$ for which

$$i_s(c) = i_s,\ i_s \leqslant \left[\frac{k}{2}\right] \text{ and } j_s(c) = j_s,\ j_s > \left[\frac{k}{2}\right],\ s = 1, 2, \ldots, \left[\gamma \frac{k}{2}\right].$$

It is clear that when the collections

$$\left(i_1, i_2, \ldots, i_{\left[\gamma \frac{k}{2}\right]}, j_1, j_2, \ldots, j_{\left[\gamma \frac{k}{2}\right]}\right)$$

and

$$\left(i_1', i_2', \ldots, i'_{\left[\gamma \frac{k}{2}\right]}, j_1', j_2', \ldots, j'_{\left[\gamma \frac{k}{2}\right]}\right)$$

are different, the sets

$$\mathfrak{M}_\gamma(\tilde{c}_0, \{i_s\}, \{j_s\},\quad s = 1, \cdots, [\gamma k/2])$$

and

$$\mathfrak{M}_\gamma(\tilde{c}_0, \{i_s'\}, \{j_s'\},\quad s = 1, \cdots [\gamma k/2])$$

do not intersect and

$$\mathfrak{M}_\gamma(\tilde{c}_0) = \bigcup_{(\{i_s\}, \{j_s\}, s=1, \cdots, [\gamma k/2])} \mathfrak{M}_\gamma(\tilde{c}_0, \{i_s\}, \{j_s\}, s = 1, \cdots, [\gamma k/2])$$

Thus

$$\tilde{\varphi}(\lambda,\kappa\,|\,\mathfrak{M}_\gamma(\tilde{c}_0)\,|\,\tilde{c}_0) = \sum_{(\{i_s\},\{j_s\},s=1,\cdots,[\gamma k/2])} \times \tilde{\varphi}(\lambda,\kappa\,|\,\mathfrak{M}_\gamma(\tilde{c}_0,\{i_s\},\{j_s\},s=1,\cdots,[\gamma k/2])\,|\,\tilde{c}_0) \times Pr(\mathfrak{M}_\gamma(\tilde{c}_0,\{i_s\},\{j_s\},s=1,\cdots,[\gamma k/2])), \tag{3.37}$$

where

$$\tilde{\varphi}(\lambda,\kappa\,|\,\mathfrak{M}_\gamma(\tilde{c}_0,\{i_s\},\{j_s\},s=1,\cdots,[\gamma k/2])\,|\,\tilde{c}_0)$$

is the conditional characteristic function of the joint distribution of the quantities $H(c'\,|\,\tilde{c}_0)$ and $N(c')$ in the conditional ensemble on the set

$$\mathfrak{M}_\gamma(\tilde{c}_0,\{i_s\},\{j_s\},s=1,\cdots,[\gamma k/2]).$$

Finally, in view of the remark made above we have

$$\tilde{\varphi}(\lambda,\kappa\,|\,\mathfrak{M}_\gamma(\tilde{c}_0,\{i_s\},\{j_s\},s=1,\cdots,[\gamma k/2])\,|\,\tilde{c}_0)$$

$$= \prod_{s=1}^{\left[\gamma\frac{k}{2}\right]} e^{2\varkappa i}\varphi_{l_{i_s}}^{(\text{dist.left})}(\lambda\,|\,\tilde{c}_0)\cdot \prod_{\substack{i<i_n,\, i\neq i_s\\ s=1,2,\dots n}} \varphi_{l_i}^{(\text{indist.left})}(\lambda,\ \varkappa\,|\,\tilde{c}_0)$$

$$\times \prod_{i_n<i\leqslant\left[\frac{k}{2}\right]} \varphi_{l_i}^{(\text{tot})}(\lambda,\ \varkappa\,|\,\tilde{c}_0)\cdot \prod_{s=1}^{\left[\gamma\frac{k}{2}\right]} \varphi_{l_{i_s}}^{(\text{dist.right})}(\lambda,\ \varkappa\,|\,\tilde{c}_0) \tag{3.38}$$

$$\times \prod_{\substack{\left[\gamma\frac{k}{2}\right]<j<i_n,\, j\neq i_s\\ s=1,2,\dots,n}} \varphi_{l_i}^{(\text{indist.right})}(\lambda,\ \varkappa\,|\,\tilde{c}_0) \prod_{j>i_n}^{k} \varphi_{l_j}^{(\text{tot})}(\lambda,\ \varkappa\,|\,\tilde{c}_0),$$

where $\varphi_{\tilde{l}_i}^{(\text{dist.left})}(\lambda\,|\,\tilde{c}_0)$ is the conditional characteristic function for the distribution of the quantity $H(c_i\,|\,\tilde{c}_0)$ in the ensemble of distinguished arrangements in C_{l_i}; it is evident that this ensemble coincides with the canonical ensemble $\{C_{\tilde{l}_i,N},\beta,N\}$ for $N=2$. The function $\varphi_{l_i}^{(\text{dist.right})}(\lambda,\kappa\,|\,\tilde{c}_0)$ has a similar meaning. Next,

$$\varphi_{l_i}^{(\text{indist.left})}(\lambda,\kappa\,|\,\tilde{c}_0)\quad(\text{and } \varphi_{l_i}^{(\text{indist.right})}(\lambda,\kappa\,|\,\tilde{c}_0))$$

is the conditional characteristic function of the joint distribution of the quantities $H(c_i\,|\,\tilde{c}_0)$ and $N(c_i)$ in the conditional grand ensemble under the additional restriction that the arrangement $c_i\subset l_i$ is indistinguished from the left (right); finally, $\varphi_{l_i}^{\text{tot}}(\lambda,\kappa\,|\,\tilde{c}_0)$ is the conditional characteristic function of the joint distribution for the quantities $H(c_i\,|\,\tilde{c}_0)$ and $N(c_i)$ in the entire ensemble $\{C_{l_i},\beta,\mu\,|\,\tilde{c}_0\}$.

We now note that the statement of Lemma 3.5, in view of (3.36) and (3.38), follows from the following lemma:

LEMMA 3.6.1. *For all* $(\lambda, \kappa) \in Q$

$$|\varphi^{(\text{dist.left})}_{\widetilde{l}_i}(\lambda \mid \widetilde{c}_0)| < \frac{a_0}{\sqrt{\lambda^2+\beta^2}} \quad \textit{for any } i, \tag{3.39}$$

where $\widetilde{a}_0$ *is an absolute constant (independent of* $\widetilde{c}_0$*).*

2. *Choose* λ_0 *such that* $\widetilde{a}_0/(\lambda^2+\beta^2) < \frac{1}{2}$ *for* $|\lambda| \geqq \lambda_0$. *Under the condition that* $(\lambda, \kappa) \in Q \backslash S$ *and* $|\Omega|^{-2/5} < |\lambda| < \lambda_0$ *we have*

$$|\varphi^{(\text{dist.left})}_{\widetilde{l}_i}(\lambda|\widetilde{c}_0)| < 1 - \widetilde{a}_1|\Omega|^{-4/5} \quad \textit{for any } i, \tag{3.40}$$

where $\widetilde{a}_1 > 0$ *is an absolute constant (independent of* Ω *and* $\widetilde{c}_0$*).*

3. *When* $|\kappa| > |\Omega|^{-2/5}$ *we have*

$$|\varphi^{(\text{dist.right})}_{\widetilde{l}_i}(\lambda, \kappa|\widetilde{c}_0)| < 1 - \widetilde{a}_2|\Omega|^{-4/5} \quad \textit{for any } i, \tag{3.41}$$

where $\widetilde{a}_2 > 0$ *is an absolute constant (independent of* Ω *and* $\widetilde{c}_0$*).*

PROOF. 1. We have

$$\varphi^{(\text{dist.left})}_{\widetilde{l}_i}(\lambda \mid \widetilde{c}_0) = Z^{-1}(\beta, 2, \widetilde{l}_i \mid \widetilde{c}_0) \int_{C_{l_i}, 2} e^{-(\beta - i\lambda) H(c|\widetilde{c}_0)}\, dc.$$

We note that, since $d(\widetilde{l}_i, \Gamma(l_i)) = R_0$, we have $H(c|\widetilde{c}_0) = H(c) = U(|x-y|)$, and the canonical partition function $Z(\beta, 2, \widetilde{l}_i|\widetilde{c}_0)$ does not depend on c_0. Consequently we have

$$\varphi^{(\text{dist.left})}_{l_i}(\lambda \mid \widetilde{c}_0) = \frac{1}{Z(\beta, 2, \widetilde{l}_i)} \iint_{r_i' \leqslant x < y \leqslant r_i''} e^{-(\beta - i\lambda) U(|x-y|)}\, dx dy, \tag{3.42}$$

where r_i' and r_i'' are the endpoints of the segment $\widetilde{l}_i$. Next we write the integral in the form

$$I = \iint_{r_i' \leqslant x < y \leqslant r_i''} e^{-(\beta - i\lambda) U(|x-y|)}\, dx dy = \int_{r_i' + \delta_0}^{r_i''} dy \int_{r_i'}^{y - \delta_0} \exp\{-(\beta - i\lambda)$$

$$\times U(|x-y|)\}\, dx dy = \int_{r_i' + \delta_0}^{r_i''} dy \int_{r_i'}^{y-\delta_0} \frac{U'(|x-y|)}{U'(|x-y|)} e^{-(\beta - i\lambda) U(|x-y|)}\, dx dy.$$

Integrating the inner integral by parts, we obtain

$$I = -\frac{1}{\beta - i\lambda} \frac{e^{-(\beta - i\lambda) U(\delta_0)} a}{U'(\delta_0)} - \frac{1}{\beta - i\lambda} \int_{r_i' + \delta_0}^{r_i''} \frac{e^{-(\beta - i\lambda) U(|r_i' - y|)}}{U'(|r_i' - y|)}\, dy$$

$$- \frac{1}{\beta - i\lambda} \int_{r_i' + \delta_0}^{r_i''} dy \int_{r_i'}^{y - \delta_0} \frac{U''(|x-y|)}{(U'(|x-y|))^2} e^{-(\beta - i\lambda) U(|x-y|)}\, dx.$$

Now, by using the fact that since

$$|r_i' - y| < \delta_0 + a,\ |x - y| < \delta_0 + a,\ |U'(r)| > \zeta_0 > 0$$

and property IV of $U(r)$, we obtain

$$(3.43)\quad \begin{aligned} |I| &< \frac{1}{\zeta_0 \sqrt{\lambda^2+\beta^2}} \Big\{ a e^{-\beta U(\delta_0)} + \int_{r_i'+\delta_0}^{r_i''} e^{-\beta U(|r_i'-y|)}\, dy \\ &+ \frac{1}{\zeta_0} \int_{r_i'+\delta_0}^{r_i''} dy \int_{r_i'}^{y-\delta_0} |U''(|x-y|)|\, e^{-\beta U(|x-y|)}\, dx \Big\} < \frac{D_1(\beta)}{\sqrt{\lambda^2+\beta^2}}, \end{aligned}$$

where $D_1(\beta)$ is an absolute constant which depends only on β. Since $Z(\beta, 2, \tilde{l}_i) > 0$, we arrive at (3.39), i.e.

$$|\varphi_{\tilde{l}_i}^{(\text{dist.left})}(\lambda \mid \tilde{c}_0)| < \frac{\tilde{a}_0}{\sqrt{\lambda^2+\beta^2}} \quad \text{for all } i,$$

where $\tilde{a}_0 = Z^{-1}(\beta, 2, \tilde{l}_i)\, D_1(\beta)$.

2. We now prove (3.40). Let $|\Omega|^{-2/5} < |\lambda| < \lambda_0$. We have

$$(3.44)\quad \varphi_{\tilde{l}_i}^{(\text{dist.left})}(\lambda \mid \tilde{c}_0) = \frac{1}{Z(\beta, 2, \tilde{l}_i)} \iint_{L_i} e^{-(\beta - i\lambda) U(|x-y|)}\, dx\, dy,$$

where L_i denotes the region of integration: $L_i = L_i(r_i' \leqq x < y \leqq r_i'')$. We consider two sets $M_1 \subset L_i$ and $M_2 \subset L_i$ defined respectively as follows:

$$M_1 = M_1\{(x, y) \in L_i : a_1 < y - x < a_2,\ r_i' + b_1 < y \leqslant r_i''\},$$
$$M_2 = M_2\{(x, y) \in L_i : d_1 + d < y - x < a_2 + d,\ r_i' + b_1 < y \leqslant r_i''\},$$

where a_1, a_2, d and b_1 are fixed constants given above (see the figure). We now rewrite the integral in (3.44) in the form

$$\begin{aligned} I' &= \iint_{L_i} e^{-(\beta - i\lambda) U(|x-y|)}\, dx dy = \iint_{M_1} e^{-(\beta - i\lambda) U(|x-y|)}\, dx dy \\ &+ \iint_{M_2} e^{-(\beta - i\lambda) U(|x-y|)}\, dx dy + \iint_{M_3} e^{-(\beta - i\lambda) U(|x-y|)}\, dx dy, \end{aligned}$$

where $M_3 = L_i \backslash (M_1 \cup M_2)$. It is easily seen that

$$(3.45)\quad \begin{aligned} |I'| &\leqslant \iint_{M_1} |\exp\{-(\beta - i\lambda) U(|y-x|)\} \\ &+ \exp\{-(\beta - i\lambda) U(|y - x + d|)\}|\, dx dy \\ &+ \iint_{M_3} \exp\{-\beta U(|y-x|)\}\, dx dy = \end{aligned}$$

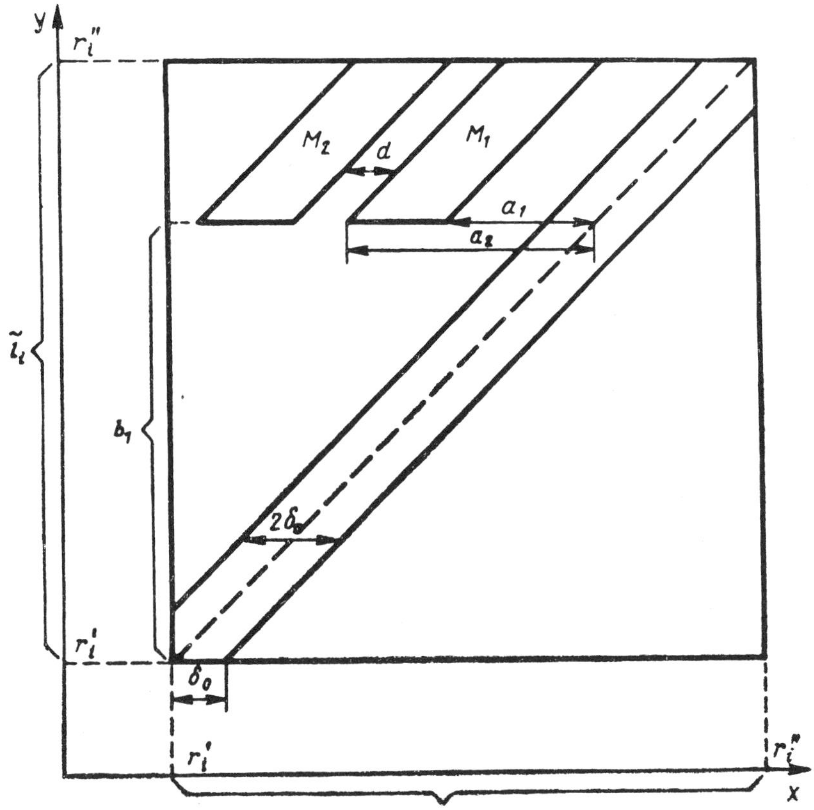

$$= \iint_{M_1} \Big| \exp\{-\beta U(|y-x|)\} + \exp\{-\beta U(|y-x+d|)\}$$

$$\times \exp\{i\lambda(U(|y-x+d|)) - U(|y-x|)\}\Big| dxdy + \iint_{M_3} \exp\{-\beta U(|y-x|)\} dxdy.$$

Next we estimate the relation

$$f(x, y) = [\exp\{-\beta U(|y-x|)\} + \exp\{-\beta U(|y-x+d|)$$
$$\times \exp\{i\lambda(U(|y-x+d|) - U(|y-x|))\}]\cdot[\exp\{-\beta U(|y-x|)\}$$
$$+ \exp\{-\beta U(|y-x+d|)\}]^{-1} = [1 + \exp\{-\beta(U(|y-x+d|) - U(|y-x|))\}$$
$$\times \exp\{i\lambda(U(|y-x+d|) - U(|y-x|))\}]\cdot[1$$
$$+ \exp\{-\beta(U(|y-x+d|) - U(|y-x|))\}]^{-1}$$

$$= P_1(x, y) + P_2(x, y)\exp\{i\lambda(U(|y-x+d|) - U(|y-x|))\},$$

where

$$P_1(x, y) = \frac{1}{1 + \exp\{-\beta(U(|y-x+d|) - U(|y-x|))\}},$$

$$P_2(x, y) = \frac{\exp\{-\beta(U(|y-x+d|) - U(|y-x|))\}}{1 + \exp\{-\beta(U(|y-x+d|) - U(|y-x|))\}}.$$

Note that

$$U(y-x+d) - U(y-x) = \int_{y-x}^{y-x+d} U'(\xi)\, d\xi.$$

Consequently for any pair $(x,y) \in M_1$ we have

$$d \cdot A_1(a_1, a_2) < |U(|y-x+d)|) - U(|y-x|)| < d \cdot A_2(a_1, a_2),$$

where

$$A_1(a_1, a_2) = \min_{a_1<r<a_2} |U'(r)| > 0, \quad A_2(a_1, a_2) = \max_{a_1<r<a_2} |U'(r)|.$$

Hence we find immediately that

$$e^{-\beta d \cdot A_2(a_1,a_2)} < \frac{P_1(x, y)}{P_2(x, y)} < e^{\beta d \cdot A_2(a_1,a_2)}.$$

Next, since

$$|\Omega|^{-\frac{2}{5}} d \cdot A_1(a_1, a_2) < |\lambda| \, |U(|y-x+d|) - U(|y-x|)| < \lambda_0 d A_2(a_1, a_2),$$

we select the constants d, a_1 and a_2 so that $\lambda_0 d \cdot A_2(a_1, a_2) < \pi$. Hence, representing the quantity $f(x,y)$ on the complex plane, by geometrical considerations it is not difficult to see that $|f(x,y)| < |z|$, where $z = 1/(1+h) + he^{i\nu}/(1+h)$ (here $h = e^{\beta d \cdot A_2(a_1,a_2)}$ and $\nu = |\Omega|^{-2/5} d A_1(a_1, a_2)$). Next, a simple calculation shows us that $|z| < 1 - h_0|\Omega|^{-4/5}$, where $h_0 = h_0(a_1, a_2, \beta, d) > 0$ is an absolute constant.

Thus

$$|f(x,y)| < 1 - h_0|\Omega|^{-4/5}. \tag{3.46}$$

Next we turn to the inequality (3.45). From (3.46) we have

$$|I'| < \iint_{M_1} (1 - h_0|\Omega|^{-\frac{4}{5}})(e^{-\beta U(|y-x|)} + e^{-\beta U(|y-x+d|)})\, dxdy$$
$$+ \iint_{M_3} e^{-\beta U(|y-x|)}\, dxdy = \left\{ \iint_{M_1} + \iint_{M_2} + \iint_{M_3} \right\} e^{-\beta U(|y-x|)}\, dxdy$$
$$- h_0|\Omega|^{-\frac{4}{5}} \left\{ \iint_{M_1} + \iint_{M_2} \right\} e^{-\beta U(|y-x|)}\, dxdy.$$

Hence from (3.44) we obtain

$$|\varphi_{\widetilde{l}_i}^{(\text{dist.left})}(\lambda/\widetilde{c}_0)| < 1 - h_0 \frac{\{\iint_{M_1} + \iint_{M_2}\} e^{-\beta U(|y-x|)}\, dxdy}{Z(\beta, 2, \widetilde{l}_i)} |\Omega|^{-\frac{4}{5}} = 1 - \widetilde{a}_1 |\Omega|^{-\frac{4}{5}},$$

where

$$\widetilde{a}_1 = h_0 Z^{-1}(\beta, 2, \widetilde{l}_i) \left\{ \iint_{M_1} + \iint_{M_2} \right\} e^{-\beta U(|y-x|)}\, dxdy.$$

This proves the inequality (3.40).

3. We next prove (3.41). We have

$$\varphi_{\widetilde{l}_i}^{(\text{dist.right})}(\lambda, \varkappa/\widetilde{c}_0) = \Xi^{-1}(\beta, \mu, \widetilde{l}_i \,|\, \widetilde{c}_0) \int_{C_{\widetilde{l}_i}} e^{-(\beta - i\lambda) H(c|\widetilde{c}_0) - (\mu - i\varkappa) N(c)}\, dc, \tag{3.47}$$

where the integral and the partition function are calculated with respect to the set of configurations distinguished from the right: it is clear that this set is precisely $C_{\widetilde{l}_i,0} \cup C_{\widetilde{l}_i,1}$. Thus, keeping in mind that $d(\widetilde{l}_i, \widetilde{c}_0) \geqq R_0$ and $H(c|\widetilde{c}_0) = 0$, we obtain

$$\varphi_{\widetilde{l}_i}^{(\text{dist.right})}(\lambda, \varkappa \,|\, \widetilde{c}_0) = \frac{1 + e^{-\mu}(\delta_0 + a)\, e^{i\varkappa}}{1 + e^{-\mu}(\delta_0 + a)}.$$

Hence, just as above, for $|\kappa| > |\Omega|^{-2/5}$ we obtain

$$|\varphi_{\widetilde{l}_i}^{(\text{dist.right})}(\lambda, \kappa|\widetilde{c}_0)| < 1 - \widetilde{a}_2 |\Omega|^{-4/5} \quad \text{for all } i,$$

where

$$\widetilde{a}_2 = \frac{1}{\pi^2} \frac{(\delta_0 + a)\, e^{-\mu}}{(1 + (\delta_0 + a)\, e^{-\mu})^2}.$$

Thus the inequality (3.41) and also Lemma 3.6 are completely proved.

IV. Next let us turn to the inequality (3.38). Let $(\lambda, \kappa) \in Q \backslash S_\Omega$ and $|\Omega|^{-2/5} < |\lambda| < \lambda_0$. Then from (3.38) with the help of (3.40) we obtain

$$|\tilde{\varphi}(\lambda, \kappa | \mathfrak{M}_\gamma(\tilde{c}_0, \{i_s\}, \{j_s\}, s = 1, \cdots, [\gamma k/2]) | \tilde{c}_0)|$$

$$\leqslant \prod_{s=1}^{\left[\gamma \frac{k}{2}\right]} |\varphi_{\widetilde{l}_{i_s}}^{(\text{от.сл.})}(\lambda/\widetilde{c}_0)| < (1 - \widetilde{a}_1 |\Omega|^{-\frac{4}{5}})^{\gamma \frac{k}{2} - 1}.$$

Since $k \sim |\Omega|$, we have

$$|\tilde{\varphi}(\lambda, \kappa | \mathfrak{M}_\gamma(c_0, \{i_s\}, \{j_s\}, s = 1, \cdots, [\gamma k/2]) | \tilde{c}_0)| < \bar{a}_0 e^{-\bar{a}_1 |\Omega|^{1/5}}, \tag{3.48}$$

where $\bar{a}_0$ and $\bar{a}_1$ are absolute constants which do not depend on Ω and $\tilde{c}_0$.

Now let $(\lambda, \kappa) \in Q \backslash S_\Omega$, $|\lambda| < |\Omega|^{-2/5}$ and $|\kappa| > |\Omega|^{-2/5}$. Then from (3.38) with the help of (3.41) we obtain

$$|\tilde{\varphi}(\lambda, \kappa | \mathfrak{M}_\gamma(c_0, \{i_s\}, \{j_s\}, s = 1, \cdots, [\gamma k/2]) | \tilde{c}_0)|$$

$$\leqslant \prod_{s=1}^{\left[\gamma \frac{k}{2}\right]} |\varphi_{l_{j_s}}^{(\text{dist.right})}(\lambda, \varkappa \,|\, \widetilde{c}_0)| < (1 - \widetilde{a}_2 |\Omega|^{-\frac{4}{5}})^{\gamma \frac{k}{2} - 1} < a_0' e^{-a_1' |\Omega|^{\frac{1}{5}}}, \tag{3.49}$$

where a_0' and a_1' are certain absolute constants which do not depend on Ω and $\tilde{c}_0$. From (3.36), (3.37), (3.48) and (3.49) we obtain the first half of Lemma 3.5, i.e. when $(\lambda, \kappa) \in Q \setminus S_\Omega$ and $|\lambda| < \lambda_0$

$$|\tilde{\varphi}(\lambda, \kappa | \mathfrak{M}_\gamma)| < a_0 e^{-a_1|\Omega|^{1/5}}.$$

When $|\lambda| > \lambda_0$ and κ is arbitrary we find from (3.38) with the help of (3.39) that

$$|\tilde{\varphi}(\lambda, \kappa | \mathfrak{M}_\gamma(\tilde{c}_0, \{i_s\}, \{j_s\}, s = 1, \cdots, [\gamma k/2])(c_0)|$$

$$< \prod_{s=1}^{\left[\gamma \frac{k}{2}\right]} |\varphi_{l_{i_s}}^{(\text{dist.left})}(\lambda | \tilde{c}_0)| < \left(\frac{\tilde{a}_0^2}{\lambda^2 + \beta^2}\right)^{\frac{1}{2}\left(\gamma\frac{k}{2} - 1\right)}$$

$$< \left(\frac{1}{2}\right)^{\frac{1}{2}\left(\gamma\frac{k}{2} - 1\right) - 2} \cdot \frac{\sigma_0'}{(\lambda^2 + \beta^2)^2} < \frac{\sigma_0}{(\lambda^2 + \beta^2)^2} e^{-b_0|\Omega|}.$$

Thus with the help of (3.36) and (3.37) we obtain the remainder of Lemma 3.5.

Now consider the integral

$$I_2 = \iint_{Q \setminus S_\Omega} \tilde{\varphi}(\lambda, \varkappa | \mathfrak{M}_\gamma) e^{i\lambda H + i\varkappa N} \, d\lambda d\varkappa$$

$$= \iint_{\substack{(\lambda,\varkappa) \in Q \setminus S_\Omega \\ |\lambda| < \lambda_0}} \tilde{\varphi}(\lambda, \varkappa | \mathfrak{M}_\gamma) e^{i\lambda H + i\varkappa N} \, d\lambda d\varkappa + \iint_{\substack{(\lambda,\varkappa) \in Q \setminus S_\Omega \\ |\lambda| > \lambda_0}} \tilde{\varphi}(\lambda, \varkappa | \mathfrak{M}_\gamma) e^{i\lambda H + i\varkappa N} \, d\lambda dx$$

From (3.34), (3.35) and Lemma 3.5 we obtain

$$|I_2| < 4\pi\lambda_0 a_0 e^{-Q_1|\Omega|^{\frac{1}{5}}} + 2\pi\sigma_0 e^{-b_0|\Omega|} \int_{|\lambda| > \lambda_0} \frac{d\lambda}{(\lambda^2 + \beta^2)^2} < \omega_0 e^{-Q_1|\Omega|^{\frac{1}{5}}},$$

where ω_0 is an absolute constant (independent of Ω).

Therefore in the expression (3.27) the integral

$$\frac{1}{\pi^2} \iint_{Q \setminus S_\Omega} \varphi(\lambda, \varkappa | \mathfrak{M}_\gamma) e^{i\lambda H + i\varkappa N} \, d\lambda d\varkappa.$$

makes a contribution not exceeding $\omega_0 e^{-Q_1|\Omega|^{1/5}}$.

Lemma 3.1 is therefore proved.

§4. Equivalence of the microcanonical, canonical and grand canonical ensembles for one-dimensional mechanical systems

In this section we shall prove the thermodynamic equivalence of the three fundamental ensembles for one-dimensional mechanical systems. Thus we shall prove the existence of limit Gibbs distributions $P_{\beta,\rho}(\cdot)$ and $P_{\rho,h}(\cdot)$ defined by means of the canonical and microcanonical ensembles, respectively, and the identity of these distributions with the limit Gibbs distribution $P_{\beta,\mu}(\cdot)$ obtained from the grand canonical ensemble, when a given relationship connects the pairs of parameters (β,μ), (ρ,h) and (β,ρ).

First, however, we shall prove the limit equivalence between the microcanonical and grand canonical ensembles. The equivalence of these ensembles to the canonical ensemble will be found as a simple consequence of this. To establish the equivalence of the microcanonical and grand canonical ensembles we must estimate the ratio of densities $\pi_U(a,E,N,\Omega)/\pi_U(a,\beta,\mu,\Omega)$ for suitable values of the parameters β and μ. To do this we make use of the local limit theorem of the preceding section. But first we must establish the relation between the parameters (β,μ) and (ρ,h) under which the equivalence of the grand and microcanonical ensembles is valid.

1. *Correspondence between the thermodynamic parameters.* On the segment $\Omega \subset R^1$ we consider the two ensembles: the grand canonical ensemble $\{\mathfrak{B}_\Omega,\beta,\mu\}$ and the microcanonical ensemble $\{\mathfrak{B}_{\Omega,N},E,N\}$. The mean particle density $\overline{N}/|\Omega|$ ($\overline{N}$ is the average particle number) and the mean energy density $\overline{E}/|\Omega|$ ($\overline{E}$ is the average energy) in $\{\mathfrak{B}_\Omega,\beta,\mu\}$ we denote by $\rho_\Omega(\beta,\mu)=\overline{N}/|\Omega|$ and $h_\Omega(\beta,\mu)=\overline{E}/|\Omega|$, respectively. From (1.5) we have

$$\rho_\Omega(\beta,\mu) = -\frac{\partial}{\partial\mu}\frac{\ln\Xi(\beta,\mu,\Omega)}{|\Omega|}, \qquad h_\Omega(\beta,\mu) = -\frac{\partial}{\partial\beta}\frac{\ln\Xi(\beta,\mu,\Omega)}{|\Omega|}. \tag{4.1}$$

We shall now select the parameters $\beta=\beta(E,N,\Omega)$ and $\mu=\mu(E,N,\Omega)$ such that the quantities $\rho_\Omega(\beta,\mu)$ and $h_\Omega(\beta,\mu)$ are respectively equal to the particle density $N/|\Omega|$ and energy density $E/|\Omega|$ in the microcanonical ensemble $\{\mathfrak{B}_{\Omega,N},E,N\}$. To determine $\beta(E,N,\Omega)$ and $\mu(E,N,\Omega)$ we have the two equations

$$\frac{E}{|\Omega|} = -\frac{\partial}{\partial\beta}\frac{\ln\Xi(\beta,\mu,\Omega)}{|\Omega|}, \qquad \frac{N}{|\Omega|} = -\frac{\partial}{\partial\mu}\frac{\ln\Xi(\beta,\mu,\Omega)}{|\Omega|}. \tag{4.2}$$

Since we shall be interested in the limiting values of the parameters (β,μ) and (ρ,h) it is natural to consider the corresponding limit equations

$$h = -\frac{\partial}{\partial\beta}\chi(\beta,\mu), \qquad \rho = -\frac{\partial}{\partial\mu}\chi(\beta,\mu), \tag{4.3}$$

where $\chi(\beta,\mu)=\lim_{|\Omega|\to\infty}\ln\Xi(\beta,\mu,\Omega)/|\Omega|$ (the so-called Gibbs potential).

Our aim is to prove the existence of a solution $\beta(E,N,\Omega)$, $\mu(E,N,\Omega)$ to the system of equations (4.2) and also a solution $\beta(\rho,h)$, $\mu(\rho,h)$ to the system (4.3), and the convergence $\beta(E,N,\Omega)\to\beta(\rho,h)$ and $\mu(E,N,\Omega)\to\mu(\rho,h)$

in the thermodynamic limit

(a) $$|\Omega| \to \infty, \qquad \frac{N}{|\Omega|} \to \rho, \qquad \frac{E}{|\Omega|} \to h$$

We first investigate what is the region of variation of the quantities $E/|\Omega|$ and $N/|\Omega|$, and also their limiting values h and ρ. It is clear that $N/|\Omega|$ can vary, taking on values which are multiples of $1/|\Omega|$ from 0 to $[|\Omega|/\delta_0]/|\Omega|$, where $[x]$ denotes the integer part of the number x. Next it is clear that for a given value of N the value of the energy can change from some value $E_0(N,\Omega)$ to $+\infty$ (the existence of a lower limit $E_0(N,\Omega) > -BN$, $B>0$, follows from the stability property of the potential $U(r)$; see (1.4)) and the corresponding energy density varies from $h_0(N,\Omega) = E_0(N,\Omega)/|\Omega|$ to $+\infty$. The limit density ρ runs over the interval $0<\rho<\rho_0 = 1/\delta_0$, and for each given ρ the value of h runs over the half line $h_0(\rho) < h < \infty$, where $h_0(\rho) = \lim_{|\Omega|\to\infty} h_0(N,\Omega)$ (the existence of the limit $h_0(\rho)$ as $N/|\Omega| \to \rho$, $0<\rho<\rho_0$, and the convexity of the function $h_0(\rho)$ are proven in Ruelle's book [**10**]).

We denote the corresponding sets on the (ρ,h)-plane by

$$G(\Omega) = \left\{(\rho,h) : \rho = \frac{N}{|\Omega|}, h_0(N,\Omega) < h < +\infty, N = 0,1,\dots,\left[\frac{|\Omega|}{\delta_0}\right]\right\}$$

and

$$G(\infty) = \{(\rho,h) : h_0(\rho) < h < +\infty, 0 < \rho < \rho_0\}.$$

We can now formulate our assertion (for the proof see part II).

THEOREM 4.1. *Let the numbers E and N (N an integer) be such that*

$$\left(\frac{E}{|\Omega|}, \frac{N}{|\Omega|}\right) \in G(\Omega).$$

Then there exists a unique solution $\{\beta(E,N,\Omega), \mu(E,N,\Omega)\}$ to the system (4.2). If $(\rho,h) \in G(\infty)$, there exists a unique solution $\{\beta(\rho,h), \mu(\rho,h)\}$ to (4.3).

In the thermodynamic limit,

$$\lim_{\substack{|\Omega|\to\infty \\ \frac{N}{|\Omega|}\to\rho,\ \frac{E}{|\Omega|}\to h}} \beta(E,N,\Omega) = \beta(\rho,h), \qquad \lim_{\substack{|\Omega|\to\infty \\ \frac{N}{|\Omega|}\to\rho,\ \frac{E}{|\Omega|}\to h}} \mu(E,N,\Omega) = \mu(\rho,h)$$

The mapping $(\rho,h) \to (\beta(\rho,h), \mu(\rho,h))$ of the region $G(\infty)$ onto the half plane $(\beta > 0, -\infty < \mu < \infty)$ is one-to-one and differentiable, as is its inverse.

2. *Limit equivalence of the microcanonical and grand canonical ensembles.*

THEOREM 4.2. *Let $(\rho,h) \in G(\infty)$. Then for any cylindrical set $S_{U,A}$ ($U \subset R^1$, $A \subset \mathfrak{B}_U$) and for any sequence of microcanonical ensembles $\{\mathfrak{B}_{\Omega_k,N_k}, E_k, N_k\}$, as $|\Omega_k| \to \infty$, $N_k/|\Omega_k| \to \rho$ and $E_k/|\Omega_k| \to h$ the limit probability*

(4.12) $$\lim_{\substack{|\Omega_k|\to\infty \\ \frac{N_k}{|\Omega_k|}\to\rho,\ \frac{E_k}{|\Omega_k|}\to h}} P_{\Omega_k, E_k, N_k}(S_{U,A}) = P_{\rho,h}(S_{U,A})$$

exists, where $P_{\Omega_k,E_k,N_k}(S_{U,A})$ *denotes the measure of the cylindrical set* $S_{U,A}$ *in the microcanonical ensemble* $\{\mathfrak{B}_{\Omega_k,N_k}, E_k, N_k\}$.

The limit probabilities $P_{\rho,h}(S_{U,A})$ *defined by* (4.12) *for all cylindrical sets* $S_{U,A}$ *generate a limit distribution* $P_{\rho,h}(\cdot)$ *on the space* $\mathfrak{B}_\infty$. *The distribution* $P_{\rho,h}(\cdot)$ *coincides with the Gibbs distribution* $P_{\beta,\mu}(\cdot)$ *for the grand canonical ensemble, where* $\beta=\beta(\rho,h)$ *and* $\mu=\mu(\rho,h)$.

PROOF. For a given segment $U\subset R^1$ we consider the density $\pi_U(a,E,N,\Omega)$ for the probability distribution on the space $\mathfrak{B}_U$ generated by the microcanonical ensemble $\{\mathfrak{B}_{\Omega,N},E,N\}$ $(U\subset\Omega)$. As we know (see (1.17)), this density has the form

$$\pi_U(a,E,N,\Omega)=\sum\nolimits^{-1}(E,N,\Omega)\sum(E-E(a),N-N(a),\Omega\diagdown U).$$

On the other hand, the similar density $\pi_U(a,\beta,\mu,\Omega)$ calculated in the grand canonical ensemble has the form (see (1.16))

$$\pi_U(a,\beta,\mu,\Omega)=\Xi^{-1}(\beta,\mu,\Omega)\,\Xi(\beta,\mu,\xi_a,\Omega\diagdown U)\,e^{-\beta E(a)-\mu N(a)},$$

where

$$\xi_a(x)=e^{-\mu}\exp\{-\beta\sum_{y\in a}U(|x-y|)\}.$$

We now consider the ratio of these densities

$$\frac{\pi_U(a,E,N,\Omega)}{\pi_U(a,\beta,\mu,\Omega)}=e^{\beta E(a)+\mu N(a)}\frac{\Xi(\beta,\mu,\Omega)}{\Sigma(E,N,\Omega)}\,\frac{\Sigma(E-E(a),N-N(a),\Omega\diagdown U)}{\Xi(\beta,\mu,\xi_a,\Omega\diagdown U)}.$$

The parameters β and μ are assumed to be chosen so that in the grand canonical ensemble the mean values $\bar{E}$ and $\bar{N}$ coincide with the values of E and N, i.e. we set $\beta=\bar{\beta}(E,N,\Omega)$ and $\mu=\bar{\mu}(E,N,\Omega)$. We estimate each of the ratios

$$\frac{\Sigma(E,N,\Omega)}{\Xi(\beta,\mu,\Omega)}\quad\text{and}\quad\frac{\Sigma(E-E(a),N-N(a),\Omega/U)}{\Xi(\beta,\mu,\xi_a,\Omega/U)}.$$

We know that

$$\frac{\Sigma(E,N,\Omega)}{\Xi(\beta,\mu,\Omega)}=e^{\mu N}P_{\text{tot}}(E,N),$$

where the quantity $p_{\text{tot}}(E,N)$ was defined in §3.

From the local limit theorem proved in the preceding section for $p_{\text{tot}}(E,N)$ and from the choice of the parameters β and μ we have

$$\frac{\Sigma(E,N,\Omega)}{\Xi(\beta,\mu,\Omega)}=e^{\mu N}\left\{\frac{1}{\pi|\Omega|\sqrt{D}}+O(|\Omega|^{-\frac{6}{5}})\right\}. \tag{4.19}$$

Next we estimate the ratio

$$\frac{\Sigma(E-E(a),N-N(a),\Omega\diagdown U)}{\Xi(\beta,\mu,\xi_a,\Omega\diagdown U)}.$$

For this purpose consider the microcanonical ensemble $\{\mathfrak{B}_{\Omega\setminus U,N}, \widetilde{E}, \widetilde{N}|a\}$ in the set $\Omega\setminus U$ under the restriction that the state $a \in \mathfrak{B}_U$ is fixed in U, where $\widetilde{N} = N - N(a)$ and $\widetilde{E} = E - E(a)$. Note that $\Sigma(E - E(a), N - N(a), \Omega\setminus U)$ is the partition function for the conditional microcanonical ensemble $\{\mathfrak{B}_{\Omega\setminus U, N-N(a)}, E - E(a), N - N(a)|a\}$. In the same way we can consider the conditional grand canonical ensemble $\{\mathfrak{B}_{\Omega\setminus U}, \beta, \mu|a\}$ for a system in the set $\Omega\setminus U$ under the same restriction; the quantity $\Xi(\beta, \mu, \xi_a, \Omega\setminus U)$ serves as the partition function of this ensemble. Thus we can write

$$(4.14) \qquad \frac{\Sigma(E - E(a), N - N(a), \Omega\setminus U)}{\Xi(\beta, \mu, \xi_a, \Omega\setminus U)} e^{\beta E(a) + \mu(N - N(a))} p_{\mathrm{tot}}(\widetilde{E}, \widetilde{N}|a),$$

where $p_{\mathrm{tot}}(\widetilde{E}, \widetilde{N}|a)$ has a similar meaning to the quantity $p_{\mathrm{tot}}(E, N)$. Just as we did in the preceding section for $p_{\mathrm{tot}}(E, N)$, we can derive a local limit theorem for $p_{\mathrm{tot}}(\widetilde{E}, \widetilde{N}|a)$:

$$(4.15) \qquad \begin{aligned} p_{\mathrm{tot}}(\widetilde{E}, \widetilde{N}|a) = \frac{1}{\pi|\Omega\setminus U|\sqrt{D}} \exp\Big\{ -\frac{1}{2}\alpha_{11}\frac{(\widetilde{E} - \bar{\widetilde{E}})^2}{|\Omega\setminus U|} - \\ - \alpha_{12}\frac{(\widetilde{E} - \bar{\widetilde{E}})(\widetilde{N} - \bar{\widetilde{N}})}{|\Omega\setminus U|} - \frac{1}{2}\alpha_{22}\frac{(\widetilde{N} - \bar{\widetilde{N}})^2}{|\Omega\setminus U|}\Big\} + \sigma(\widetilde{E}, \widetilde{N}, a), \end{aligned}$$

where $\sigma(\widetilde{E}, \widetilde{N}, a) < O(|\Omega\setminus U|^{-6/5})$, and $\bar{\widetilde{E}}$ and $\bar{\widetilde{N}}$ are the mean values of the quantities $\widetilde{E}$ and $\widetilde{N}$, respectively, evaluated in the conditional ensemble $\{\mathfrak{B}_{\Omega\setminus U}, \beta, \mu|a\}$, and the quantities $\alpha_{11}, \alpha_{12}, \alpha_{22}$ and D were defined in the preceding section. The proof of this local limit theorem is similar to the proof of the local limit theorem of §3. The only thing here that requires more careful consideration is the verification of estimates of the form (2.15), (2.20) and (2.23). To avoid complicating this exposition we have put off this verification to part II.

We note that for our choice of the parameters β and μ the differences $\widetilde{E} - \bar{\widetilde{E}}$ and $\widetilde{N} - \bar{\widetilde{N}}$ are bounded by some constant K which depends only on U, β, μ and a. Indeed, $\widetilde{N} - \bar{\widetilde{N}} = \bar{N} - N(a) - \bar{\widetilde{N}}$, where $\bar{N}$ is the mean value of the particle number in the grand canonical ensemble. For the difference $\bar{N} - \bar{\widetilde{N}}$ we have

$$\bar{N} - \bar{\widetilde{N}} = \int_{\Omega} r_1(x, \beta, \mu, \Omega)\, dx - \int_{\Omega\setminus U} r_1(x, \beta, \mu, \xi_a, \Omega\setminus U)\, dx.$$

In the same way, the difference $\widetilde{E} - \bar{\widetilde{E}} = \bar{E} - E(a) - \bar{\widetilde{E}}$. Next, we have

$$\begin{aligned} \bar{E} - \bar{\widetilde{E}} &= \int_{\Omega}\int_{\Omega} U(|x - y|)\, r_2(x, y, \beta, \mu, \Omega)\, dx dy \\ &- \int_{\Omega\setminus U}\int_{\Omega\setminus U} U(|x - y|)\, r_2(x, y, \beta, \mu, \xi_a, \Omega\setminus U)\, dx dy. \end{aligned}$$

By using the estimates of §2 we can show that the difference $\bar{E} - \bar{\widetilde{E}}$ is bounded:

$$|\bar{E} - \bar{\widetilde{E}}| < K_1(\beta, \mu, U, a),$$

where the constant $K_1(\beta,\mu,U,a)$ does not depend on Ω. Hence we have

$$|\widetilde{E}-\widetilde{\widetilde{E}}| < K_1(\beta,\mu,U,a)+E(a). \tag{4.21}$$

By a similar argument we derive the inequality

$$|\widetilde{N}-\widetilde{\widetilde{N}}| < K_2(\beta,\mu,U,a)+N(a), \tag{4.22}$$

where $K_2(\beta,\mu,U,a)$ does not depend on Ω. From the inequalities (4.21) and (4.22) we find that

$$\begin{gathered} P_{\text{tot}}\,(\widetilde{E},\widetilde{N}/a)=\frac{1}{\pi\,|\,\Omega\diagdown U\,|\,\sqrt{D}}+\sigma\,(\widetilde{E},\widetilde{N},\,a) \\ (\sigma\,(\widetilde{E},\widetilde{N},\,a)<0\,(\,|\,\Omega\diagdown U\,|^{-\frac{5}{6}})). \end{gathered} \tag{4.23}$$

As follows from (4.21) and (4.22), this asymptotic behavior is uniform on any set $T_{E_0,N_0}\subset\mathfrak{B}_U$ of the form $T_{E_0,N_0}\{a\in T_{E_0,N_0}: E(a)<E_0 \text{ and } N(a)<N_0\}$, where E_0 and N_0 are arbitrary fixed constants. By using (4.13), (4.14) and (4.23) we obtain

$$\frac{\pi_U\,(a,E,N,\Omega)}{\pi_U\,(a,\bar{\beta},\,\bar{\mu},\Omega)}=1+O(\,|\,\Omega\,|^{-\frac{6}{5}}), \tag{4.24}$$

where this asymptotic relation is valid uniformly on any set T_{E_0,N_0}. It follows from the results of [**1, 2, 15, 16, 20**] that in the thermodynamic limit $|\Omega|\to\infty$ the density $\pi_U(a,\beta,\mu,\Omega)$ converges uniformly with respect to $a\in\mathfrak{B}_U$ to the limit density $\pi_U(a,\beta,\mu,\infty)$. Moreover, this limit density depends continuously on β and μ. Hence it follows that in the thermodynamic limit (a)

$$\pi_U(a,\bar{\beta}(E,N,\Omega),\bar{\mu}(E,N,\Omega),\Omega)\to\pi_U(a,\beta(\rho,h),\mu(\rho,h),\,\infty). \tag{4.25}$$

Thus we find from (4.24) and (4.25) that for each $a\in\mathfrak{B}_U$

$$\lim_{\substack{|\Omega|\to\infty \\ \frac{N}{|\Omega|}\to\rho,\ \frac{E}{|\Omega|}\to h}} \pi_U\,(a,E,N,\Omega)=\pi_U\,(a,\beta\,(\rho,h),\mu\,(\rho,h),\infty), \tag{4.26}$$

where this convergence is uniform for every set $T_{E_0,N_0}\subset\mathfrak{B}_U$. Thus

$$\int_{T_{E_0,N_0}}\pi_U\,(a,E,N,\Omega)\,da\to\int_{T_{E_0,N_0}}\pi_U\,(a,\beta\,(\rho,h),\mu\,(\rho,h),\infty)\,da$$

for any E_0 and N_0. However, in view of the fact that

$$\int_{\mathfrak{B}_U}\pi_U\,(a,E,N,\Omega)\,da=\int_{\mathfrak{B}_U}\pi_U\,(a,\beta\,(\rho,h),\,\mu\,(\rho,h),\infty)\,da=1$$

as $|\Omega|\to\infty$, we have

$$\int_{\mathfrak{B}_U\setminus T_{E_0,N_0}}\pi_U\,(a,E,N,\Omega)\,da\to\int_{\mathfrak{B}_U\setminus T_{E_0,N_0}}\pi_U\,(a,\beta\,(\rho,h)\,\mu\,(\rho,h),\infty)da\,.$$

Hence we finally conclude that for any cylindrical set $S_{U,A}$ the limit

$$\lim_{\substack{|\Omega|\to\infty \\ \frac{E}{|\Omega|}\to h,\ \frac{N}{|\Omega|}\to\rho}} P_{\Omega,E,N}(S_{U,A}) = P_{\rho,h}(S_{U,A})$$

$$= \int_{A\in\mathfrak{B}_U} \pi_U(a, \beta(\rho, h), \mu(\rho, h), \infty)\, da = P_{\beta,\mu}(S_{U,A})$$

exists. For each fixed $U \subset \Omega$ this convergence is uniform with respect to all $A \subset \mathfrak{B}_U$.

Therefore the limit distribution $P_{\rho,h}(\cdot)$ generated by the microcanonical ensemble exists and coincides with the limit Gibbs distribution $P_{\rho,h}(\cdot)$ under the condition that β and μ are related to ρ and h by the relation (4.3).

3. *The limit equivalence of the canonical and grand canonical ensembles for one-dimensional mechanical systems.* We shall now establish the equivalence of the canonical and grand canonical ensembles. To do this we first look at what the relation must be between the thermodynamic parameters. Consider the canonical ensemble $\{\mathfrak{B}_{\Omega,N}, \beta, N\}$ and the grand canonical ensemble $\{\mathfrak{B}_\Omega, \beta, \mu\}$ (having the same value of the parameter β). As we saw in the preceding subsection, for each given β we can find $\bar{\mu} = \bar{\mu}(N, \Omega, \beta)$ such that $\rho_\Omega(\beta, \bar{\mu}) = N/|\Omega|$, where $\rho_\Omega(\beta,\mu)$ is the mean particle density in the grand canonical ensemble. Moreover, for the interval $0 < \rho < \rho_0$ and given β there exists a $\mu = \mu(\beta, \rho)$ such that $\rho_\infty(\beta,\mu) = \rho$, where $\rho_\infty(\beta,\mu)$ is the limit particle density in the grand canonical ensemble. In addition, $\bar{\mu}(N,\Omega,\beta) \to \mu(\beta,\rho)$ as $|\Omega| \to \infty$ and $N/|\Omega| \to \rho$.

Theorem 4.3. *Let the parameters β and ρ $(0 < \rho < \rho_0)$ be fixed. Then for any cylindrical set $S_{U,A}$ $(U \subset R^1$, $A \subset \mathfrak{B}_U)$ and for any sequence of canonical ensembles $\{\mathfrak{B}_{\Omega_k,N_k}, \beta, N_k\}$ as $|\Omega_k| \to \infty$ and $N_k/|\Omega_k| \to \rho$ the limit probability*

$$\lim_{|\Omega_k|\to\infty} P_{\Omega_k,N_k,\beta}(S_{U,A}) = P_{\beta,\rho}(S_{U,A}) \tag{4.32}$$

exists (here $P_{\Omega_k,N_k}(S_{U,A})$ is the measure of the cylindrical set in the canonical ensemble $\{\mathfrak{B}_{\Omega_k,N_k}, \beta, N_k\}$). The limit probabilities $P_{\beta,\rho}(S_{U,A})$ defined by (4.32) *for all cylindrical sets generate a limit distribution $P_{\beta,\rho}(\cdot)$ on the space $\mathfrak{B}_\infty$. The distribution $P_{\beta,\rho}(\cdot)$ coincides with the limit Gibbs distribution $P_{\beta,\mu}(\cdot)$ obtained from the grand canonical ensemble, where $\mu = \mu(\rho)$.*

Fundamentally, the proof of this theorem, just like that of the preceding one, consists in proving a local limit theorem for the particle number. By our proof of Theorem 4.2, for the distribution of both energy and particle number we have a local limit theorem for the particle number alone. By repeating word for word the argument leading to Theorem 4.2, we have a proof for Theorem 4.3. In the case of a small number of particles (in any number of dimensions) this theorem has been proven by Halfina [**17**].

The identity of the limit distributions $P_{\rho,h}(\cdot)$ and $P_{\beta,\rho}(\cdot)$ obtained from the microcanonical and canonical ensembles respectively follows from

Theorems 4.3 and 4.2, where the parameters are related as follows: ρ is the same in both cases and $\beta = \beta(\rho, h)$, where the function $\beta(\rho, h)$ was defined in subsection 1 of this section.

Received 25/NOV/71

Bibliography

[1] R. L. Dobrušin, *Gibbsian random fields. The general case,* Funkcional. Anal. i Priložen. **3** (1969), no. 1, 27-35 = Functional Anal. Appl. **3** (1969), 22-28. MR **39** # 1151.

[2] ———, *Definition of a system of random variables by means of conditional distributions,* Teor. Verojatnost. i Primenen. **15** (1970), 469-497 = Theor. Probability Appl. **15** (1970), 458-486. MR **45** # 7765.

[3] R. L. Dobrušin and R. A. Minlos, *Existence and continuity of pressure in classical statistical physics,* Teor. Verojatnost. i Primenen. **12** (1967), 595-618 = Theor. Probability Appl. **12** (1967), 535-559. MR **36** # 7394.

[4] I. A. Ibragimov and Ju. V. Linnik, *Independent and stationarily connected variables,* "Nauka", Moscow, 1965; English transl., Noordhoff, Groningen, 1971. MR **34** # 2049.

[5] M. A. Krasnosel'skiĭ, *Positive solutions of operator equations,* Fizmatgiz, Moscow, 1962; English transl., Noordhoff, Groningen, 1964. MR **26** # 2862; **31** # 6107.

[6] R. A. Minlos, *Gibbs' limit distribution,* Funkcional. Anal. i Priložen. **1** (1967), no. 2, 60-73 = Functional Anal. Appl. **1** (1967), 141-150. MR **35** # 5202.

[7] R. A. Minlos and A. M. Halfina, *Two-dimensional limit theorem for the particle number and energy in the grand canonical ensemble,* Izv. Akad. Nauk SSSR Ser. Mat. **34** (1970), 1173-1191 = Math. USSR Izv. **4** (1970), 1183-1202. MR **42** # 8847.

[8] R. A. Minlos and A. Haitov, *Limiting equivalence of various ensembles for one-dimensional classical systems of statistical physics (infinitely distributed potential)* (to appear). (Russian)

[9] S. V. Nagaev, *Some limit theorems for stationary Markov chains,* Teor. Verojatnost. i Primenen. **2** (1957), 389-416 = Theor. Probability Appl. **2** (1957), 378-406. MR **20** # 1355.

[10] D. Ruelle, *Statistical mechanics: Rigorous results,* Benjamin, New York, 1969; Russian transl., "Mir", Moscow, 1971. MR **44** # 6279.

[11] F. Riesz and B. Sz.-Nagy, *Leçons d'analyse fonctionnelle,* Akad. Kiadó, Budapest, 1952; 2nd ed., 1953; Russian transl. of 1st ed., IL, Moscow, 1954; English transl. of 2nd ed., Ungar, New York, 1955. MR **14**, 286; **15**, 132; **17**, 175.

[12] Ja. G. Sinaĭ, *Construction of the dynamics in one-dimensional systems of statistical mechanics,* Teoret. Mat. Fiz. **11** (1972), no. 2, 248-258 = Theoret. Math. Phys. **11** (1972), 487-494.

[13] S. H. Siraždinov, *Limit theorems for stationary Markov chains,* Izdat. Akad. Nauk Uzbek. SSR, Tashkent, 1955. (Russian) MR **18**, 944.

[14] V. A. Statulevičius, *Limit theorems for sums of random variables that are connected in a Markov chain.* I, II, III, Litovsk. Mat. Sb. **9** (1969), 345-362, 635-672; ibid. **10** (1970), 161-170. (Russian) MR **42** # 1188.

[15] Ju. M. Suhov, *An application of the matrix method to continuous systems of classical statistical mechanics,* Uspehi Mat. Nauk **25** (1970), no. 2 (152), 277-278. (Russian) MR **41** # 6292.

[16] ———, *The matrix method for continuous systems in classical statistical mechanics,* Trudy Moskov. Mat. Obšč. = Trans. Moscow Math. Soc. (to appear).

[17] A. M. Halfina, *The limiting equivalence of the canonical and grand canonical ensembles (low density case),* Mat. Sb. **80 (122)** (1969), 3-51 = Math. USSR Sb. **9** (1969), 1-52.

[18] A. Ja. Hinčin, *Statistical mechanics as a problem of the theory of probability,* Uspehi Mat. Nauk **5** (1950), no. 3 (37), 3-46. (Russian) MR **12**, 68.

[19] L. Van-Hove, *Sur l'intégrale de configuration pour les systèmes de particules à une dimension,* Physica **16** (1950), 137-143. MR **12**, 68.

[20] G. Gallavotti and S. Miracle-Sole, *Absence of phase transitions in hard-core one-dimensional systems with long-range interactions,* J. Mathematical Phys. **11** (1970), 147-154. MR **43** # 1573.

[21] T. Kato, *Perturbation theory for linear operators,* Die Grundlehren der math. Wissenschaften, Band 132, Springer-Verlag, New York, 1966. MR **34** # 3324.

[22] N. N. Bogoljubov, O. Ja. Petrina and B. I. Hacet, *Mathematical description of the equilibrium state of classical systems on the basis of the canonical ensemble formalism*, Teoret. i Mat. Fiz. **1** (1969), 251-274 = Theoret. Math. Phys. **1** (1969), 194-212.

[23] R. A. Minlos and Ja. G. Sinaĭ, *The phenomenon of "phase separation" at low temperatures in some lattice models of a gas*. I, Mat. Sb. **73 (115)** (1967), 375-448 = Math. USSR Sb. **2** (1967), 335-396. MR **35** # 3998.

[24] Ī. L. Simjatickiĭ, *Remarks on the paper by N. N. Bogoljubov, O. Ja. Petrina and B. I. Hacet, "Mathematical description of the equilibrium state of classical systems on the basis of the canonical ensemble formalism"*, Teoret. i Mat. Fiz. **6** (1971), 230-237 = Theoret. Math. Phys. **6** (1971), 169-174.

Translated by
M. L. GLASSER